Springer Series in
OPTICAL SCIENCES 165

founded by H.K.V. Lotsch

Editor-in-Chief: W. T. Rhodes, Atlanta

Editorial Board: A. Adibi, Atlanta
T. Asakura, Sapporo
T. W. Hänsch, Garching
T. Kamiya, Tokyo
F. Krausz, Garching
B. Monemar, Linköping
H. Venghaus, Berlin
H. Weber, Berlin
H. Weinfurter, München

Springer Series in
OPTICAL SCIENCES

The Springer Series in Optical Sciences, under the leadership of Editor-in-Chief *William T. Rhodes*, Georgia Institute of Technology, USA, provides an expanding selection of research monographs in all major areas of optics: lasers and quantum optics, ultrafast phenomena, optical spectroscopy techniques, optoelectronics, quantum information, information optics, applied laser technology, industrial applications, and other topics of contemporary interest.

With this broad coverage of topics, the series is of use to all research scientists and engineers who need up-to-date reference books.

The editors encourage prospective authors to correspond with them in advance of submitting a manuscript. Submission of manuscripts should be made to the Editor-in-Chief or one of the Editors. See also www.springer.com/series/624

Editor-in-Chief
William T. Rhodes
Georgia Institute of Technology
School of Electrical and Computer Engineering
Atlanta, GA 30332-0250, USA
E-mail: bill.rhodes@ece.gatech.edu

Editorial Board
Ali Adibi
Georgia Institute of Technology
School of Electrical and Computer Engineering
Atlanta, GA 30332-0250, USA
E-mail: adibi@ee.gatech.edu

Toshimitsu Asakura
Hokkai-Gakuen University
Faculty of Engineering
1-1, Minami-26, Nishi 11, Chuo-ku
Sapporo, Hokkaido 064-0926, Japan
E-mail: asakura@eli.hokkai-s-u.ac.jp

Theodor W. Hänsch
Max-Planck-Institut für Quantenoptik
Hans-Kopfermann-Straße 1
85748 Garching, Germany
E-mail: t.w.haensch@physik.uni-muenchen.de

Takeshi Kamiya
Ministry of Education, Culture, Sports
Science and Technology
National Institution for Academic Degrees
3-29-1 Otsuka, Bunkyo-ku
Tokyo 112-0012, Japan
E-mail: kamiyatk@niad.ac.jp

Ferenc Krausz
Ludwig-Maximilians-Universität München
Lehrstuhl für Experimentelle Physik
Am Coulombwall 1
85748 Garching, Germany   and
Max-Planck-Institut für Quantenoptik
Hans-Kopfermann-Straße 1
85748 Garching, Germany
E-mail: ferenc.krausz@mpq.mpg.de

Bo Monemar
Department of Physics
and Measurement Technology
Materials Science Division
Linköping University
58183 Linköping, Sweden
E-mail: bom@ifm.liu.se

Herbert Venghaus
Fraunhofer Institut für Nachrichtentechnik
Heinrich-Hertz-Institut
Einsteinufer 37
10587 Berlin, Germany
E-mail: venghaus@hhi.de

Horst Weber
Technische Universität Berlin
Optisches Institut
Straße des 17. Juni 135
10623 Berlin, Germany
E-mail: weber@physik.tu-berlin.de

Harald Weinfurter
Ludwig-Maximilians-Universität München
Sektion Physik
Schellingstraße 4/III
80799 München, Germany
E-mail: harald.weinfurter@physik.uni-muenchen.de

Please view available titles in *Springer Series in Optical Sciences*
on series homepage http://www.springer.com/series/624

Ana Belén Cristóbal López
Antonio Martí Vega
Antonio Luque López

Editors

Next Generation of Photovoltaics

New Concepts

With 204 Figures

Editors
Ana Belén Cristóbal López
Antonio Martí Vega
Antonio Luque López
Universidad Politécnica de Madrid
Instituto de Energía Solar. E.T.S.I. Telecomunicación
Madrid, Spain

Springer Series in Optical Sciences   ISSN 0342-4111   e-ISSN 1556-1534
ISBN 978-3-642-23368-5   e-ISBN 978-3-642-23369-2
DOI 10.1007/978-3-642-23369-2
Springer Heidelberg Dordrecht London New York

Library of Congress Control Number: 2012931939

© Springer-Verlag Berlin Heidelberg 2012
This work is subject to copyright. All rights are reserved, whether the whole or part of the material is concerned, specifically the rights of translation, reprinting, reuse of illustrations, recitation, broadcasting, reproduction on microfilm or in any other way, and storage in data banks. Duplication of this publication or parts thereof is permitted only under the provisions of the German Copyright Law of September 9, 1965, in its current version, and permission for use must always be obtained from Springer. Violations are liable to prosecution under the German Copyright Law.
The use of general descriptive names, registered names, trademarks, etc. in this publication does not imply, even in the absence of a specific statement, that such names are exempt from the relevant protective laws and regulations and therefore free for general use.

Printed on acid-free paper

Springer is part of Springer Science+Business Media (www.springer.com)

Preface

This book is the result of a productive meeting that took place in Cercedilla (Spain) in 2008 to discuss about new concepts that could be used to improve the efficiency of photovoltaic (PV) devices. Speakers were encouraged to write their contributions and this book is the result of that effort.

The first three chapters review *multijunction* solar cells and the use of *concentrated* light. It is not long ago that multi-junction solar cells were considered as new concepts and now hold the record efficiency over any other photovoltaic device. Proposed by Jackson in 1955 (E. D. Jackson, Trans. Conf. on the Use of Solar Energy, Tucson, 1955, University of Arizona Press, Tucson, vol. 5, pp. 122–126, 1958) they were considered difficult to achieve in practise in 1960 because the use of tunnel junctions was still not considered (M. Wolf, Proc. IRE; Vol/Issue: 48:7, pp. Pages: 1247–1263, 1960). Nowadays, multi-junction solar cells and concentration systems are progressing fast toward industrialization.

Chapters 4 and 5 deal with a topic of common interest for any photovoltaic device as it is *light management*. In particular, Chapter 5 reviews the potential of *plasmon polaritons*. *Light management*, in general, pursues how to absorb more photons in less material volume. This is of interest, not only because the use of less material usually leads to a lower cost of the device, but also because the use of less material reduces non-radiative recombination and therefore increases the efficiency of the device. On the other hand, many of the novel photovoltaic devices, as the intermediate band solar cell, can be regarded as multicolored devices and mastering *light management* is useful to assist photon absorption at those wavelengths where absorption is weak. This mastering often demands dealing with light at electromagnetic level as it is the case of *plasmon polaritons*.

Chapter 6 and 7 deal with *up-conversion* in organic systems and *multiple exciton generation* solar cells, respectively. The first approach pursues converting the absorption of several below bandgap energy photons, not useful for PV conversion, and convert them into a useful high energy photon. The second approach, on the other hand, pursues the conversion of a high energy photon in several lower energy photons. In this way the energy in excess of the photon can be used to create not one but several electron-hole pairs.

Chapters 8–13 deal with the *intermediate band* solar cell concept that pursues the conversion of below bandgap energy photons into solar cell current without voltage degradation. Chapter 8 introduces the concept and the rest of the chapters expand it along different frameworks: modeling (Chap. 9), quantum dots (Chap. 10), thin films (Chap. 11), GaInN (Chap. 12), and ion implantation on silicon (Chap. 13).

Finally, we would like to acknowledge our contributors for their work and patience with the edition of this book, to Springer Verlag for the careful edition of this book, and to all of our families, also in the name of the authors of this book, for the time they have sacrificed of being with their beloved ones.

Madrid

Ana Belén Cristóbal López
Antonio Martí Vega
Antonio Luque López

Contents

1 **Present Status in the Development of III–V Multi-Junction Solar Cells** 1
  Simon P. Philipps, Wolfgang Guter, Elke Welser, Jan Schöne,
  Marc Steiner, Frank Dimroth, and Andreas W. Bett
  1.1 Introduction 1
  1.2 Research Status 4
       1.2.1 Numerical Simulation 4
       1.2.2 Device Fabrication 9
  1.3 Next Steps 14
  References 19

2 **The Interest and Potential of Ultra-High Concentration** 23
  Carlos Algora and Ignacio Rey-Stolle
  2.1 Motivation of Ultra-High Concentration 23
  2.2 Cost Calculation Update 28
  2.3 Series Resistance 29
  2.4 Tunnel Junctions for UHCPV 35
  2.5 Increase of the MJSC Efficiency at Ultra-High Concentration 39
  2.6 Advanced Characterisation 45
       2.6.1 Quantum Efficiency Under Concentration 45
       2.6.2 Electroluminescence Spectroscopy 47
       2.6.3 Electroluminescence Intensity Mapping 49
  2.7 Reliability 52
  References 56

3 **The Sun Tracker in Concentrator Photovoltaics** 61
  Ignacio Luque-Heredia, Goulven Quéméré, Rafael Cervantes,
  Olivier Laurent, Emmanuele Chiappori, and Jing Ying Chong
  3.1 Introduction 61
  3.2 Requirements and Specifications 63
  3.3 The Tracker Structure 67

	3.4	Sun Tracking Control	74
		3.4.1 Background	74
		3.4.2 The Autocalibrated Sun Tracking Control Unit	79
	3.5	Sun Tracking Accuracy Monitoring	83
		3.5.1 The Tracking Accuracy Sensor	84
		3.5.2 The Monitoring System	85
		3.5.3 Accuracy Assessment: Example of the Autocalibrated Tracking Strategy	87
	References		91
4	**Light Management in Thin-Film Solar Cell**		**95**
	Janez Krč, Benjamin Lipovšek, and Marko Topič		
	4.1	Introduction	95
	4.2	Optical Properties of Thin-Film Photovoltaic Devices	97
		4.2.1 Optical Modelling and Simulation	100
		4.2.2 Importance of Light Management in Thin-Film Solar Cells	101
	4.3	Advanced Light Management in Thin-Film Solar Cells	106
		4.3.1 Efficient Light Scattering	106
		4.3.2 Minimisation of Optical Losses	113
		4.3.3 Enhanced Utilisation of the Solar Spectrum	121
	4.4	Conclusion	124
	References		126
5	**Surface Plasmon Polaritons in Metallic Nanostructures: Fundamentals and Their Application to Thin-Film Solar Cells**		**131**
	Carsten Rockstuhl, Stephan Fahr, and Falk Lederer		
	5.1	Introduction	131
	5.2	Plasmon Polaritons and Their Use in Solar Cells	132
	5.3	Fundamentals	134
		5.3.1 Localized Surface Plasmon Polaritons	134
		5.3.2 Propagating Surface Plasmon Polaritons	138
		5.3.3 Simulation Techniques	142
	5.4	Use of Localized Surface Plasmon Polaritons in Solar Cells	143
		5.4.1 Thin-Film Solar Cells	145
		5.4.2 Tandem Solar Cells	147
	5.5	Use of Propagating Surface Plasmon Polaritons in Solar Cells	149
		5.5.1 Thin-Film Solar Cells	150
	5.6	Conclusions	152
	References		153
6	**Non-Coherent Up-Conversion in Multi-Component Organic Systems**		**157**
	Stanislav Baluschev and Tzenka Miteva		
	6.1	Comparison Between Up-Conversion Processes	157
	6.2	Requirements of an Organic System for Observation of Efficient TTA-UC	165

	6.3	Characterization of the TTA-UC	169
		6.3.1 Replacement of the Volatile Organic Solvent	170
		6.3.2 Materials and Methods	173
		6.3.3 Transparency Window	173
		6.3.4 Triplet Harvesting	175
		6.3.5 Intensity Depence	176
		6.3.6 Dependence on the Beam Diameter and Viscosity of the Optically Inactive Solvent	177
		6.3.7 Molar Concentration Dependence	179
		6.3.8 Molar Concentration Dependence	180
		6.3.9 Temperature Dependence in Solid-State Films	181
	6.4	Sunlight UC and Its Application for Organic Solar Cells	183
		6.4.1 Active UC-Media: The Molecular Couples	185
		6.4.2 Active UC-Media: The Molecular Couples	186
	6.5	Conclusion	188
	References	189
7	**Next Generation Photovoltaics Based on Multiple Exciton Generation in Quantum Dot Solar Cells**	191
	Arthur J. Nozik	
	7.1	Introduction	192
	7.2	Solar Cells Utilizing Hot Carriers for Enhanced Conversion Efficiency	193
	7.3	Quantum Dots, Multiple Exciton Generation, and Third Generation Solar Cells	194
	7.4	Configurations of QD Solar Cells	199
		7.4.1 Photoelectrodes Composed of Quantum Dot Arrays	200
		7.4.2 Quantum Dot-Sensitized Nanocrystalline TiO_2 Solar Cells	201
		7.4.3 Quantum Dots Dispersed in Organic Semiconductor Polymer Matrices	201
	7.5	Schottky Junction and p–n Junction Solar Cells Based on Films of QD Arrays	202
	7.6	Conclusions	203
	References	204
8	**Fundamentals of Intermediate Band Solar Cells**	209
	Antonio Martí and Antonio Luque	
	8.1	Introduction	209
	8.2	Intermediate Band Solar Cell Model	213
	8.3	Quantum Dot Intermediate Band Solar Cells	216
	8.4	Thin-Film Intermediate Band Solar Cells	219
	8.5	InGaN Intermediate Band Solar Cells	220
	8.6	Highly Mismatched Alloys	221
	8.7	ZnTe:O Intermediate Band Solar Cells	222
	8.8	Molecular Approach	223

	8.9	In$_2$S$_3$:V Intermediate Band Material	223
	8.10	Other Intermediate Band Solar Cell Systems	225
	8.11	Conclusions	225
	References	226
9	**Modelling of Quantum Dots for Intermediate Band Solar Cells**	229
	Stanko Tomić	
	9.1	Introduction	229
	9.2	Theory of QD Array Electronic Structure	232
		9.2.1 The k·p Theory	232
		9.2.2 Plane Waves Implementation of QD Array Electronic Structure Solver	234
		9.2.3 Model QD Arrays	234
		9.2.4 Wavefunction Delocalization and Intermediate Band Formation in QD Arrays	235
		9.2.5 Electronic Structure of QD Arrays	237
		9.2.6 Absorption Characteristics of QD Arrays	239
	9.3	Radiative and Non-Radiative Processes in QD Array	242
		9.3.1 Radiative Processes in QD Arrays	242
		9.3.2 Electron–Phonon Interaction in QD Arrays	244
		9.3.3 Auger-Related Nonradiative Times in QD Arrays	246
	9.4	Conclusions	247
	References	249
10	**The Quantum Dot Intermediate Band Solar Cell**	251
	Colin R. Stanley, Corrie D. Farmer, Elisa Antolín,	
	Antonio Martí, and Antonio Luque	
	10.1	Introduction	251
	10.2	Molecular Beam Epitaxy of III–V Semiconductors	253
		10.2.1 Introduction	253
		10.2.2 MBE Growth Kinetics	254
		10.2.3 The Utility of Reflection High Energy Electron Diffraction in MBE	255
	10.3	MBE Growth of InAs Quantum Dots on (100)-GaAs	256
		10.3.1 Introduction	256
		10.3.2 MBE Growth of InAs Quantum Dots	256
		10.3.3 Capping of InAs QDs	257
		10.3.4 "Seeded" InAs QD Growth	259
		10.3.5 Multi-Layer Stacks of InAs QDs	260
	10.4	MBE Growth of InAs Quantum Dots for IBSCs	261
		10.4.1 Introduction	261
		10.4.2 Optimising the CB→IB Energy Separation	262
		10.4.3 Multiple QD Layers	264
		10.4.4 Populating the IB with Electrons	265

	10.5	Manufacture of QD-IBSC Die	265
		10.5.1 Introduction	265
		10.5.2 Low Resistance Ohmic Metallisations to n- and p-Type Emitters	268
		10.5.3 Definition of Diode Area	269
		10.5.4 Exposure of AlGaAs Window Layer	269
		10.5.5 Broad Band Anti-Reflection Coating	269
	10.6	Characterisation of Quantum Dot Intermediate Band Solar Cells	270
	10.7	Conclusions	272
	References		274

11 Thin-Film Technology in Intermediate Band Solar Cells: Advanced Concepts for Chalcopyrite Solar Cells 277
David Fuertes Marrón
	11.1	Why Not Thin-Film Intermediate Band Solar Cells? Theoretical Aspects	278
	11.2	The Impurity Approach	287
	11.3	The Nanostructuring Approach	294
	11.4	Conclusions and Outlook	298
	References		304

12 InGaN Technology for IBSC Applications 309
C. Thomas Foxon, Sergei V. Novikov, and Richard P. Campion
	12.1	Introduction	309
	12.2	Motivation and Theoretical Consideration for InGaN:Mn in IBSC Applications	310
	12.3	Growth of Mn Doped In-Rich InGaN	312
	12.4	Proposed IBSC Structures and Progress So Far	317
	12.5	Future Plans and Options	318
	12.6	Summary and Conclusions	318
	References		319

13 Ion Implant Technology for Intermediate Band Solar Cells 321
Javier Olea, David Pastor, María Toledano Luque,
Ignacio Mártil, and Germán González Díaz
	13.1	Introduction	321
	13.2	Experimental	323
	13.3	Results	324
		13.3.1 Structural Characterization	324
		13.3.2 Electrical Transport Properties	329
	13.4	Discussion	343
	13.5	Conclusions	344
	References		345

Index 347

Contributors

Carlos Algora Instituto de Energía Solar, Universidad Politécnica de Madrid, Madrid, Spain

Elisa Antolín Instituto de Energía Solar, Universidad Politécnica de Madrid, Madrid, Spain

Stanislav Baluschev Max-Planck-Institute for Polymer Research, Mainz, Germany

Optics and Spectroscopy Department, Sofia University "St. Kliment Ochridski", Sofia, Bulgaria

Andreas W. Bett Fraunhofer Institute for Solar Energy Systems, Freiburg, Germany

Richard P. Campion School of Physics and Astronomy, University of Nottingham, Nottingham, UK

Rafael Cervantes BSQ Solar, Madrid, Spain

Emmanuele Chiappori BSQ Solar, Madrid, Spain

Jing Ying Chong BSQ Solar, Madrid, Spain

Germán González Díaz Departamento de Física Aplicada III, Electricidad y Electrónica, Universidad Complutense de Madrid, Madrid, Spain

Frank Dimroth Fraunhofer Institute for Solar Energy Systems, Freiburg, Germany

Stephan Fahr Institute of Condensed Matter Theory and Optics, Friedrich-Schiller-Universität Jena, Jena, Germany

Corrie D. Farmer School of Engineering, University of Glasgow, Glasgow, UK

C. Thomas Foxon School of Physics and Astronomy, University of Nottingham, Nottingham, UK

Wolfgang Guter Fraunhofer Institute for Solar Energy Systems, Freiburg, Germany

Janez Krč University of Ljubljana, Ljubljana, Slovenia

Olivier Laurent BSQ Solar, Madrid, Spain

Falk Lederer Institute of Condensed Matter Theory and Optics, Friedrich-Schiller-Universität Jena, Jena, Germany

Benjamin Lipovšek University of Ljubljana, Ljubljana, Slovenia

Antonio Luque Instituto de Energía Solar, Universidad Politécnica de Madrid, Madrid, Spain

María Toledano Luque Departamento de Física Aplicada III, Electricidad y Electrónica, Universidad Complutense de Madrid, Madrid, Spain

Ignacio Luque-Heredia BSQ Solar, Madrid, Spain

David Fuertes Marrón Instituto de Energía Solar, Universidad Politécnica de Madrid, Madrid, Spain

Antonio Martí Instituto de Energía Solar, Universidad Politécnica de Madrid, Madrid, Spain

Ignacio Mártil Departamento de Física Aplicada III, Electricidad y Electrónica, Universidad Complutense de Madrid, Madrid, Spain

Tzenka Miteva Sony Deutschland GmbH, Materials Science Laboratory, Stuttgart, Germany

Sergei V. Novikov School of Physics and Astronomy, University of Nottingham, Nottingham, UK

Arthur J. Nozik National Renewable Energy Laboratory Golden, Golden, CO, USA
and
University of Colorado Boulder, Boulder, CO, USA

Javier Olea Departamento de Física Aplicada III, Electricidad y Electrónica, Universidad Complutense de Madrid, Madrid, Spain

David Pastor Departamento de Física Aplicada III, Electricidad y Electrónica, Universidad Complutense de Madrid, Madrid, Spain

Simon P. Philipps Fraunhofer Institute for Solar Energy Systems, Freiburg, Germany

Goulven Quéméré BSQ Solar, Madrid, Spain

Ignacio Rey-Stolle Instituto de Energía Solar, Universidad Politécnica de Madrid, Madrid, Spain

Carsten Rockstuhl Institute of Condensed Matter Theory and Optics, Friedrich-Schiller-Universität Jena, Jena, Germany

Jan Schöne Fraunhofer Institute for Solar Energy Systems, Freiburg, Germany

Elke Welser Fraunhofer Institute for Solar Energy Systems, Freiburg, Germany

Colin R. Stanley School of Engineering, University of Glasgow, Glasgow, UK

Marc Steiner Fraunhofer Institute for Solar Energy Systems, Freiburg, Germany

Stanko Tomić Computational Science and Engineering Department, STFC Daresbury Laboratory, Cheshire, UK

Marko Topič University of Ljubljana, Ljubljana, Slovenia

Chapter 1
Present Status in the Development of III–V Multi-Junction Solar Cells

Simon P. Philipps, Wolfgang Guter, Elke Welser, Jan Schöne, Marc Steiner, Frank Dimroth, and Andreas W. Bett

Abstract During the last years high-concentration photovoltaics (HCPV) technology has gained growing attention. Excellent operating AC-system efficiencies of up to 25% have been reported. One of the driving forces for this high system efficiency has been the continuous improvement of III–V multi-junction solar cell efficiencies. In consequence, the demand for these solar cells has risen, and strong efforts are undertaken to further increase the solar cell efficiency as well as the volume of cell output. The production capacity for multi-junction solar cells does not constitute a limitation. Already now several tens of MWp per year can be produced and the capacities can easily be increased. The state-of-the art approach for highly efficient photovoltaic energy conversion is marked by the $Ga_{0.50}In_{0.50}P/Ga_{0.99}In_{0.01}As/Ge$ structure. This photovoltaic device is today well established in space applications and recently has entered the terrestrial market. The following chapter presents an overview about the present research status in III–V multi-junction solar cells at Fraunhofer ISE regarding cell design, expected performance, numerical simulation tools, adaptation of devices to different incident spectra and the fabrication of these devices. Finally, an outlook on future developments of III–V multi-junction solar cells is given.

1.1 Introduction

There are different approaches to reduce the levelized costs of electricity from photovoltaics. On one hand, module costs decrease due to economies of scale, less material and energy consumption, or the use of cheap materials. On the other

---

S.P. Philipps (✉) · A.W. Bett (✉)
Fraunhofer Institute for Solar Energy Systems, Heidenhofstr. 2, 79110 Freiburg, Germany
e-mail: simon.philipps@ise.fraunhofer.de; andreas.bett@ise.fraunhofer.de

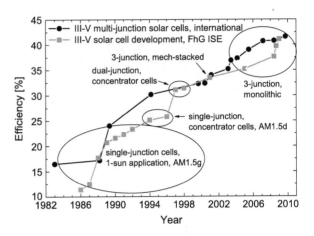

Fig. 1.1 Development of record efficiencies at Fraunhofer ISE (*squares*) and elsewhere (*circles*) for different kinds of III–V multi-junction solar cells from the 1980s to autumn 2009

hand, system costs can be reduced by an increase in module efficiency, which also provides the advantage of smaller systems and less use of area. Hence, all components of a HCPV system have to be further developed to reach highest efficiencies. Large progress can be observed, especially in the field of III–V multi-junction solar cells, where record efficiencies above 41% have been reported by different groups in 2009 [1, 2]. However, despite the high concentration levels, the solar cell still represents up to 20% of the overall costs of a HCPV system [3]. Therefore, a key element for further energy cost reduction is a highly efficient multi-junction solar cell. Multiple stacking of solar cells with growing bandgap energies increases the efficiency of the overall device since the solar spectrum is exploited more profitably. This becomes obvious when looking at the cell development of Fraunhofer ISE and other institutions from the 1980s to autumn 2009, which is summarized in Fig. 1.1. Until the middle of the 1990s, single-junction solar cells were investigated and achieved efficiencies above 25%. Then monolithic dual-junction solar cells have boosted the efficiency records above 30%. Monolithic triple-junction structures have finally surpassed the 40% mark and are still heading for higher efficiencies. In 2009 a GaAs single-junction concentrator solar cell made at Fraunhofer ISE reached a record efficiency of 29.1% under the AM1.5d ASTM G173-03 spectrum (in the following: AM1.5d) and a concentration of 117 suns (1 sun corresponds to $1\,\mathrm{kW\,m^{-2}}$). For a monolithic III–V dual-junction solar cell, a record value of 32.6% under 1,000 suns (AM1.5d) was achieved at the UPM Madrid using $Ga_{0.51}In_{0.49}P$ and GaAs subcells [4].

However, the state-of-the-art device is a lattice-matched triple-junction solar cell consisting of monolithically stacked $Ga_{0.50}In_{0.50}P$-, $Ga_{0.99}In_{0.01}As$-, and Ge junctions. It has reached conversion efficiencies of 41.6% at concentrations of 364 suns under the AM1.5d spectrum [2]. Yet, detailed balance calculations [5] show that the bandgap combination of the lattice-matched design is not optimally adjusted to the solar spectrum. The GaInAs middle cell uses the smallest part of the spectrum and hence produces the lowest current $J_{SC,GaInAs}$ compared to the GaInP top cell with about 11% more current and the Ge bottom cell with about twice $J_{SC,GaInAs}$.

1 Present Status in the Development of III–V Multi-Junction Solar Cells

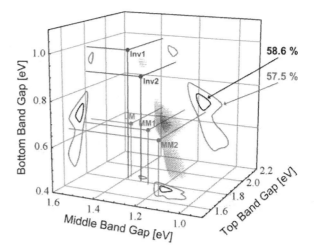

Fig. 1.2 Ideal efficiencies of triple-junction solar cell structures calculated with etaOpt [9] under the AM1.5d ASTM G173-03 spectrum with a concentration ratio of 500 suns and a cell temperature of 65°C

This is why different approaches for achieving current-matching conditions have been suggested [6–8]. Presently, the use of metamorphic structures as discussed below proves to be the most successful strategy.

Figure 1.2 shows how the ideal efficiency of a triple-junction solar cell design varies with the bandgap energies of the three individual junctions. This diagram has been calculated with the model etaOpt [9] (for download see [10]), which is based on the detailed balance method first introduced by Shockley and Queisser [5]. It is assumed that radiative recombination is the only recombination mechanism. Furthermore, an external quantum efficiency (EQE) of unity is assumed for each subcell. In order to improve current matching under a certain spectrum, photocurrent from upper subcells can be transferred to lower ones. In reality this is achieved by thinning the absorbing layers. In order to model this effect, each subcell has an individual degree of transparency, which can be adjusted to improve current-matching. The efficiency then is calculated according to the one-diode model. The calculations are carried out under the AM1.5d spectrum with a concentration of 500 suns and a cell temperature of 65°C. These operating conditions represent a reasonable average for today's concentrator systems. As Fig. 1.2 illustrates, the global maximum for AM1.5d lies at a bandgap combination of 1.75, 1.18, and 0.70 eV. A separated local maximum with a 2.4% (rel.) lower efficiency is found for the relatively high bandgap combination 1.86, 1.34, and 0.93 eV. This twofold maximum results from the absorption band of atmospheric water and carbon dioxide around 1,400 nm, which significantly deteriorates the efficiency of bandgap combinations in between the two maxima. Accordingly, under the extraterrestrial AM0 spectrum only a single maximum exists. Assuming that – as a rule of thumb – about 70–80% of the theoretical efficiency can be achieved in practice, the theoretical model provides a reasonable guideline to assess the potential of a solar cell design. The bandgap combinations of five specific triple-junction solar cell structures, for which efficiencies of over 40% under the concentrated AM1.5d spectrum have already been experimentally realized, are indicated: lattice-matched $Ga_{0.50}In_{0.50}P/Ga_{0.99}In_{0.01}As/Ge$ (LM) [11–13]; metamorphic $Ga_{0.44}In_{0.56}P/Ga_{0.92}In_{0.08}As/Ge$ (MM1) [11]; metamorphic $Ga_{0.35}In_{0.65}P/Ga_{0.83}In_{0.17}As/Ge$ (MM2) [1]; inverted metamorphic $Ga_{0.50}In_{0.50}P/GaAs/Ga_{0.73}In_{0.27}As$ (Inv1); inverted (double) metamorphic device $Ga_{0.63}In_{0.37}As/Ga_{0.96}In_{0.04}As/GaAs$ (Inv2) [14].

Today, the industry standard is still the lattice-matched triple-junction solar cell, as similar structures have been developed, produced, and successfully tested for space applications [15–17]. Metamorphic or inverted concepts have not been produced in large quantities yet. Standards for the long-term stability test of such highly strained solar cell structures have to be developed and the devices have to be qualified before mass production. As the space-market is comparatively small, up to now the lattice-matched triple-junction solar cell can be purchased from only a few global suppliers, among these are AZUR SPACE Solar Power GmbH (Germany), Emcore Inc. (USA), and Spectrolab Inc. (USA).

The following section presents an overview about the actual research status on III–V-based multi-junction solar cells at Fraunhofer ISE regarding numerical device simulation as well as device fabrication. Subsequently, the possible next steps in cell design are outlined.

1.2 Research Status

In order to optimize such complex solar cell structures as multi-junction cells, numerical modeling of these devices is indispensible to reduce the number of expensive and time-consuming experiments. At Fraunhofer ISE sophisticated numerical modeling tools are used in the optimization process. The simulation is closely linked to the experimental optimization, concerning feasibility of the semiconductor structures, material quality, and evaluation of the models.

1.2.1 Numerical Simulation

The high number of layers in III–V multi-junction solar cell structures makes a pure experimental optimization very expensive and protracted. An accurate and reliable modeling is desirable to accelerate the optimization procedure considerably. However, a predictive modeling of these sophisticated structures is challenging due to the complex electrical and optical interactions between the different layers and the high number of material parameters and physical phenomena that need to be considered. In recent years, the capabilities of various approaches and tools for the simulation of III–V multi-junction solar cells have greatly improved [18–23]. In the III–V group at Fraunhofer ISE, three different approaches are used. The optimal number of bandgaps and the ideal bandgap combination is evaluated with etaOpt (see Sect. 1.1). We analyze the semiconductor layer structure with the commercially available semiconductor simulation environment Sentaurus TCAD from Synopsys. The grid design is optimized with the circuit simulator LTSpice from Linear Technology Corporation [24]. In the following, a short overview of the status of our modeling capabilities is presented.

Two prerequisites have to be fulfilled to enable realistic simulations with Sentaurus TCAD: first, the necessary models describing the occurring physical phenomena need to be implemented and validated. Of particular importance for III–V multi-junction solar cells are optical interference effects, optical generation [25] and recombination of minority carriers, tunneling effects [20] and carrier transport at hetero-interfaces [26]. Second, material parameters such as optical constants, carrier mobilities, bandgap energies, electron affinity and parameters for radiative, Auger, Shockley-Read-Hall as well as interface recombination are required for each semiconductor layer in the structure. Both prerequisites are satisfactory fulfilled for the materials used in our GaAs single-junction solar cells, as well as in our lattice-matched $Ga_{0.51}In_{0.49}P$/GaAs dual-junction solar cells. However, for other materials, especially those in metamorphic III–V multi-junction solar cells, the lack of material data limits the modeling capabilities [27]. To keep the computational effort within tolerable limits, the smallest two-dimensional symmetry element of the solar cell is modeled, which is constructed by a cut through the layers from cap to substrate perpendicular to the grid fingers. The element covers a width corresponding to half of the finger spacing to ensure that series resistance effects caused by lateral current flow in the device are taken into account.

Figure 1.3 shows a comparison between measured and simulated EQE, reflection and I–V curve of two GaAs solar cells with different material for the front surface field (FSF) layer. The model and material parameters are based on [26]. The good agreement between measurement and simulation proves the validity of the numerical model. Note that all material parameters of the solar cell except for the FSF layer have been identical. The GaInP FSF layer leads to significant absorption in the short wavelength range between 300 nm and therefore reduces the EQE. This underlines the importance of a high bandgap material for the FSF layer.

An additional challenge for the modeling of multi-junction solar is the requirement of a proper and numerically stable model for the tunnel diode, which connects the subcells in series. It was found that nonlocal interband tunnel diode models reproduce measured tunnel diode I–V curves very well in a large voltage range [20, 28]. These models cover the full nonlinearity of the tunneling mechanism and enable the simulation of multi-junction solar cells within semiconductor simulation environments. However, detailed quantum mechanical calculations propose that interband tunneling cannot explain the high currents of typical tunnel diodes for multi-junction solar cells [29]. Rather resonant tunneling through defects homogeneously distributed in the junction is identified as possible tunneling mechanism. Thus, the phenomena of tunneling in III–V multi-junction solar cells need to be investigated further in the future.

Yet, with the nonlocal interband tunnel diode model, we were able to model a lattice-matched dual-junction solar cell with a top cell of $Ga_{0.51}In_{0.49}P$ and a bottom cell of GaAs [21]. The sophisticated device contains an anti-reflection

Fig. 1.3 Comparison between measured and simulated EQE and reflection (**a**) and I–V curve under AM1.5g (**b**) for two GaAs solar cells with different front surface field (FSF) layer material. The devices with an area of 1 cm^2 were designed for operation under AM1.5g ((**a**) reprinted with permission from[27]. Copyright 2010 MIDEM Society)

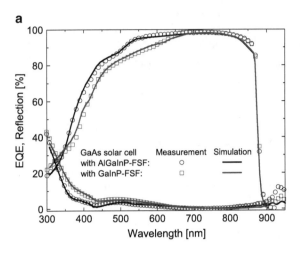

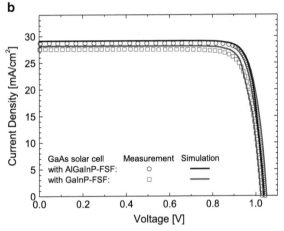

coating of MgF_2/TiO_2, a p-GaAs/n-GaAs Esaki interband tunnel diode, as well as a distributed Bragg reflector composed of 20 alternating layers of $Al_{0.80}Ga_{0.20}As$ and $Al_{0.10}Ga_{0.90}As$. As shown in Fig. 1.4, a good agreement is achieved between experimental and simulated EQE of top and bottom cell, reflection and I–V curve under the AM1.5g spectrum. The slight deviations in the modeled reflection for wavelengths higher than the bandgap of the bottom cell are caused by a minor inaccuracy of the transfer matrix method [25] used for the description of the optical processes. A deviation is also observable in the I–V curve in the range of 0.7 V and about 1.7 V. In contrast to the slightly increasing measured current, the simulated value remains constant. In the real device, such a current decrease can either be caused by a distributed series resistance effect along the grid fingers or by a current leakage at the cell edge. Both effects are not covered in our two-dimensional modeling approach.

The sophisticated numerical model constitutes a quick and cost-efficient tool to study the effect of structural changes on the cell performance. As illustrated in

Fig. 1.4 A good agreement is achieved between simulated and measured EQE and reflection (**a**) and I–V curve (**b**) for the investigated $Ga_{0.51}In_{0.49}P/GaAs$ dual-junction solar cell (reprinted with permission from [21]. Copyright 2008. Wiley)

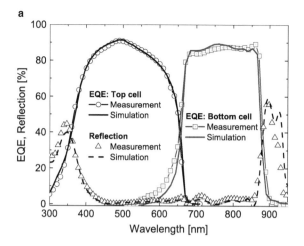

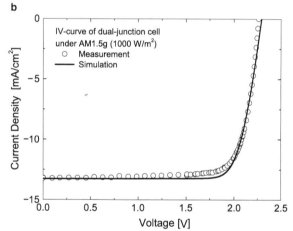

Fig. 1.5, doubling of the tunnel diode thickness strongly reduces the bottom cell's EQE. We explain this by absorption in the GaAs material of the tunnel diode. Due to the stronger absorption of high energy photons, the decrease of the EQE is more pronounced for lower wavelengths. The simulations underline that it is very important to make the tunnel diode as thin as possible if it consists of the same material as the absorber of the lower cell. In most cases, a better option is to use higher bandgap materials for the tunnel diode [30]. A further application of the recently developed numerical solar cell models is design optimization [21, 22, 27, 31].

As shown above, the semiconductor layer structure can be very well modeled with a two-dimensional symmetry element. Yet, for the optimization of the front contact grid such a model is not sufficient. In principle it would be possible to model and simulate a complete solar cell in all three dimensions within the Sentaurus TCAD simulation environment. However, due to the high number of mesh points

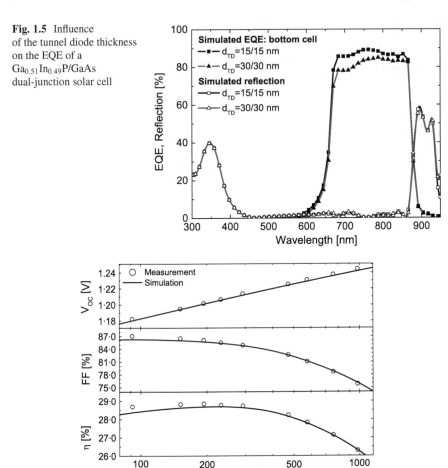

Fig. 1.5 Influence of the tunnel diode thickness on the EQE of a Ga$_{0.51}$In$_{0.49}$P/GaAs dual-junction solar cell

Fig. 1.6 Comparison between measured and simulated fill factor and efficiency for a GaAs concentrator solar cell with an active area of 5 mm². For the simulation the SPICE network model presented by Steiner et al. [23] was used

necessary for a realistic model, the computational effort would be enormous, leading to intolerable computing time of weeks or even months. Therefore, we optimize the front contact separately with an electrical network model. The solar cell is modeled as a network of elementary cells consisting of diodes, resistances, and current sources to model the saturation currents and the photo-generated current. The elementary cells are connected in parallel through ohmic resistances representing, for instance, the lateral conducting emitter layer or the metal fingers. Thereby, a network of electrical components is created, which describes the whole solar cell. The IV-characteristic is calculated with the circuit simulator LTSpice, which uses a Simulation Program with Integrated Circuit Engineering (SPICE) approach. More details about our network model can be found in [23]. Figure 1.6 shows a comparison between measured and simulated I–V parameters as a function of the concentration ratio for a GaAs concentrator solar cell. Measurement and simulation agree well. Note that the experimental cell reaches a high efficiency of 28.8 ± 1.2% at 230 suns under the reference spectrum AM1.5d [32]. The network modeling approach is highly predictive and has successfully been applied for the design of contact grids for different cell structures and various illumination conditions. Together with the etaOpt approach and the simulations with Sentaurus TCAD, a very powerful set of modeling tools is available. These numerical modeling techniques are now well established for supporting the design process of multi-junction solar cells.

1.2.2 Device Fabrication

In order to avoid strain and defects in the crystal structure of a multi-junction solar cell, all III–V compounds are usually grown lattice-matched to the substrate. Consequently, from the commercially available semiconductor substrates, GaAs and Ge are the most suitable for the further growth of (Al)GaInAs- and (Al)GaInP-based compounds. Considering the diagram shown in Fig. 1.2, Ge with its bandgap of 0.66 eV is the obvious choice as bottom cell in a triple-junction structure. Hence, the straightforward lattice-matched approach makes use of a $Ga_{0.50}In_{0.50}P$ top cell, a $Ga_{0.99}In_{0.01}As$ middle cell, and a Ge bottom cell. While the upper cells are commonly deposited via epitaxy, the Ge subcell is established via diffusion of group-V atoms into a p-doped Ge substrate. This kind of structure has been leading to 41.6% conversion efficiency [23]. However, this lattice-matched structure appears to be quite far away from the optimum bandgap configuration (see Fig. 1.2) and is highly current-mismatched. Lower bandgap materials are required for the upper subcells. This can be achieved by increasing the indium content in the top and middle cell. With the almost ideal bandgaps of 1.67 eV for GaInP and 1.18 eV for GaInAs, the metamorphic triple-junction solar cell approach approximates the optimum configuration. However, the higher In-content in the upper subcells also increases the lattice-constant of these materials. The combination of $Ga_{0.35}In_{0.65}P$, $Ga_{0.83}In_{0.17}As$, and Ge causes a high lattice-mismatch of 1.1% between the Ge and the upper two subcells, while the top and middle subcell are lattice-matched to each other. This metamorphic cell structure (Fig. 1.2 , MM2) was developed at Fraunhofer ISE [1, 8, 12, 33].

Figure 1.7 illustrates the difference between the lattice-matched and the metamorphic concept. The larger lattice constant of the upper subcells requires a transition region from the lattice constant of Ge to the lattice constant of $Ga_{0.83}In_{0.17}As$. This transition is realized by a metamorphic buffer structure.

We have developed an optimized step-graded buffer structure made from $Ga_{1-y}In_yAs$ to overcome the high mismatch in the lattices. This buffer increases the lattice constant as required, reduces the amount of residual strain to a minimum of only 6–9%, and avoids the penetration of threading dislocations into the middle and top cell. The degree of strain relaxation of an epitaxial semiconductor layer with cubic lattice is defined as the ratio of the in-plane lattice constant to the lattice

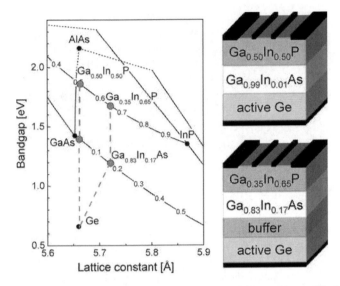

Fig. 1.7 Relation between bandgap and lattice constant for different binary (*black dots*) and ternary (*lines*) semiconductor materials as well as for germanium (*left*). The triple-junction solar cell lattice-matched to Ge features a Ga$_{0.99}$In$_{0.01}$As middle cell and a sGa$_{0.50}$In$_{0.50}$P top cell (*top right* structure). The Ga$_{0.83}$In$_{0.17}$As and the Ga$_{0.35}$In$_{0.65}$P subcell of a metamorphic triple-junction solar cell (*bottom right* structure) are lattice-matched to each other, but have a 1.1% mismatch to the substrate. This mismatch is managed by a buffer structure

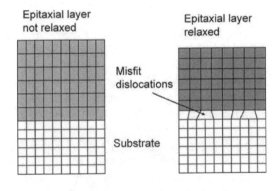

Fig. 1.8 Schematic illustration of a pseudo-morphically strained layer (*left*) grown on a substrate with smaller lattice constant. As a critical thickness is exceeded, the strained epitaxial layer starts to relax by the generation of misfit dislocation (*right*)

constant of the relaxed lattice normalized by the lattice constant of the substrate. Figure 1.8 schematically illustrates the process of strain relaxation of a strained epitaxial layer. Up to a critical thickness, first defined by Matthews et al. [34], the layer is pseudo-morphically strained. As the layer thickness increases beyond the critical thickness, first misfit dislocations are generated within the interface due to bending of pre-existing substrate dislocations. With a further increase of the layer thickness, several generation mechanisms for misfit dislocation set in, which finally lead to a significant strain relaxation [35, 36].

1 Present Status in the Development of III–V Multi-Junction Solar Cells   11

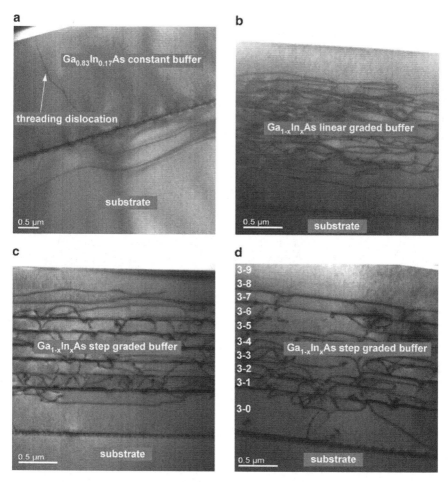

Fig. 1.9 TEM cross-section images from a constant buffer with abrupt change in lattice constant (**a**), a linearly graded buffer with linear change in lattice constant (**b**), a step graded buffer with about seven grading steps (**c**), and a step graded metamorphic $Ga_{1-x}In_x$As buffer with increasing In content from 1% to 17% (3–1 to 3–7) and an overshooting layer (3–8) with 23% In in (**d**). (TEM images measured at the Christian-Albrechts-University in Kiel, Germany)

Figure 1.9 shows TEM cross-section images of different buffer concepts which we have investigated: a constant buffer (a), a linearly graded buffer (b), and two step graded buffers (c and d). In contrast to the graded structures, the constant buffer growth leads to the generation of a large number of threading dislocations which may act as recombination centers in the photoactive region of the multi-junction solar cells. For high performance devices threading dislocation densities below 10^6 cm^{-2} are necessary. According to our experience, this can only be achieved with the graded buffers. Especially the step graded buffer localizes the dislocations generated within the interface regions between the steps and prevents the generation of threading dislocations, penetrating into the photoactive regions of the solar cell structure. This is the reason for the success of these buffer structures.

A more detailed development of the step graded buffers enables the reduction of remaining strain to a minimum of only 6–9%. This is achieved by triggering the strain-relaxation of the upper part of the buffer by introducing an overshooting $Ga_{1-x}In_xAs$ layer with even larger lattice constant than originally intended (x > 0.17). The subsequent growth continues with the lattice constant of $Ga_{0.83}In_{0.17}As$. The degree of strain relaxation can be adjusted by the thickness of the overshooting layer and by the strength of overshooting. Figure 1.9d shows a TEM cross-section image of a buffer consisting of seven $Ga_{1-x}In_xAs$ grading steps from 1% In to 17% In followed by an overshooting layer with 23% In. With this structure, the remaining strain in the following layers is reduced to about 6–9%.

The remaining strain of a metamorphic triple-junction solar cell structure can be reduced considerably. However, in situ collected data show that the 300 μm thick Ge-wafers bend during growth. Figure 1.10 shows the curvature transients from in situ measurement during the growth process for different buffer structures with relaxations between 81% and 94%. The first highly strained layers of the buffer cause a strong increase in curvature. As relaxation sets in, the curvature remains almost constant. The remaining value is proportional to the remaining strain in the structure. Experiments also show that thinner wafers in the range of 150 μm are unsuitable for metamorphic structures, since the even stronger bending of the substrate affects growth on the wafer. Today, different extensions to the step graded buffer concepts are under investigation. Additional blocking layers made from hard materials such as GaInNAs can be deposited over the buffer structure to bend the threading dislocations from growth direction into the growth plane [37]. This allows even less dislocations to penetrate into the photoactive regions of the structure. Adding aluminium to the $Ga_{1-x}In_xAs$ alloy increases its bandgap and hence renders the buffer more transparent. This may be important to increase the current generated by the Ge subcell.

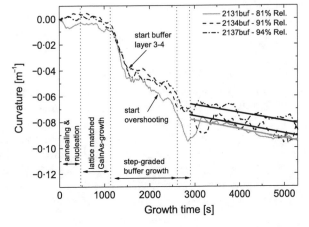

Fig. 1.10 In situ measured curvature of the wafer during growth of different metamorphic triple-junction solar cell structures. The different buffers result in remaining strain between 19% and 6%. This correlates with the remaining values in curvature

In lattice-matched as well as in metamorphic multi-junction solar cells, the individual subcells are interconnected via interband tunnel diodes, which provide a low electric resistance and can be designed transparent for light converted in the following subcell. Tunnel diodes essentially consist of two degenerately doped layers with different polarity. High n-doping levels in the range of 10^{19} cm^{-3} have been achieved with tellurium or silicon in GaInAs or GaInP. Very high p-doping in the range of 10^{20} cm^{-3} is achieved with carbon in AlGaAs [38]. Tunnel current densities above 100 A cm^{-2} have been measured. However, tunnel diodes in metamorphic solar cells need larger lattice constants, which can be realized by adding In to the mentioned compounds, as already done for the subcells. But the carbon doping of such AlGaInAs layers turns out to be more difficult. The commonly used carbon precursor CBr$_4$ cannot be used anymore as it etches indium [38]. Intrinsic C-doping from the alkyl groups of TMAl had been applied, but the doping levels achieved are one order of magnitude lower than with In-free AlGaAs. The use of lattice-mismatched layers in the tunnel diode, such as AlGaAs, generates dislocations, which degrade the solar cell structure. This is why we designed strain compensated tunnel diodes. These structures make use of a highly doped AlGaAs layer, which has a too small lattice constant and causes tensile strain in the tunnel diode structure. In order to reduce this strain, a neighboring layer such as GaInAs or GaInP is grown with a too large lattice constant. This compensates the strain in the structure and avoids the generation of dislocations. Figure 1.11 illustrates this concept with a p-AlGaAs/n-GaInAs tunnel diode. Due to strain compensation, the rest of the structure is not affected by the highly strained tunnel diodes.

As explained above, the subcells of a metamorphic triple-junction solar cell are almost ideally adapted to the AM1.5d spectrum. Figure 1.12 compares the EQEs of such a structure with a lattice-matched triple-junction cell. The lower bandgaps in the metamorphic cell shift the EQEs to higher wavelengths. The favorable bandgap combination together with the improvements regarding the metamorphic buffer and the tunnel diodes are the key points that lead to an efficiency of 41.1% with our metamorphic triple-junction solar cell (Fig. 1.12).

The performance of a concentrator solar cell also strongly depends on the cell design including size, grid structure, and intended operation illumination intensity. Figure 1.13 illustrates how different front side grid layouts with decreasing amount of metallization shift the maximum efficiency to lower concentrations as shading is reduced. Furthermore, the absolute efficiency maximum rises with decreased

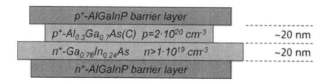

Fig. 1.11 Schematic illustration of a strain compensated tunnel diode. The width of the boxes represents the lattice constant. A compressively strained GaInAs layer is compensated with a tensile AlGaAs layer

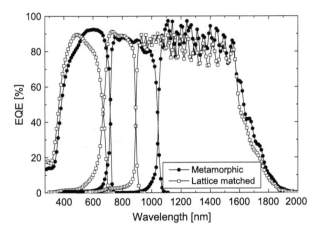

Fig. 1.12 External quantum efficiencies of a lattice-matched $Ga_{0.50}In_{0.50}P/Ga_{0.99}In_{0.01}As/Ge$ as well as a metamorphic $Ga_{0.35}In_{0.65}P/Ga_{0.83}In_{0.17}As/Ge$ triple-junction solar cell

shading. Note that the logarithmic increase of efficiency with illumination intensity is caused by an increase in open circuit voltage and in fill factor. For higher concentration ratios, more metallization is required to carry the high current densities generated. At Fraunhofer ISE, we have developed a network simulation tool to compare and optimize the size and front contact layout for various solar cell structures and different illumination intensities (see Sect. 1.2.1) [23]. Generally a small cell requires less metallization for the same resistive losses as a larger cell and thus achieves higher theoretical efficiencies because of less shading losses. In summary, this section shows that lattice-matched and metamorphic triple-junction solar cell concepts have surpassed the 41% mark and offer even further potential to increase the conversion efficiency toward 45%.

1.3 Next Steps

In recent years, research efforts for the development of III–V multi-junction solar cells with more than three subcells have significantly grown as an increase in cell efficiency is expected [2, 39–41]. It is a well-known fact that the ideal efficiency of a multi-junction solar cell under a certain spectrum increases with the number of pn-junctions [40, 42]. However, multi-junction solar cells are also known to be highly sensitive to changes in the solar spectrum [43, 44]. In terrestrial applications, the spectral distribution of the incident light strongly varies throughout day and year. In addition, the irradiance conditions can change significantly with the intended place of operation. As the sensitivity to spectral variations increases with the number of subcells, it is important to investigate the possible gain in energy production before a decision is made on which number of subcells and which material combination to realize.

Fig. 1.13 Fill factor and efficiency of two metamorphic triple-junction solar cells as a function of the concentration. Different front side metallization, which is shown in the photographs, leads to different maximum positions. The cell in (**a**) is suitable for high concentration ratios well above 1,000 suns, the cell in (**b**) is adapted for 500 suns

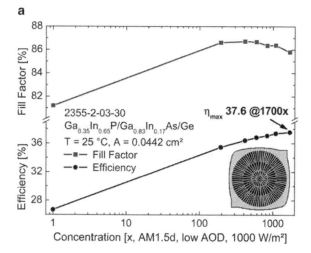

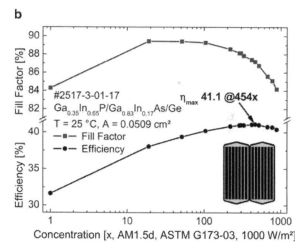

For this purpose, we have developed a theoretical energy harvesting model [45], in which the solar cell is modeled according to the detailed balance model etaOpt (see Sect. 1.1). Here, we investigate the energy harvesting at two locations on Earth with distinct spectral conditions: Solar Village in Saudi Arabia with rather red-rich incident light and La Parguera in Puerto Rico with a higher share of blue light. For these places, measured atmosphere parameters from the AERONET database [46] are used to calculate direct solar spectra with the computer code SMARTS 2.9.5 [47–49]. A discretization of one spectrum per hour was chosen resulting in more than 4,400 spectra for each geographical location. Concerning the operating conditions, we assume a concentration factor of 1,000 suns, which is expected to become standard for future concentrator systems. In addition, a constant cell temperature of 338 K is assumed. From our experiences, this value represents a reasonable average, which is also used by other groups [50]. Based on the simulated spectra the annual sum of the produced energy, which is computed from the produced energy for each day in the year, is calculated and is then referred to the irradiated energy. The overall annual incident energies are 2,599 kWh m^{-2} a^{-1} at Solar Village and 2,849 kWh m^{-2} a^{-1} at La Parguera. The ratio of the energy produced to the irradiated energy defines the energy harvesting efficiency.

III–V multi-junction concentrator solar cells are usually optimized and rated under the reference spectrum AM1.5d. Following this approach, we first calculated the ideal bandgap combination under the reference spectrum (1,000 suns, 338 K) for solar cells with 1–6 subcells. Then the energy harvesting efficiencies of the resulting bandgap designs at Solar Village (Fig. 1.14a) and La Parguera (Fig. 1.14b) were calculated. The results correspond to the left bars in each figure. The right bars indicate the energy harvesting efficiencies that could be realized with the optimal bandgap combination for the intended place of operation. These were determined

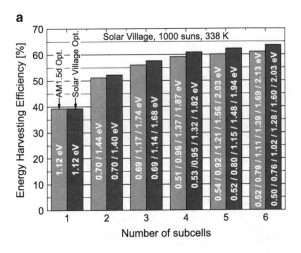

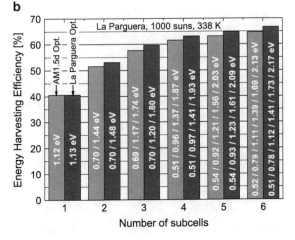

Fig. 1.14 Comparison of the maximum energy harvesting efficiency versus the number of subcells at Solar Village, Saudi Arabia (**a**) and La Parguera, Puerto Rico (**b**). The indicated bandgap combinations result from the optimization of the cell efficiency under the reference spectrum AM1.5d ASTM G173-03 (*left bars*), and of the energy harvesting efficiency for the intended place of operation (*right bars*)

by maximizing the yearly produced energy based on the more than 4,400 spectra for Solar Village or La Parguera, respectively.

It is noteworthy that the energy harvesting efficiency increases with the number of subcells in all cases despite the considered spectral variations throughout day and year. However, the relative gain of adding additional junctions is small for more than four junctions. Compared to the bandgap combinations resulting from the optimization of the cell efficiency under the reference spectrum AM1.5d, a rather strong increase in energy harvesting efficiency can be realized with the optimal bandgap combinations for each location. However, these bandgap combinations go into different directions. At Solar Village, slightly lower bandgaps are favorable due to the low share of blue light at this location. In contrast, higher bandgaps are favorable under the blue-rich spectral conditions of La Parguera. It should be noted that the detailed balance model assumes ideal solar cells. As a rule of thumb 70–80% of the theoretical cell efficiencies can be achieved in practice. However, the realization of good material quality usually becomes more challenging for solar cells with a higher number of subcells due to the greater complexity of the structure and the novel materials involved (see discussion below). Thus, it is still an open question if solar cells with more than four pn-junctions can be realized with sufficient material quality to harvest the small relative gains predicted by our energy harvesting model.

Figure 1.15 summarizes the development roadmap for III–V multi junction solar cell concepts investigated at Fraunhofer ISE. Apart from the already discussed lattice-matched (a) and metamorphic (d) triple-junction solar cells, different

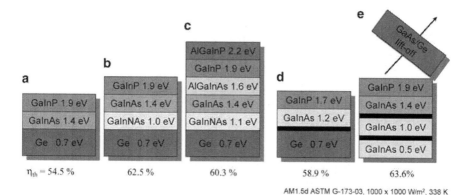

Fig. 1.15 Roadmap for the development of III–V multi-junction solar cells at Fraunhofer ISE and corresponding maximum efficiencies calculated with etaOpt (AM1.5d, 1,000 suns, 338 K). The number of junctions and the efficiency of the lattice-matched GaInP/GaInAs/Ge approach (**a**) can be increased by adding a GaInNAs subcell (**b**). As the performance of this device is limited by the diffusion length in GaInNAs, a six-junction cell has been designed (**c**). An almost ideal bandgap combination is achieved with a metamorphic triple-junction solar cell (**d**). Adding a second buffer leads to a metamorphic quadruple-junction solar cell (**e**). In order to grow as many junctions as possible lattice-matched to the substrate, the structure is grown inverted

Fig. 1.16 Measured internal quantum efficiency of a four-junction (**a**) and a six-junction solar cell (**b**), which were produced at Fraunhofer ISE. The subcells' short circuit current densities are also indicated

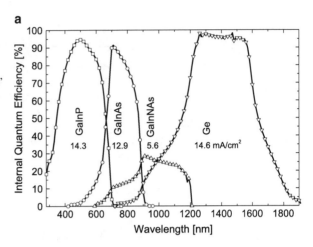

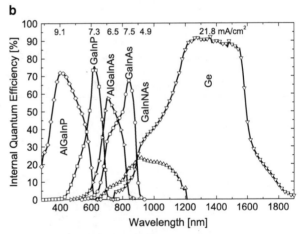

concepts with more than three subcells are under investigation. The concepts can be divided into two groups. The first group uses materials that can be grown lattice-matched on germanium substrates (a–c). The second group includes buffer structures to enable the use of metamorphic materials (d–e). For a lattice-matched approach on a Ge substrate, a strong increase in efficiency is expected from the inclusion of a fourth junction with a bandgap of about 1 eV (b), which may be realized with the quaternary alloy GaInNAs [51]. The (a) diagram in Fig. 1.16 shows the internal quantum efficiency (IQE) of such a device produced at Fraunhofer ISE. The GaInNAs subcell limits the current due to the quite low minority diffusion lengths, which are very likely caused by the formation of nitride chains. A lot of work is done to solve this deficiency. One promising method is to use thermal annealing to dissolve the nitride chains [52].

Another way to get around the low diffusion length is to increase the number of subcells. In a six-junction device, each subcell produces a smaller current than in a quadruple cell and hence the subcells are thinner. Thus, shorter diffusion lengths are acceptable. A six-junction cell may be realized by adding AlGaInP and AlGaInAs subcells to the stack (Fig. 1.15c). The graph (b) in Fig. 1.16 shows the IQE of such a six-junction solar cell. The Ge bottom cell has a quite high current due to the not ideal bandgap combination of this device. Smaller bandgaps for the lower cells would be favorable. A way to realize such combinations is to use more than one lattice-transition buffer and to grow the stack inversely (Fig. 1.15e).

This chapter showed that III–V triple-junction solar cells have reached efficiencies of 41.6% under concentrated sunlight [2]. We expect that the record efficiency of these devices will soon reach above 43%. For even higher efficiencies toward 50% III–V multi-junction solar cells with more than three subcells are promising.

References

1. W. Guter, J. Schöne, S.P. Philipps, M. Steiner, G. Siefer, A. Wekkeli, E. Welser, E. Oliva, A.W. Bett, F. Dimroth, Appl. Phys. Lett. **94**(22), 223504 (2009)
2. R.R. King, A. Boca, W. Hong, D. Larrabee, K.M. Edmondson, D.C. Law, C.M. Fetzer, S. Mesropian, N.H. Karam, in *Conference Record of the 24th European Photovoltaic Solar Energy Conference and Exhibition*, Hamburg, Germany, 21–25 Sep 2009, pp. 55–61
3. H. Lerchenmüller, A.W. Bett, J. Jaus, G. Willeke, in *Conference Record of the 3rd International Conference on Solar Concentrators for the Generation of Electricity or Hydrogen*, Scottsdale, AZ, USA, 1–5 May 2005, p. 6.
4. I. García, I. Rey-Stolle, B. Galiana, C. Algora, Appl. Phys. Lett. **94**(5), 053509 (2009)
5. W. Shockley, H.J. Queisser, J. Appl. Phys. **32**(3), 510 (1961)
6. D.J. Friedman, J.F. Geisz, S.R. Kurtz, J.M. Olson, J. Cryst. Growth **195**(1–4), 409 (1998)
7. K.W.J. Barnham, I. Ballard, J.P. Connolly, N.J. Ekins-Daukes, B.G. Kluftinger, J. Nelson, C. Rohr, Phys. E-Low-Dimensional Syst. Nanostructures **14**(1–2), 27 (2002)
8. A.W. Bett, C. Baur, F. Dimroth, J. Schöne, Mater. Photovoltaics **836**, 223 (2005)
9. G. Létay, A.W. Bett, in *Conference Record of the 17th European Photovoltaic Solar Energy Conference and Exhibition*. WIP-Renewable Energies, Munich, Germany, 22–26 Oct 2001, pp. 178–180
10. G. Létay, A.W. Bett, etaOpt, www.ise.fraunhofer.de/downloads/software/etaOpt.zip/view (2001)
11. R.R. King, D.C. Law, K.M. Edmondson, C.M. Fetzer, G.S. Kinsey, H. Yoon, R.A. Sherif, N.H. Karam, Appl. Phys. Lett. **90**(18), 183516 (2007)
12. F. Dimroth, Phys. Status Solidi C – Curr. Top. Solid State Phys. **3**(3), 373 (2006)
13. M. Yamaguchi, T. Takamoto, K. Araki, Sol. Energ. Mater. Sol. Cell. **90**(18–19), 3068 (2006)
14. J.F. Geisz, D.J. Friedman, J.S. Ward, A. Duda, W.J. Olavarria, T.E. Moriarty, J.T. Kiehl, M.J. Romero, A.G. Norman, K.M. Jones, Appl. Phys. Lett. **93**(12), 123505 (2008)
15. R.R. King, C.M. Fetzer, D.C. Law, K.M. Edmondson, H. Yoon, G.S. Kinsey, D.D. Krut, J.H. Ermer, P. Hebert, B.T. Cavicchi, N.H. Karam, in *Conference Record of the 2006 IEEE 4th World Conference on Photovoltaic Energy Conversion*, Waikoloa, HI, 17–12 May 2006, pp. 1757–1762
16. P.R. Sharps, A. Comfeld, M. Stan, A. Korostyshevsky, V. Ley, B. Cho, T. Varghese, J. Diaz, D. Aiken, in *Conference Record of the PVSC: 2008 33rd IEEE Photovoltaic Specialist Conference*, San Diego, USA, 11–16 May 2008, pp. 2046–2051
17. A.W. Bett, F. Dimroth, W. Guter, R. Hoheisel, E. Oliva, S.P. Philipps, J. Schöne, G. Siefer, M. Steiner, A. Wekkeli, E. Welser, M. Meusel, W. Köstler, G. Strobl, in *Conference Record of the 24th European Photovoltaic Solar Energy Conference and Exhibition*, Hamburg, Germany, 21–25 Sep 2009, pp. 1–6

18. S. Michael, Sol. Energ. Mater. Sol. Cell. **87**(1–4), 771 (2005)
19. Z. Li, G. Xiao, Z. Li, High Low Concentration Sol. Electric Appl. **6339**, 633909 (2006)
20. M. Hermle, G. Létay, S.P. Philipps, A.W. Bett, Progr. Photovoltaics **16**(5), 409 (2008)
21. S.P. Philipps, M. Hermle, G. Létay, F. Dimroth, B.M. George, A.W. Bett, Phys. Status Solidi-Rapid Res. Lett. **2**(4), 166 (2008)
22. M. Baudrit, C. Algora, Phys. Status Solidi a-Appl. Mater. Sci. **207**(2), 474 (2010)
23. M. Steiner, S.P. Philipps, M. Hermle, A.W. Bett, F. Dimroth, Prog. Photovoltaics **19**(1), 73 (2011)
24. LTSpice, Switcher CAD III/LT Spice. Tech. rep., Linear Technology Coorporation (2007)
25. G. Létay, M. Breselge, A.W. Bett, in *Conference Record of the 3rd World Conference on Photovoltaic Energy Conversion*, Osaka, Japan, 11–18 May 2003, pp. 741–744
26. G. Létay, M. Hermle, A.W. Bett, Prog. Photovoltaics **14**(8), 683 (2006)
27. S.P. Philipps, W. Guter, M. Steiner, E. Oliva, G. Siefer, E. Welser, B. George, M. Hermle, F. Dimroth, A.W. Bett, Informacije MIDEM- J. Microelectron. Electron. Compon. Mater. **39**(4), 201 (2010)
28. M. Baudrit, C. Algora, in *Conference Record of the PVSC: 2008 33rd IEEE Photovoltaic Specialist Conference*, San Diego, USA, 11–16 May 2008, pp. 576–580
29. K. Jandieri, S.D. Baranovskii, O. Rubel, W. Stolz, F. Gebhard, W. Guter, M. Hermle, A.W. Bett, J. Appl. Phys. **104**(9), 094506 (2008)
30. T. Takamoto, M. Kaneiwa, M. Imaizumi, M. Yamaguchi, Progr. Photovoltaics **13**(6), 495 (2005)
31. S.P. Philipps, M. Hermle, G. Létay, W. Guter, B.M. George, F. Dimroth, A.W. Bett, in *Conference Record of the 23rd European Photovoltaic Solar Energy Conference and Exhibition*, Valencia, Spain, 1–5 Sep 2008, pp. 90–94
32. M.A. Green, K. Emery, Y. Hishikawa, W. Warta, Progr. Photovoltaics **17**(5), 320 (2009)
33. F. Dimroth, R. Beckert, M. Meusel, U. Schubert, A.W. Bett, Prog. Photovoltaics **9**(3), 165 (2001)
34. J.W. Mattews, A.E. Blakeslee, J. Cryst. Growth **27**, 118 (1974)
35. D.D. Perovic, D.C. Houghton, in *Microscopy of Semiconducting Materials* ed. by A.G. Cullis, A.E. StatonBevan. Conference on Microscopy of Semiconducting Materials, 20–23 Mar 1995, pp. 117–134 (Oxford, England, 1995)
36. F.K. Legoues, B.S. Meyerson, J.F. Morar, P. Kirchner, J. Appl. Phys. **71**(9), 4230 (1992)
37. J. Schöne, E. Spiecker, F. Dimroth, A.W. Bett, W. Jäger, Appl. Phys. Lett. **92**(8), 081905 (2008)
38. W. Guter, F. Dimroth, J. Schöne, S.P. Philipps, A.W. Bett, in *Conference Record of the 22th European Photovoltaic Solar Energy Conference and Exhibition*, Milan, Italy, 3–7 Sep 2007, pp. 122–125
39. D.J. Friedman, J.F. Geisz, A.G. Norman, M.W. Wanlass, S.R. Kurtz, in *Conference Record of the 2006 IEEE 4th World Conference on Photovoltaic Energy Conversion*, Waikoloa, Hi, 07–02 May 2006, pp. 598–602
40. F. Dimroth, S. Kurtz, MRS Bull. **32**(3), 230 (2007)
41. M. Yamaguchi, K.I. Nishimura, T. Sasaki, H. Suzuki, K. Arafune, N. Kojima, Y. Ohsita, Y. Okada, A. Yamamoto, T. Takamoto, K. Araki, Sol. Energ. **82**(2), 173 (2008)
42. S. Kurtz, D. Myers, W.E. McMahon, J. Geisz, M. Steiner, Prog. Photovoltaics **16**(6), 537 (2008)
43. M. Meusel, R. Adelhelm, F. Dimroth, A.W. Bett, W. Warta, Progr. Photovoltaics **10**(4), 243 (2002)
44. S. Kurtz, J.M. Olson, P. Faine, Sol. Cell. **30**(1–4), 501 (1991)
45. S.P. Philipps, G. Perharz, R. Hoheisel, T. Hornung, N.M. Al-Abbadi, F. Dimroth, A.W. Bett, Sol. Energ. Mater. Sol. Cell. **94**(5), 869 (2010)
46. International AERONET Federation, AERONET (AErosol RObotic NETwork) program, aeronet.gsfc.nasa.gov (2009)
47. C. Gueymard, Sol. Energ. **71**(5), 325 (2001)
48. C. Gueymard. Simple Model of the Atmospheric Radiative Transfer of Sunshine (SMARTS), Version 2.9.5, www.nrel.gov/rredc/smarts (2009)

49. C. Gueymard, SMARTS: Simple Model of the Atmospheric Radiative Transfer of Sunshine: Algorithms and Performance Assessment. Tech. Rep. FSEC-PF-270-95, Florida Solar Energy Center, 1679 Clearlake Rd., Cocoa, FL, USA (1995)
50. G.S. Kinsey, K.M. Edmondson, Prog. Photovoltaics **17**(5), 279 (2009)
51. M. Kondow, K. Uomi, A. Niwa, T. Kitatani, S. Watahiki, Y. Yazawa, Jpn. J. Appl. Phys. Part1 Regular Papers Short Notes Rev. Papers **35**(2B), 1273 (1996)
52. K. Volz, T. Torunski, O. Rubel, W. Stolz, J. Appl. Phys. **104**(5), 053504 (2008)

Chapter 2
The Interest and Potential of Ultra-High Concentration

Carlos Algora and Ignacio Rey-Stolle

Abstract The benefits of the ultra-high concentration (>1,000 suns) are shown in terms of cost reduction, raw material availability and efficiency increase. The main challenges for the operation at such high concentrations, namely, (a) to keep a low series resistance; (b) to manufacture tunnel junctions with high peak currents and low series resistance; (c) to design and build advanced characterisation methods and tools; and (d) to increase the reliability of the devices are addressed and solutions for them are proposed. As examples of success in manufacturing multijunction solar cells at ultra-high concentration, we have developed and manufactured a GaInP/GaAs dual-junction device, which exhibits an efficiency of 32.6% for a concentration range going from 499 to 1,026 suns. This efficiency is the world record efficiency for a dual-junction solar cell. Besides, the efficiency is still as high as 31% at 3,000 suns. We have extended this strategy to lattice-matched GaInP/Ga(In)As/Ge triple junction solar cells. At the current state of development, such cells show an efficiency of 36.2% at 700 suns. The theoretical optimisation shows that an efficiency well over 40% at 1,000 suns is achievable.

2.1 Motivation of Ultra-High Concentration

III–V multijunction solar cells (from now on, MJSCs) have achieved the highest conversion efficiency of any photovoltaic device and have much room for improvement. However, MJSCs are very expensive. In order to compensate this disadvantage, concentrator optical systems are used to illuminate the solar cells. By concentrating the light, smaller solar cells are required so their high cost is counterbalanced by the low cost of optics making the resulting technology

---

C. Algora (✉)
Instituto de Energía Solar, Universidad Politécnica de Madrid, Avenida Complutense 30, 28040 Madrid, Spain
e-mail: algora@ies-def.upm.es

economically feasible. For this starting technology, one of the open aspects is the determination of the most recommendable concentration range. Currently, most companies developing MJSC-based concentrator systems have chosen a concentration level of around 500 suns. However, in many cases the selection of the concentration level does not derive from a deep analysis of the implicit technological trade-offs but from the direct integration of standard available solar cells, optics, etc. For example, the more widespread concentrator optics is the Fresnel lens (one-stage optics), which concentrates light at levels around 500 suns. The operation at concentrations much higher, namely 1,000 suns or more requires the use of two-stage optics (the so-called primary and secondary optic elements) whose availability is less extended. Simultaneously, the record efficiencies of triple junction solar cells have been historically achieved within the range going from 200–500 suns: 40.8% at 326× by NREL [1], 41.1% at 454× by Fraunhofer Institute [2] and 41.6% at 364× by Spectrolab [3]. In this sense, the most recent efficiency record achieved by Spire although uses a different approach (bifacial epitaxy) again peaks its efficiency 42.3% at 406 suns.[1] Besides, in most cases, the efficiency dramatically drops at concentrations of 1,000 suns or higher. As a consequence, the solar cell manufacturers have developed their triple junction solar cells for optimum performance within the same concentration range (see for example [4–6]).

It is important to remark that the need of a good performance of MJSCs at concentrations about 1,000 suns (or higher) is not only a strategic issue (as we will show in this section) but it is a need derived from the fact that real optical concentrators do not produce uniform illumination spots on the cell. On the contrary, most optical concentrators exhibit peak irradiances significantly higher than the nominal concentration [7]. Therefore, it is required that concentrator MJSCs have a soft efficiency decrease after their maximum efficiencies. In the opposite case, the portion of the cell illuminated with an intensity higher than the nominal will have a much lower efficiency resulting in a poor performance of the whole cell. Figure 2.1 shows champion cell efficiencies for single, dual and triple junction solar cells. It is clear the very different performance of cells after their peak efficiency. For example, the record dual-junction cell exhibits better performance than several record triple junction cells beyond 1,000 suns. So, this dual-junction cell could result in a better efficiency in several concentrator modules for operation at nominal concentrations close to 1,000 suns.

Besides that need, the main widespread reasons for using concentrations higher than 500× are: (a) concentrator systems based on available silicon cells operating at 200–400× would have a much lower cost. Even silicon flat panel-based installations have already reached a price below €2.0/Wp which clearly beats the €3/Wp expected cost of concentration photovoltaic (CPV) systems operating at 500×; (b) material availability whose first consequence has been the increase in the raw material costs for the manufacture of solar cells. The most severe limitation of

---

[1] Nowadays, the new efficiency record has been achieved by Solar Junction (April 2011). Its cell measured a peak efficiency of 43.5% at 418 suns.

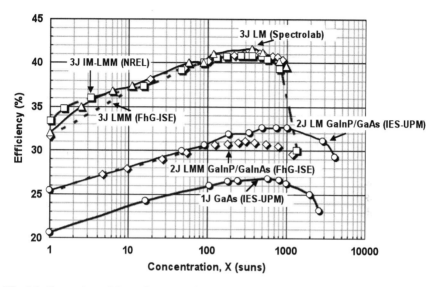

Fig. 2.1 Comparison of the performance of the best III–V concentrator solar cells. The number of junctions (from 1J to 3J), the type of structure (Lattice Matched LM, Lattice Mismatched LMM or Inverted Metamorphic IM) and the lab or company which manufactured the cell are also indicated

these raw materials could be the availability of germanium substrates, which is currently the preferred one for MJSCs [8, 9], though this is still under discussion. Other materials used in the MJSC manufacturing such as TMGa for the MOVPE growth has increased its price as a consequence of the rapid penetration of LED backlighting modules in LCD TVs, which absorbs most of the production of MOVPE consumables to date [10]. Therefore, an increase in the concentration level would relax these aspects; (c) a reduced cost of the whole concentration system because of the size reduction of the most expensive element, the MJSC.

On the other hand, the reasons argued by many people for not going to higher concentrations than 500× are: (a) such high concentrations usually lead to chromatic aberration in the optics [11]; (b) such high concentrations raise doubts about cooling, series resistance and stability of MJSC [11]; and (c) although the cost impact of the MJSCs is reduced, an increase on cost of other elements such optics and tracking appears because of the need of a higher pointing accuracy. Because the discussion on these subjects is not over yet, it is necessary to add more arguments and to clarify several topics which are wrongly assumed though widely used.

Solar cells, optics, modules and trackers are the main components of CPV systems. In our opinion, the pursued characteristics of these components should be:

- Solar cells: Very efficient although expensive.
- Optics: Very efficient and with high acceptance angle.
- Module: Efficient heat extraction and cheap assembling process.
- Tracker: Precise pointing, low acceptance angle losses and low cost.

Perhaps, the most surprising, among the aforementioned characteristics, is the possibility of using very efficient and expensive solar cells as far as they operate at high enough concentration. If the above characteristics are met, our cost calculations showed that an electricity cost below c €3/kWh for a place like Madrid (Spain) and a nominal cost for the whole grid-connected installation below €1/Wp could be attained [12]. If these cost figures were achieved, CPV would become a real breakthrough. Following the cost forecast of Luque [13], PV could supply an important part of the world electricity in this century (23% by 2025 and 34% by 2050) provided that this long sought-for breakthrough occurs. The influence of concentration on cost was shown in [7]. From that reference, many cases can be derived. For example, a complete installation based on 30% efficient solar cells operating at 1,000 suns would have a cost somewhat higher than €3/Wp after a cumulated production of 10 MWp, which would be the same cost than that of an installation based on 40% efficient MJSCs operating at 400 suns. Although these costs are now rather obsolete (and will be updated below), the general trend shown in [7] is that the hierarchy of factors on cost are first, learning, second, concentration and third, efficiency. Therefore, the key factor for reducing the cost of CPV will be by means of learning and production optimisation but while a significant cumulated production is achieved, the running CPV installations should take into account the efficiency-concentration pair because concentration appears as a factor governing cost more severely than efficiency.

Another reason for the operation at 1,000 suns also derives from our experience in close proximity to the manufacture of complete CPV prototype modules which started in 1996. First, we did the proof of concept of proper operation at 1,000 suns by using an RXI optical concentrator and a GaAs solar cell [14]. After this experience, in 1999, we participated in the industrialisation of the concept. The work was supported by the European Commission within the project called INFLATCOM: INdustrialisation of ultra-FLAT COncentrator Module of high efficiency. At the beginning of the INFLATCOM project, a fully commercial photovoltaic system with an objective cost of €2.8/Wp assuming a nominal concentration of 1,000 suns and 10 MWp of cumulated production was stated. Nevertheless, at the end of the project, a cost of €4.38/Wp was envisaged for the same production volume. Several unpredicted additional costs arose and the operation at 1,000 suns appeared to be a key factor in order to partially offset the unexpected extra costs. If the designed concentration had been 500 suns, the envisaged cost would result around €7–10/Wp. Unfortunately, we think this has been and is being a common situation for many companies entering the CPV market. From now on, we will call Ultra High Concentration Photovoltaic (UHCPV) to concentrations about 1,000 suns and higher in opposition to High Concentration Photovoltaic (HCPV) typically assigned to concentrations about 500 suns and something higher and to the more general acronym CPV. But there are not only strategic (and then, subjective) reasons for the operation at UHCPV but also incontestable ones such as the MJSC efficiency increase. The detailed balance theory shows that the efficiency of a solar cell increases as a function of concentration as a result of the variation of its open circuit voltage [15]:

$$V_{oc} = \frac{E_G}{q} - \frac{kT}{q} \cdot \ln\left[\frac{A_0 E_G^2 E(W)}{J_L A(W)}\right] + \frac{kT}{q} \cdot \ln\left[\frac{X}{r(\theta)}\right], \quad (2.1)$$

where X is the concentration level, $r(\theta)$ is the reflectivity for the cone of light with an angle, θ, which impinges the cell from the concentrator, $A(W)$ and $E(W)$ are the average values for the thickness (W) of absorbance and emittance, respectively, and A_0 is a constant. The rest of symbols have their usual meaning. $A(W) > E(W)$ because of the different temperatures of the equivalent black-body spectra for incident and emitted radiation. Thus, V_{oc} can be higher than for $A = E = 1$, usually referred as the limiting efficiency case. An interesting remark is that because V_{oc} depends on the $X/r(\theta)$ ratio, the same effect can be obtained by increasing X than by decreasing $r(\theta)$. This means that the maximum efficiency (i.e, V_{oc}) can be reached at one-sun operation with a highly angularly selective solar cell or with an isotropic cell at a concentration of 46,050 (for a refraction index unity) or with any other combination in between keeping $X/r(\theta) = 46{,}050$ [15].

This is the theory in the ideal case of a solar cell with only radiative recombination, infinite mobility, zero series resistance, etc. However, it is well known that the main limitation of solar cells under concentration is the series resistance, which is the responsible of the efficiency decrease once the increase of open circuit voltage with concentration cannot counterbalance the decrease in fill factor. Therefore, the consideration of series resistance and the ways to minimise its deleterious effect are of key importance and will be analyzed in Sect. 2.3.

Section 2.4 will be devoted to a key structure of the MJSC for operation at UHCPV – the tunnel junction – and its performance when the impinging illumination on the cell is not uniform. Because at UHCPV the light power density impinging the MJSC is about 1 MWp m^{-2} or higher, an important aspect is the heat extraction. This subject has been extensively considered elsewhere [7]. Basically, the heat removal at UHCPV (>1 MWp m^{-2}) is a question of the solar cell size. As it was shown, for solar cell sizes below about 10 mm^2 no active cooling is required and a temperature increase well below 50°C over the ambient temperature can be obtained.

Once the solutions for the challenges related to series resistance, tunnel junction (TJ) and heat extraction are stated, the ways for increasing the MJSC efficiency will be addressed in Sect. 2.5 (Increase of the MJSC efficiency at ultra-high concentration). The accurate illumination I-V characterisation of the MJSC is a tricky subject, which is properly carried out by only a few specialised laboratories and research centres. However, the illumination I–V curve supplies information about the performance of the whole solar cell. Just the quantum efficiency measurement relies on some aspects of the individual performance of each sub-cell within the whole MJSC. These limitations are of key importance, because the R&D process for improving the MJSC efficiency requires to know the complete behaviour of each sub-cell. In order to face this challenge, several characterisation methods are described in Sect. 2.6 (Advanced characterisation).

Finally, the challenge of the proper operation of MJSCs at UHCPV for the scheduled period of time, i.e., their reliability is analyzed in Sect. 2.7 (Reliability). Preliminary results indicate that III-V solar cells are robust devices.

2.2 Cost Calculation Update

Figure 2.2 shows the cost of a complete PV installation (including modules, tracking, inverter, land, etc.) as a function of the MJSC efficiency and concentration. As it was stated in the former section, the cost analysis carried out in [7] is now updated in order to include some rapidly changing costs. As can be seen, the cost reduction resulting from the increase in the solar cell efficiency when going from 35 to 50% at a given concentration is about €0.5/Wp. Both the thicker lines only consider the efficiency increase for the cost reduction. Such high efficiencies can be achieved by three or more junctions of III–V semiconductors grown on germanium substrates. The analysis assumes a cost of €80 for 4-in. Ge substrates. The cost reduction thanks to the operation at 1,000 suns is also shown. As can be seen, a complete installation based on MJSCs of a given efficiency operating at 1,000 suns will result in the same cost than an installation with MJSCs with an efficiency 6–8%

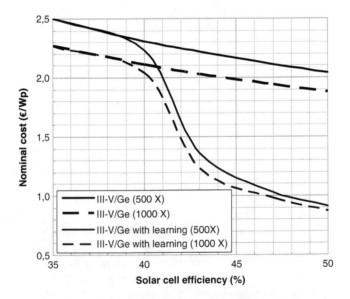

Fig. 2.2 Nominal cost of a complete MJSC-based PV plant as a function of the solar cell efficiency operating at concentrations of 500 and 1,000 suns. III–V MJSCs on Ge substrates are considered. *Thicker lines* represent the evolution of cost as a function of the solar cell efficiency only. *Thinner lines* assume an additional reduction of cost because of "learning" in the mass-production of all components of the CPV installation. In this last case, the evolution of the curve is not representative (transient period) and only the initial and final costs must be taken into account

(absolute) higher but operating at 500 suns. However, the evolution presented by both thicker lines in Fig. 2.2 is not fully realistic because for the achievement of commercial MJSC efficiencies close to 50%, a time of about 5–10 years is required. In such period, it is expected that additional cost reductions in all elements of the CPV system will occur because of learning. For example, in the 2008–2010 period, the cost of the inverters has been reduced from about €1/Wp to about a third. Correspondingly, the tracking cost has decreased from about €2/Wp to less than 1. These huge cost reductions are the consequence of the tremendous growth of the photovoltaic market in the last years [16]. Therefore and in order to consider this type of cost decrease, a new cost calculation is presented in Fig. 2.2 by means of the thinner lines. As can be seen, during the time required to increase the MJSC efficiency from about 40–50%, the cost reduction by learning would allow to reach a cost close to about €0.9/Wp. The optoelectronic learning coefficient, 0.52, has been assumed for related components and processes such as solar cells, encapsulating of cells while the photovoltaic learning coefficient, 0.27, has been assumed for the corresponding components and processes such as structure, tracking, inverter [12]. It must be remarked that the evolution of the curve (transient period) is not accurate and only the initial and final costs must be taken into account. This is because there are many unknowns about the evolution of several aspects. For example, an important cost decrease will be related to the transition in the manufacturing of MJSCs from 4-in. Ge substrates to 6-in. It is expected that this transition to 6-in. wafers of the main manufacturers will be completed by 2013 but delays can appear.

2.3 Series Resistance

The most limiting parameter of a solar cell for its proper operation at UHCPV is its series resistance. The case of MJSCs is especially complex since in addition to the usual contributions derived from the front metal grid, semiconductor layers and specific contact resistances, the impact of TJs must be carefully taken into account. Therefore, it is compulsory to have a suitable modelling of the whole MJSC able to guide the subsequent technological development.

Traditionally, series resistance has been evaluated or simulated by means of models based on lumped parameters, which consider one or two diodes for the recombination, one series resistance, one shunt resistance and one current source. Because of the high current densities achieved in concentrator solar cells and the actual concentrator set-up characteristics, three-dimensional (3-D) or at least distributed models are required [18]. In the mid-1980s, one of the first attempts for simulating solar cells with a distributed model was presented [19]. Since then, several authors have presented different distributed models to explain the different series resistance losses occurring in solar cells but none of them linked together all of the main effects, till the proposal of a 3-D distributed model for high-concentrator solar cells based on elemental units made up of electrical circuits [20]. In that 3-D

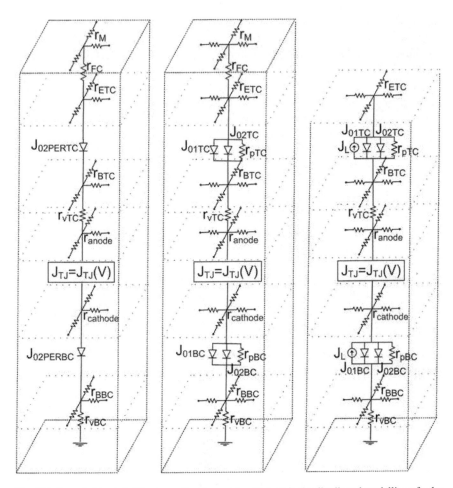

Fig. 2.3 From *left* to *right* it is shown the elemental units used in the distributed modelling of solar cells corresponding to illuminated areas, dark areas, both being inner regions; and perimeter of a dual-junction solar cell containing a TJ [17]

distribution model, the recombination mechanisms are dealt with in detail, paying special attention to the perimeter properties. No ohmic effect is omitted and shunt resistances are also included (see Fig. 2.3). The model also allows the simulation of the external connections made on the solar cell bus bar. The main result of these studies is that for concentration operation the models using lumped parameters were not able to accurately reproduce the solar cell performance. For example, the parameter extraction from the traditional two-diode lumped model in [20] revealed mistakes leading to an overestimation of J_{01} and series resistance components and also its unavailability to simulate the solar cell for a wide range of concentrations. With the distributed model presented in [20], the advantages of treating each part of the solar cell independently without lumped parameters was shown and a correct parameter extraction was carried out to a concentrator single junction GaAs solar cell within a wide concentration range (1–2,000×).

The use of the 3D distributed model together with the one-diode lumped model and the numerical simulation (TCAD software, ATLAS© from Silvaco) has been applied to a GaInP/GaAs dual junction solar cell in order to compare the three different models [21]. The reproduction of the I–V curves was achieved with reasonable accuracy by the one-diode model in a limited concentration range. The calculation of the I–V curves at high concentration with numerical models showed the need of improvement since it exhibited a discrepancy between experimental and modeled data. On the contrary, within the whole concentration range investigated (1–4,100 suns) the 3-D distributed model yielded excellent results.

The power of the distributed models in the determination of the impact of the different series resistance components has been also demonstrated. For example, the origin of the differences between the dark and illumination ohmic losses has been presented in [22], showing that the most important series resistance components for the dark I–V curve are those related to the vertical current flow, such as the vertical resistance and the front contact-specific resistance, while the emitter sheet resistance and the metal sheet resistance have little impact on the dark I–V measurements, but they have a key role under concentrated illumination. The distributed model is also required for the optimisation of the grid design. For example, in [23] once a given front grid geometry is decided, efficiency was calculated by varying the front contact specific resistance and the metal sheet resistance for 500×, 1,000× and 2,000× as a function of the shadowing factor. The optimum number of fingers was found for each situation. Again, differences were found between the optimum design produced by the lumped two-diode model and the distributed one. However, when the need of the distributed model appears as essential is when real operation of MJSCs is considered. The most important situations appearing in real operation are [18]: (a) different illumination spectra from the standard ones, as a consequence of light going through a given optics; (b) inhomogeneous illumination on the solar cell because of the focusing of light; (c) light impinging the cell within a cone, as a consequence of the different areas of the optics and solar cell; (d) chromatic aberration and (e) temperature gradients: at the horizontal plane, they appear as a consequence of the inhomogeneous illumination while at the vertical plane, as a consequence of the different light absorption at the different semiconductor layers together with the joule effect produced by the vertical current flow. All of these situations also affect the series resistance effects.

The 3D distributed models have already shown their capabilities in some of these aspects. For example, the effect of non-uniform light profiles, produced by a practical concentrator optics, on the performance of a concentrator GaInP top cell has been studied (see Fig. 2.4) [24]. Its front grid design was also optimised even taking into account the asymmetries of the light distribution that can appear due to a possible misalignment of the tracker. Finally, the potentialities of this modeling for the analysis of distributed effects on MJSCs were briefly explored by analyzing the impact of the chromatic aberration on the performance of a double junction GaInP/GaAs concentrator solar cell .

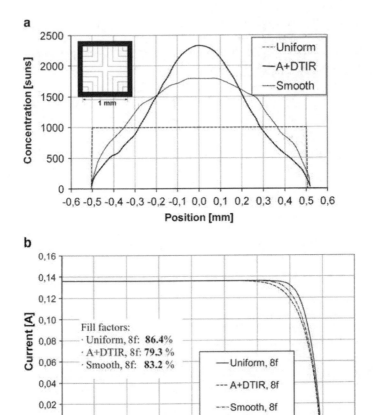

Fig. 2.4 Uniform, A+DTIR and smooth Light profiles studied (*top*) and the resulting I–V curves of a single junction GaInP concentrator solar cell illuminated with them (*bottom*). The integrated light power is kept constant to 1,000× (Reprinted with permission from [24]. Copyright 2008. IEEE)

In the achievement of the world record efficiency dual-junction solar cell (whose analysis will be carried out in Sect. 2.5 of this chapter), the control of the series resistance (achieved in a great extent by the use of the 3D distributed model) was of key importance [26] An efficiency of 32.6% at 1,000 suns was obtained and it still remained at 31% at 3,000 suns with a shadowing factor for the front grid of only 2.7%. From its use for the optimisation of record dual junction cell, the distributed model has been improved in order to allow the analysis, among others, of the influence of the busbar width (also considering the influence of the perimeter recombination) and the post-growth technological imperfections [25]. Figure 2.5 represents an example of this kind of studies where the impact of the number of external connections (wirebonds) placed on the busbar is analyzed. It was found

2 The Interest and Potential of Ultra-High Concentration

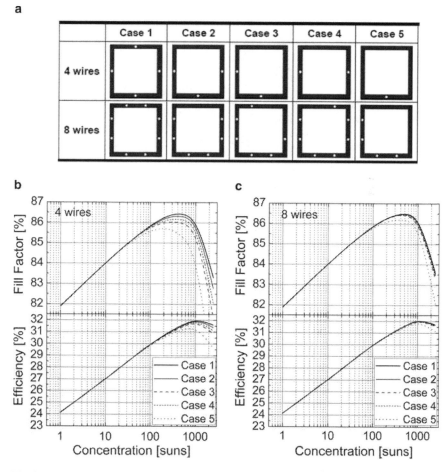

Fig. 2.5 (a) Wire-bonding defect cases studied, for the 4- and 8-wire configurations. (b) and (c) Simulation results of the GaInP/GaAs dual-junction solar cell concentration response for the different cases of wire-bonding defects analyzed [25]

that, for the current level managed by the GaInP/GaAs dual-junction solar cell, the efficiency obtained at 1,000× with the 4- and 8-wire approaches is virtually identical (see Fig. 2.5). Moreover, in both cases, defects in which one wire is missing do not produce an important efficiency drop. If defects involving more wires are to be expected, then an 8-wire scheme should be chosen. The impact of non-uniform light intensity profiles in the series resistance, and then in the efficiency of a dual-junction solar cell has been also analyzed in [25] as an extension of the work carried out in [24] for single junction cells. Drops below 1 and 5% absolute are obtained for aspect ratios of the Gaussian light profile as high as 4 and 10, respectively, and for an average concentration of 1,000×, mainly because of a reduction of the FF and also in the V_{oc} (see Fig. 2.6). The non-homogeneity of the resistive paths throughout

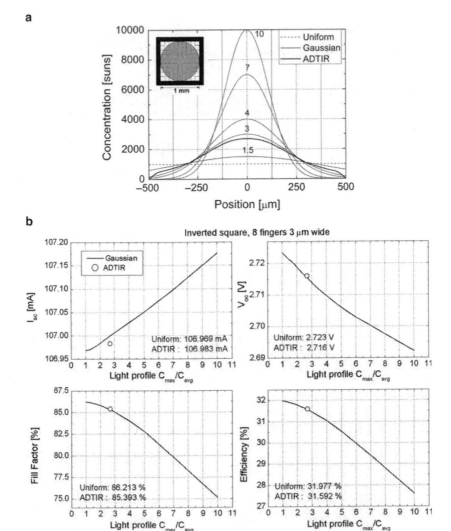

Fig. 2.6 (a) Light intensity profiles used to study the effect of the non-uniformity of light on the performance of 1 mm² dual junction solar cell. The average concentration is 1,000× in all cases. The peak concentration to average concentration ratio (C_{max}/C_{avg}) is modified from 1 to 10. (b) Simulated standard parameters against the (C_{max}/C_{avg}) ratio for the GaInP/GaAs dual junction solar cell with inverted square grid and shadowing factor of 2.7% [25]

the solar cell active area with inverted square front grid with constant finger pitch [27], is one of the origins of the FF drop. A solution for this problem is obtained by means of variable finger pitch configurations, which can be optimised for a given irradiance distribution [25].

2.4 Tunnel Junctions for UHCPV

In the preceding section, there is no mention to the TJ and its impact in series resistance or its performance at the real operation conditions described in [18]. However, its role is key as will be shown below. TJ is one of the most critical parts of the MJSC in order to attain high efficiencies under very high concentrations. High optic transmittance together with a high peak current and a low voltage drop across the TJ (i.e. low series resistance) are required. This fact has a special relevance in the case of concentrator photovoltaic systems, since as has been shown above, the concentrator optics forms a non-uniform illumination profile on the cell with a maximum irradiance significantly higher than the nominal (average) concentration level. Consequently, it is important to develop TJs that can assure peak currents well over the maximum level of concentrated light on the MJSC together with negligible voltage drops (series resistance). The simultaneous combination of all these characteristics is complicated, and it is in the origin of the failure of many MJSCs to properly operate at UHCPV.

The highest peak tunneling currents are obtained using low band gap materials, since the lower barrier height generated with them in the junction favors the tunneling process. In fact, a peak current density as high as 19,904 A cm^{-2} was published for an InGaAs/GaAsSb TJ [28]. For GaAs, a complete tunnel junction I–V curve with peak current densities higher than this value has not been published so far. A 340 A cm^{-2} peak current density was described in [29]. Recently, a GaAs-based TJ was presented showing a peak current density of 2,000 A cm^{-2}, which is beyond any value published before for a same type TJ in the MJSCs field [25]. In terms of series resistance, the voltage drop through that TJ was lower than 50 mV for concentrations below 6,000 suns and smaller than 100 mV for concentrations up to 13,000 suns (see Fig. 2.7) after the annealing. With these characteristics, the performance of the MJSC working inside any of the current concentrators [30] will not be affected and also broaden the field for UHCPV MJSC research. As we have already seen, 3D distributed model based on elemental circuit units constitutes an excellent tool to simulate the MJSC under concentration conditions. However, to date and to the best of our knowledge, in these models the TJ has been modelled as a short circuit [31] or as a resistor [21], which only takes into account the tunnelling region of the I–V curve. This simplified approach is not accurate at high current densities, and it fails in situations that produce currents in the solar cell higher than the peak current of the TJ. In order to address these limitations, reference [17] presents a new method that operates without restriction throughout the entire J-V curve of the TJ. In that method, the TJ is modelled using the analytical expressions that reproduce its three main operating regimes (ohmic region, excess current region and thermal diffusion region) together with several resistances that allow lateral current spreading in the TJ layers.

This new simulation approach has been used to simulate MJSCs at the operational limit of their tunnel junctions to revisit some typical assumptions made when interpreting the results of solar cells operating at ultra-high concentration [32].

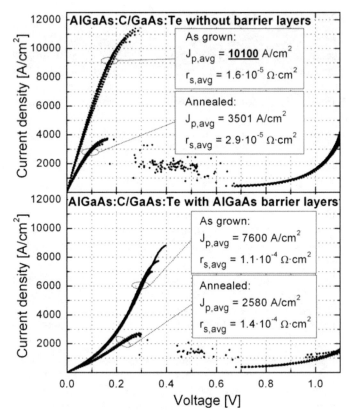

Fig. 2.7 J-V curves of TJs as-grown and after an annealing process at 675°C under AsH$_3$ atmosphere during 30 min, in order to emulate the growth of a GaInP top cell on the tunnel junction described in [25]

For instance, it is typically assumed that to determine the peak current of the limiting tunnel junction in an MJSC it is enough to increase the irradiance on the device and monitor the I–V curve until a dip in the curve appears [33]. At this point, it can be assumed that the short circuit current of the MJSC equals the TJ peak current. To test this assumption, a dual-junction solar cell with a tunnel junction described in Fig. 2.8 left and a short circuit current density of 13.5 mA cm−2 was simulated (i.e. for this device a dip in the I–V curve should appear just above 3,000 suns) [32]. The bottom part of Fig. 2.8 depicts the result of such simulations at two different concentration levels: 3,090× and 3,100× carried out as presented in [32]. As expected, a dip in the I–V curve becomes apparent in the case of 3,100×. However, a surprising result is obtained at 3,090×, since no effects in the I–V curve are detectable. Therefore, if the simulations are taken as correct, this dual-junction solar cell can operate with illumination currents beyond the TJ peak current without exhibiting the dramatic dip-related effects, up to at least 3,090×. Greyscale

Fig. 2.8 (**a**) J-V curve of the TJ used in the simulations presented in [17]. $J_p = 40.5\,\text{A cm}^{-2}$ corresponds to the photogenerated current density at 3,000 suns for that solar cell. (**b**) I–V curves for a dual-junction solar cell with a TJ whose J-V characteristic is that of the top and operating under uniform irradiance conditions at different concentration levels [17]

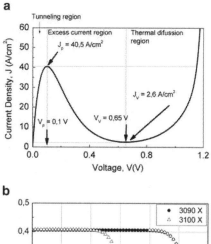

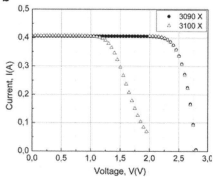

maps at 1.5 V (in the dip) for the two irradiance levels, 3,090× and 3,100×, have been depicted in Fig. 2.9, in order to understand the origin of this surprising result. Figure 2.9 shows that at 3,090× the voltage drop in the TJ is around 0.1 V, then it is working in the tunneling region (see Fig. 2.8 top). However, at 3,100× the voltage drop in the TJ is around 1.12 V, then the TJ is working in the thermal diffusion region.

The results shown in Figs. 2.8 and 2.9 can be understood by considering the actual current flow in the MJSC. Figure 2.10 aids this by plotting the lateral current density along the tunnel junction (i.e. along r_{anode} of Fig. 2.3) versus half the distance between two adjacent fingers. At 3,090×, the TJ operates in the tunneling region with a voltage drop close to 0.1 V (Fig. 2.9 left), then the current through the junction is virtually equal to its peak current. This phenomenon is possible, despite the fact the photogenerated current is higher than the peak current because there is a lateral current density that diverts the excess towards the dark areas where it can flow vertically. At 3,100×, the TJ is operating in the thermal diffusion region (Fig. 2.9 right), the voltage drop is around 1.12 V and thus the current through the TJ decreases sharply and the dip in the I–V curve is observed (Fig. 2.10).

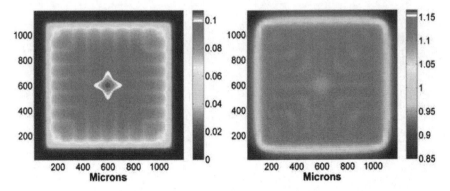

Fig. 2.9 Greyscale maps of the voltage drop (vertical scale in volts) in the TJ when the solar cell of Fig. 2.8 is biased at 1.5 V. (*Left*) Uniform irradiance of 3,090× and (*right*) uniform irradiance of 3,100× (Reprinted with permission from [32]. Copyright 2010. American Institute of Physics)

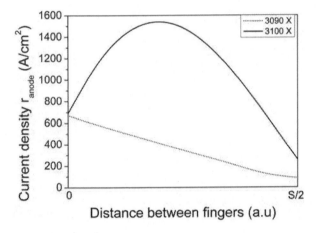

Fig. 2.10 Lateral current through the tunnel junction anode, ranode, versus half the distance between two adjacent fingers (where S is the finger pitch) calculated in [32]

This surprising result reveals that it is not strictly necessary that the current through the solar cell is lower than the peak current of the TJ, in order to avoid the appearance of the dip in the solar cell I–V curve has been experimentally observed [34]. This phenomenon can be explained thanks to the method described in [17], as follows. In practice, MJSCs comprise areas covered with metal and exposed areas. In the areas covered with metal no photogenerated current is produced, since they are in the dark. Thus, they can be considered as regions where the short circuit current is 0. These regions act as current sinks which somehow relax the requirement of the TJ peak current being higher than the whole solar cell short circuit current. In the simulations, these current sinks cannot be effective if the lateral current flow is not allowed. This is why this effect cannot be explained with the traditional models for the TJ consisting on a simple resistor or short-circuit. On the contrary, the new method [17] contemplates the lateral current flow, which, though being low, is enough to allow that the MJSC I–V curve does not exhibit the dip for current values slightly above the TJ peak current, as was observed experimentally. In the same sense, a recent work [35] arrives also to the same conclusions (the need of current spreading in the anode of the TJ) by using numerical simulations. This powerful method of modeling the TJ allows also the simulation of MJSC under chromatic aberration conditions [32]. Numerical simulations based on commercial software are also a useful way to evaluate series resistance effects and TJ performance. In fact, a good knowledge of the dual junction solar cells has been recently achieved [36, 37], which could be easily extended to triple junction solar cells. However, because of the present high computing cost of this approach, it is only applied to a trench of the MJSC, i.e., it is a 2D analysis, so no results of the series resistance or of the operation under real conditions of the whole MJSC have been achieved to date.

2.5 Increase of the MJSC Efficiency at Ultra-High Concentration

As stated in the Introduction of this chapter, tremendous progress has been made in recent years in increasing the efficiency of multijunction concentrator solar cells. Several labs, including NREL [1], Fraunhofer ISE [2] and Spectrolab [3], have reported devices with efficiencies over 40% for concentrations below 500 suns. However, the performance of these devices at concentrations above 1,000 suns is either much more modest or even not reported. This fact is the result of the steep efficiency roll-off that these devices suffer after peaking efficiency. Consequently, it seems that the development of an MJSC technology able to provide high efficiency for UHCPV is still a challenging issue. Taking advantage of the advances on the control of the series resistance and TJ performance described in Sects. 2.3 and 2.4 of this chapter, in this section we go a step forward by presenting the efforts of IES-UPM around this topic, by conducting a detailed discussion of the design and performance of record MJSCs above 1,000 suns.

The manufacture of highly efficient MJSCs for UHCPV requires two main technological steps. First, the growth of a highly efficient semiconductor structure has to be accomplished (typically by using the MOVPE technique) and, second, these structures have to be processed into concentrator solar cells minimising the deleterious effects of the series resistance. Figure 2.11 summarises these two steps for a record performing dual-junction solar cell developed at IES-UPM [25, 26]. As shown by the left part of the figure, the structure corresponds to a GaInP/GaAs dual junction solar cell with an AlGaAs/GaAs tunnel junction. Essentially, this structure involves the growth of about 15 layers of five different materials using four different dopant sources. Thicknesses have to be controlled to the nanometre scale, compositions of ternary and quaternary alloys have to be finely adjusted to

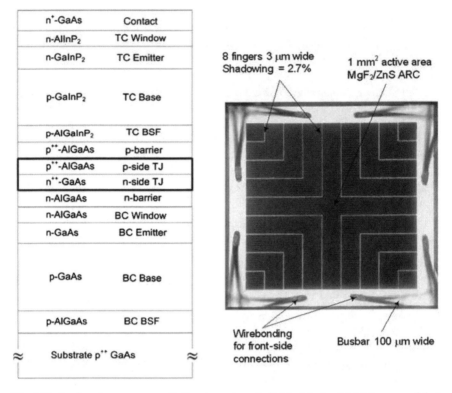

Fig. 2.11 Semiconductor structure (*left*) and top view of a finished device (*right*) for a record dual junction solar cell [26]

within 100 ppm and dopant concentrations covering three orders of magnitude have to be repeatedly controlled. The key aspects of this structure in terms of UHCPV operation are: (a) the design of the top cell emitter; (b) the TJ and (c) a careful tuning of all heterointerfaces to avoid additional (and sometimes unexpected) contributions to the series resistance. Regarding the top cell emitter, a special emphasis was put to maximise its conductivity at the cost of impacting the photogeneration at low wavelengths, as will be discussed when dealing with the quantum efficiency of this design. The importance and performance of the TJ has already been discussed in Sect. 2.4 of this chapter. In addition, special attention was paid to many of the heterointerfaces present in the device in order to minimise their impact in the series resistance [38].

As mentioned above, the front grid metallisation plays a major role in minimising the series resistance when working under concentration. The main parameters that influence the design of the optimum front grid are the top-cell emitter sheet resistance, the technological characteristics of the grid (maximum attainable sheet conductance and minimum semiconductor/metal specific contact resistance) and the short-circuit current of the device. Thus, to optimise the grid design, it is necessary to perform a multivariable optimisation for which the 3D models based on distributed circuit units described in Sect. 2.3 were used [25]. The right part of Fig. 2.11 depicts a finished device, which shows the final result of this optimisation process to define the front grid. As seen in Fig. 2.11, the finger pitch was kept constant because this approach has been demonstrated to give similar results as those obtained with the current-balanced design for uniform light profiles and is clearly advantageous for non-uniform light profiles [24]. The solar-cell active area used is 1 mm^2, which, in accordance to the compromise situation imposed by the series resistance, perimeter recombination, packaging cost and heat dissipation for an operation of the device at 1,000 suns, is close to the optimum value. The resulting geometrical dimensions of the device make it appropriate to use the "LED-like approach" [7] for the post-growth processing of the semiconductor structure. Consequently, the grid was formed by photolithography and Physical Vapour Deposition (PVD) techniques using the well-known AuGe/Ni/Au system [39]. As thousands of millimetre-sized devices fit in a single wafer, wet mesa etching consisting of an optimised sequence of acidic and basic solutions was applied to isolate individual solar cells and minimize the perimeter recombination. Finally, a MgF$_2$/ZnS anti-reflecting coating (ARC) was deposited onto the devices.

Figure 2.12 presents the external quantum efficiency (hereinafter QE for brief) of these dual-junction concentrator solar cells as measured by the *Calibration Laboratory of the Fraunhofer Institute for Solar Energy Systems* (FhG-ISE, Freiburg, Germany). The reflectivity of the ARC-coated devices is also plotted. The apparently poor blue response of the GaInP top cell is a consequence of the emitter design, which is optimised to maximize the efficiency at 1,000 suns by reducing its series resistance component. This leads to the necessity of using a thickness and a doping level in this layer that produce such a reduced QE in the 300–500 nm range. The short-circuit current densities (J$_{SC}$) under the solar spectrum AM1.5D low AOD show values of 14.3 and 13.5 mA cm^{-2} for the top and bottom cells, respectively,

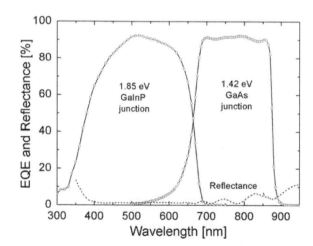

Fig. 2.12 External quantum efficiency of the solar cell in Fig. 2.11

meaning that there is a current mismatch of 6%. This result is partly due to the deposited ARC layer, which is not negligibly reflective in the near-IR region, as shown in Fig. 2.12. Although this current mismatch gives rise to an increased fill factor, the loss in short-circuit current produces a net conversion efficiency that is lower than that in a current-matched device. This means that there is still room for improvement in this dual-junction solar cell, by simply adjusting the current matching via fine-tuning of the thickness of the top-cell base layer and by using an appropriate ARC coating thickness and material quality. An additional way of improving the efficiency is raising the band-gap of the top cell – E_G in the top cell is 1.85 eV as marked in Fig. 2.12 – by disordering the GaInP by means of the use of a surfactant during the MOVPE growth [40].

Figure 2.13 summarises the certified concentration measurements which were performed on these devices at the *Calibration Laboratory of the FhG-ISE* (Freiburg, Germany). The solar spectrum used was the AM1.5D low AOD, with an irradiance of 1,000 W m^{-2}. The calibration at 1 sun gave an open-circuit voltage of 2.216 V, a J_{SC} value of 13.5 mA cm^{-2}, and an FF of 85%. The relatively low V_{oc} value at one sun, compared to other GaInP/GaAs devices, is mostly due to the high perimeter-to-area ratio in this solar cell, which causes a notable influence of the perimeter recombination on the value of V_{oc} at low currents [21]. Regarding efficiency, it can be observed that the maximum value achieved is 32.6%, remaining

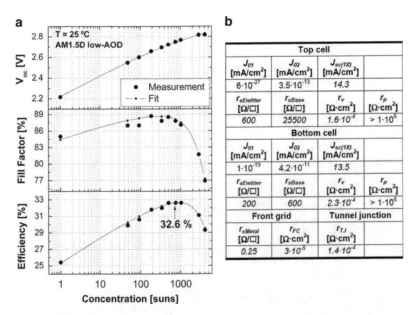

Fig. 2.13 (a) Efficiency, fill factor and open circuit voltage vs. concentration of the ultra-high concentrator dual-junction solar cell described in Fig. 2.11. (b) The fit obtained with the 3D distributed model described in Sect. 2.3 using the parameters of the table is also plotted as a thin line [25]

virtually constant for concentrations ranging from 400 to 1,000 suns. Moreover, the efficiency is above 30% up to concentrations higher than 3,500 suns. Therefore, the multivariable-design approach applied herein, which was employed with the aim of obtaining the maximum efficiency at 1,000 suns has been fruitful. Fig. 2.13 shows that, as concentration goes up, the open circuit voltage recovers rapidly as the perimeter recombination becomes less influential, and reaches a value of 2.76 V at 1,000 suns. The FF has its peak value of 88.7% at around 400 suns and decreases steadily at higher concentrations. This behaviour suggests that for concentrations up to 4,000 suns, the TJ is still working in the ohmic region, as expected from the discussions in Sect. 2.4 of this chapter.

In summary, a design approach oriented toward achieving a peak conversion efficiency at 1,000 suns has been demonstrated for a dual-junction solar cell. All the actions taken – from the optimisation of the TJ to the accurate design of the front grid to minimise the series resistance – have resulted in a device with a record conversion efficiency for any dual-junction solar cell and that even outperforms many record triple-junction designs at concentrations above 1,000 suns. However, despite these excellent results it is clear that a dual-junction device – though having interesting applications in approaches that aim substrate reuse or need substrate removal [41] – has a limited efficiency potential when compared to state-of-the-art triple-junction solar cells. Accordingly, this methodology for UHCPV design is being applied to the development of GaInP/GaInAs/Ge triple-junction devices, with the goal of achieving efficiencies in excess of 40% at 1,000 suns and above. Such efficiencies are, according to our simulations, reachable [42]. In this respect, Fig. 2.14 summarises the results of the current design of triple-junction solar cells under development at IES-UPM. The left part of the figure shows a sketch of the semiconductor structure of these devices where it can be noticed that the transition to a triple-junction device forces key changes as compared to the dual-junction structure shown in Fig. 2.11. The main change is of course related to having to deal with a new substrate – a germanium wafer – which will act both as the mechanical support for the whole device and as the third subcell. In addition, there are more subtle changes such as: (a) the need to modify the composition of virtually all the layers in the middle and top subcells to cope with the slight difference in lattice constant between Ge and GaAs – this is the reason why GaInAs with ~1% indium has to be used for the midlle subcell instead of pure GaAs– and (b) a second tunnel junction has to be added between the bottom and middle subcell, which has to be fine tuned to withstand the long thermal load associated with the growth of the middle and top subcell without severe degradation of the tunnelling characteristics.

However, by far the most challenging task in this design is related to the heteroepitaxial growth of III–V layers on germanium, which has essentially two main requirements: (1) III–V epilayers on Ge have to provide a defect-free template for subsequent epitaxial growth; and (2) during the initial stages of III–V growth on Ge, diffusion of group-V atoms into the substrate has to be controlled to form a third junction with the required characteristics for producing a high efficiency solar cell. The attainment of these two requirements is usually accomplished by the growth of two specific layers, namely, the nucleation layer grown at conditions which

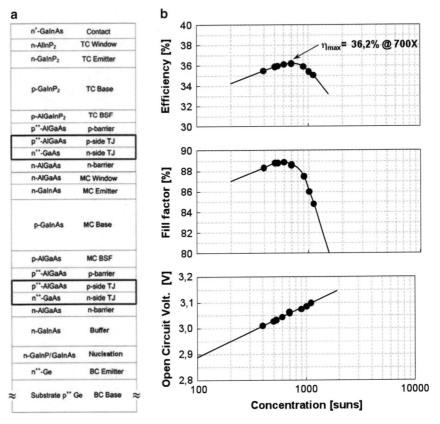

Fig. 2.14 (a) Semiconductor structure for the current design of triple-junction solar cells developed at IES-UPM. (b) Results for efficiency, fill factor and open circuit voltage vs. concentration for these cells

optimise the III–V/IV interface and the buffer layer grown at conditions that hinder the propagation of any crystallographic defect that might occur during the growth of the nucleation layer and thus provide an optimum template for further epitaxial growth. The growth of such layers of a given III–V material on germanium with the aforementioned features is not a straight-forward task. In fact, the growth of a polar III–V compound – such as GaAs or GaInP – on non-polar Germanium typically produces a variety of problems, such as of antiphase domains (APD) [43], misfit dislocations, hillocks [44, 45], uncontrolled etching [46], lack of charge neutrality across the heterointerface and diffusions from the layer into the substrate and vice versa [47]. In summary, a lot of material science has to be put into this interface to build on it a high efficiency solar cell.

The right part of Fig. 2.14 summarises the concentrator results of the current design of triple-junction solar cells under development at IES-UPM. As this figure shows, efficiency is over 36% for concentrations between 550 and 850 suns, peaking at 700× with a 36.2%. After peaking, the decrease in efficiency is much steeper than in the case of the record dual-junction solar cell (see Fig. 2.11) as a result of the rapid decrase of the fill factor after 500×. Preliminary investigations indicate that this behaviour is related to a processing issue (front contact problem) as the new series resistance components associated with this structure (i.e. the germanium wafer and the additional tunnel junction) should only add minor contributions to the overall series resistance as compared to the dual-junction design for UHCPV.

Simulations indicate that the way to improve the results of this triple junction solar cell is pretty much the same way to follow with the dual-junction [42]. A high bandgap cathode for the top tunnel junction would increase the short circuit current, and raising the bandgap of the GaInP in the top cell would also add up to the open circuit voltage. These two changes, together with a successful processing of the contacts and antireflection coatings, should bring efficiency above 40% at 1,000×.

2.6 Advanced Characterisation

The characterisation of MJSCs designed for UHCPV is a challenging task. On the one hand, MJSCs are inherently complex devices – with several subcells and TJs – with only two terminals in typical configurations. The latter fact implies that, when measuring the device, we get the overall combined response of the solar cell and have no access (typically) to each individual component of the MJSC. On the other hand, the nominal operating conditions are extremely far from the one-sun conventional characterisation set-ups used in photovoltaics. These two facts cause the portfolio of available techniques for cell characterisation to be ample and slightly divergent from conventional photovoltaics (both in terms of equipment and measurement procedures). A good review of the main techniques involved in MJSC characterisation can be found in [48]. Also of capital interest are the techniques used to get the concentrator response of these devices (i.e. the evolution of efficiency, fill factor and open circuit voltage versus concentration), which for MJSCs are particularly intricate [49]. It is not the purpose of this section to review these conventional techniques but to discuss some advanced characterisation techniques useful for achieving an accurate diagnosis of the performance of MJSCs, particularly for UHCPV. Here, the term advanced refers to the fact that these techniques are not of widespread use and their set-ups are not readily available as commercial products.

2.6.1 Quantum Efficiency Under Concentration

The spectral response or external quantum efficiency of a photovoltaic device is one of the most revealing measurements, which can be taken on a solar cell. This magnitude provides a picture of the internal performance of the device layer by layer. According to ASTM and IEC standards [50, 51], the spectral measurement of a solar cell has to be performed using *a continuous white light beam (bias light) to illuminate the entire device at an irradiance approximately equal to normal end use operating conditions intended for the cell.* Therefore, the quantum efficiency of high concentrator solar cells should be measured, in principle, using a bias light source with irradiance equal to their nominal concentration which, in the case of MJSCs, may range from some 100 suns to more than 1,000 suns.

Additionally, there are other interesting applications of this kind of QE measurement. For instance, when characterising concentrator MJSCs, the linearity between irradiance and concentration is often assumed. However, there are not many studies dealing with this subject [52, 53]. Bias-dependent QE measurements can provide some insight when trying to understand the lack of linearity with concentration of the photocurrent of a solar cell since this must be the result of a variation of the minority carrier collection parameters (lifetimes, diffusion lengths, etc.) of some of the layers in the device with irradiance.

The set-up needed to do quantum efficiency measurements under high irradiance bias light is schematically shown in Fig. 2.15. This set-up consists of the following parts: (a) a high intensity Xenon lamp, a chopper and a grating monochromator to obtain a monochromatic spot of high intensity; (b) a temperature controlled monitor cell connected to its own lock-in amplifier to take into account fluctuations of the

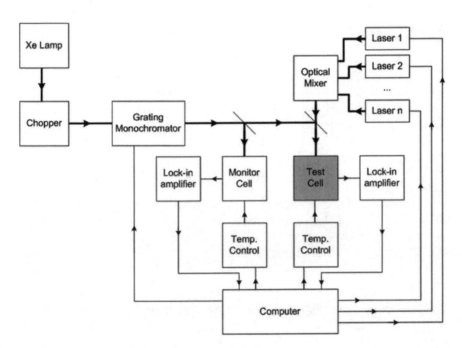

Fig. 2.15 Set-up for quantum efficiency measurements with high irradiance bias light. The bias light is produced by a set of lasers and an optical mixer

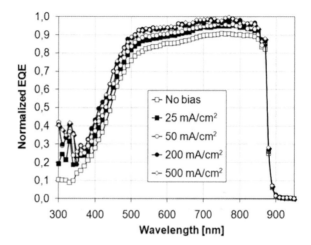

Fig. 2.16 External quantum efficiency measurements taken with different irradiance levels of the bias light on a single junction GaAs solar cell

intensity produced by the Xe lamp; (c) a set of lasers (with the corresponding power sources) and an optical mixer to produce a high intensity bias light with the required spectral content; (d) the device under test, placed on a temperature controlled chuck and connected to a lock-in amplifier; and (e) a computer to control the different elements, and to process and display the results of the measurement.

Figure 2.16 shows the evolution of the QE of a GaAs concentrator solar cell depending on the irradiance of the bias light. In this case, the bias light covers a range from 1 to 20 suns, since larger concentrations produced no change in the QE [54]. This solar cell was designed to work at 1,000 suns so the region where the lack of linearity occurs is far below the nominal concentration. However, if an insufficient bias light (or no bias light at all) was used to measure the quantum efficiency and then the QE was used to calculate the short circuit current at the nominal concentration then this J_{SC} will be clearly underestimated.

2.6.2 Electroluminescence Spectroscopy

Characterisation techniques based on the spectroscopic analysis of the light emitted by a semiconductor device, namely photoluminescence (PL) and electroluminescence (EL), are known to be one of the most sensitive ways to detect defects and impurities in semiconductors. These techniques have the advantages of being non-destructive, having high spatial resolution, and not needing special sample preparation, thus enabling to evaluate the cells as they are manufactured [55–60]. However, in terms of time consumed and required set-up, PL and EL are quite different. PL spectral analysis is usually performed at very low temperatures (ranging from 4 to 20 K) to maximise the radiative to non-radiative recombination ratio. Therefore, time-consuming cooling down cycles have to be implemented, before measuring. In compensation, in complex devices – as MJSCs – PL provides the ability to easily discriminate the response of every part of the device. By using a laser of adequate wavelength, the different cells in the stack can be individually excited and the unwanted crossed influence in others (if any) can be easily filtered. On the other hand, the set-up for EL characterisation is extremely simple and this analysis is usually carried out near room temperature; but (as biasing is applied to the whole device, not to a sole junction) it is more difficult to control the individual excitation of each subcell. The latter of course applies to two-terminal devices, which is the case for the vast majority of current state-of-the-art MJSCs.

The main goal of this section is to present some possibilities of EL spectroscopy as a fast characterisation technique able to provide valuable information on the manufacturing process, as for instance: (a) information on top cell composition and ordering (E_G of GaInP); (b) compare minority carrier properties of different runs; (c) photon recycling and coupling studies (radiative recombination of one subcell stimulates other subcells); (d) information on the thermal performance of encapsulated devices. Moreover, in terms of quantitative information, the analysis of the QE combined with the EL produced under different injection currents can be used for recovering the individual I–V curve of each subcell (and the corresponding diode quality factors) [61] using the reciprocity theorem between electroluminescent emission and external quantum efficiency of solar cells [62].

The set-up needed for EL spectroscopy is schematically shown in Fig. 2.17. This set-up consists of the following parts: (a) a programmable power supply to bias the solar cell; in most cases, this would be used as a high precision current source; (b) the device under test, placed on a temperature controlled chuck; (c) a grating monochromator for the spectral decomposition of the emitted light; (d) a detector (a calibrated photodiode) endowed with suitable collecting optics; and (f) a computer to process and display the results.

Figure 2.18 shows normalised EL spectra of a GaInP/GaAs dual junction solar cell at 25°C for different bias currents. Important information which can be extracted from this plot is: (a) position of the EL maximum of each peak and cut-off frequency, in the low energy side, which relates to the bandgap; (b) FWHM for each peak and its deviation from the expected thermal broadening, which gives

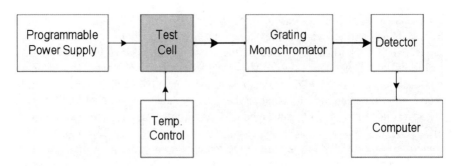

Fig. 2.17 Set-up for electroluminescence spectroscopy

Fig. 2.18 Normalised EL spectra of a dual junction solar cell at different bias currents

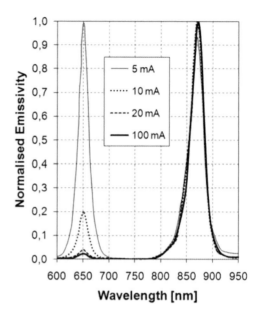

information of the non-thermal broadening mechanisms. In Fig. 2.18, the maximum of the EL peak corresponding to the GaAs subcell is significantly more intense than the one associated with the GaInP top cell. However, as the forward current is varied, the relative height of the peaks changes. For low bias currents (I <5 mA → J_{SC}<350 mA cm^{-2}), the high energy EL peak (top cell) is somewhat more intense than that of the GaAs cell. For any other bias current, the GaAs sub-cell peak is clearly more intense. This evolution can be explained taking into account the radiative efficiency of each subcell and its dependence on bias [55].

An experiment similar to that summarised in Fig. 2.18 can be used to implement a thermal analysis of the device since any change in the temperature will cause a shift in the maximum of the EL peak (as a result of the change in E_G). So, if the temperature control is disabled and the device is biased at different currents, any shift detected in the maximum of EL will reflect changes in the temperature of the junctions, giving valuable information about how the heat removal is working [63].

2.6.3 Electroluminescence Intensity Mapping

Electroluminescence intensity mapping [60,64–66] is an interesting characterisation tool to ascertain the spatial uniformity of a solar cell, although it is far less extended than PL mapping. In general, PL mapping is a fast and quite widespread technique to assess the uniformity of a semiconductor structure, at wafer level, in most cases just after the epitaxial growth. It is a fast measurement, taken at room temperature, which can be even done as a routine quality control in-line. On the other hand,

EL mapping is a device oriented measurement, as it involves the use of direct biasing. Therefore, it is more time consuming and just can be done when the device is finished. On the other hand, it can provide information related to the electrical performance of the solar cell, which is not accessible by PL. Taking into account that electrical measurements have to be made, EL mapping is a complementary tool that can be easily integrated into electrical characterisation tools and provides valuable additional information about the solar cell.

The signal detected by EL intensity mappers can be roughly seen as proportional to the integral of one of the spectra presented in Fig. 2.18. Therefore, any sort of spectral dependent information is lost (as the bandgap of the cells, or even the cell which is mostly contributing to the EL at a given bias), if no selective filtering is used. Thus, in most cases, EL intensity signals can be interpreted in terms of:

1. *Defect maps*: Some regions emit less than others as the result of a higher concentration of defects in those areas. Defects provide non-radiative recombination possibilities so, despite being at the same voltage, the light emission of those regions is weaker. Normally, the intensity distribution in these maps is somewhat random and does not follow the symmetry of the device.
2. *Voltage maps*: Some regions emit less than others because of the lower voltage in those areas. The lower the voltage, the lower the injected current and thus the lower the amount of carriers recombined radiatively. Normally, the intensity distribution in these maps is symmetrical in respect to any of the symmetry axes of the device.

The set-up needed for EL intensity mapping is schematically shown in Fig. 2.19. This set-up consists of the following parts: (a) a programmable power supply to bias the solar cell; in most cases this would be used as a high precision current source; (b) the device under test, placed on a temperature controlled holder; (c) a filter wheel to be able to eliminate the EL coming from any of the subcells; (d) a focusing optics to adjust the image to be generated; and (e) a CCD detector connected to a computer to process the image and display the results.

With this set-up, it is also possible to take pictures of the unbiased solar cell. For this purpose only an extra illumination source is needed (i.e. an external lamp). In this situation, the CCD camera just captures the conventional image of the solar

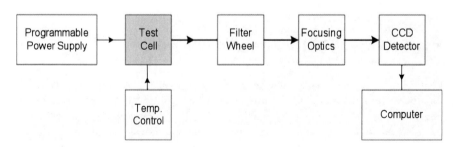

Fig. 2.19 Set-up for electroluminescence intensity mapping

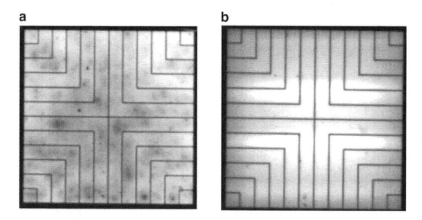

Fig. 2.20 Comparison between EL intensity maps produced by solar cells with a defective (**a**) and a good crystal structure (**b**). *Bright* (*clear*) areas indicate good quality material and *dark* areas indicate defective regions

cell and not the EL image. It is interesting to compare the conventional image and the EL image to interpret the EL results. For instance, a dark spot in the EL intensity map may be the result of a defective area or just the result of a speck of dust or any other kind of deposit produced during manufacture. In the first case, the affected area will look normal in the conventional image, while in the second case the speck of dust will appear in the conventional picture with the same shape as it appears in the EL intensity map.

Figure 2.20 represents a typical example of the use of EL intensity maps to compare the crystal quality of two different devices. Both figures are grey scale pictures of the solar cell active area where dark areas correspond to poor EL emission while bright areas are points with strong EL. Of course, the metallised regions are completely dark since no emitted light can go through the metal layer. In the picture on the left, greyish areas and dark spots are present all over the active area of the device, revealing an almost ubiquitous presence of crystal defects. On the other hand, in the picture on the right the bright areas are uniformly distributed following the symmetry of the grid. In this case, the EL intensity map is interpreted as a defect map. In this respect, EL mapping is a powerful tool to actually see and map recombination centers [67].

Figure 2.21 shows EL intensity maps for other couple of solar cells, this time revealing problems during the manufacturing process. In this figure, EL intensity maps are presented using a gray scale, where dark areas represent high intensities and clear areas low intensities. In the picture on the left, there is a clear (light gray) area on the left and a dark gray on the right, indicating that the right part of the solar cell is emitting more than the left part, as a result of the higher voltage. This uneven voltage distribution is caused by a defective formation of the wire bonding connections placed at the left side of the device; because of this

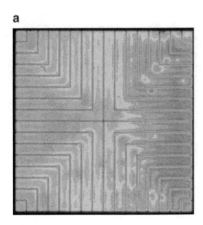 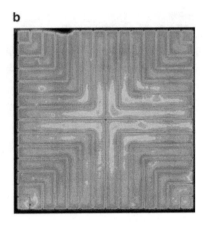

Fig. 2.21 Comparison between EL intensity maps produced by solar cells with a defective wire bonding (**a**) and a good wire bonding (**b**). In both cases wires were placed at corners. *Bright* (*clear*) areas indicate regions with lower voltage, while *dark* areas indicate regions biased at higher voltage. See discussions in the text

defective connection, during forward bias the current in the device is flowing from the connections at the right side causing this right-left voltage/EL drop pattern. On the other hand, the picture on the right shows a perfectly symmetrical EL pattern (i.e. voltage distribution) showing no problems during connection formation. In this case, the EL intensity map is interpreted as a voltage distribution map in a relatively defect free device.

2.7 Reliability

The satisfactory efficiency record results described at the beginning of this chapter have crystallised into the first commercial ventures of MJSC-based CPV systems. Consequently, a new qualification standard, namely the IEC-62108, has been developed [68] in which the procedure for qualifying CPV systems and assemblies is described. However, such standard does not consider specifically concentrator solar cells and there is no other that defines their failure condition or power degradation. However, before reaching a well-defined CPV module with a warranty to the customer, it is necessary to demonstrate the reliability of III–V high concentrator MJSCs. It is well known that silicon flat plate modules are very reliable systems capable of withstanding 25 years in field operation [69]. Nowadays, warranties offered by the manufacturers are precisely of about 25 years, but this value is expected to increase until 30 years in the near future [70]. If CPV systems are expected to be competitive with silicon flat plate modules, they must be capable of reaching similar warranties. Nevertheless, there are still many open questions regarding the reliability of MJSCs operating at high concentration that should be answered in order to give such warranty.

Although there are some preliminary results regarding the degradation of these solar cells [71–74], there is not enough accumulated experience to evaluate their reliability. The study of degradation is of great importance, but it is important to keep in mind that reliability is a completely different issue [75]. For instance, in terms of degradation, a typical study could consist in a set of solar cells introduced in a climatic chamber, the temperature is increased and the cells are biased to a specific current level, the number of failures is registered and the Mean Time to Failure (MTTF) is calculated with the following expression:

$$\text{MTTF} = \sum_{i=1}^{N} \frac{t_{Fi}}{N_F}, \qquad (2.2)$$

where t_{Fi} is the time of every failure and N_F is the number of failures. In this case, the test is just a degradation limited in time study. The problem is that this kind of tests does not establish a correlation between the test time and the lifetime of the device. With degradation limited in time, it is not possible to determine in which periods of the classical bath-tube curve (decreasing, constant or increasing failure rate) is the device working at [75].

In order to clarify how a reliability test should be carried out, it is important to consider the following points:

1. To define what reliability is. A good definition is that *reliability is the probability that a component part, equipment, or system will satisfactorily perform its intended function under given circumstances, such as environmental conditions, limitations as to operating time, and frequency and thoroughness of maintenance for a specified period* [76].
2. To establish a test procedure in which the following tasks should be taken into account: (a) to define a failure criterion; (b) to apply a stress factor; (c) to carry out statistics, with the aim of getting the activation energy, the acceleration factor and the main reliability functions (reliability $(R(t))$, failure rate $(\lambda(t))$ and MTTF).

Summarising, reliability testing should be directed to determine not only how long devices are going to live, but also the way in which these devices are going to live. In other words, reliability is also interested in knowing the probability distribution of failure in nominal conditions of operation [75].

Considering the similarities between light emitting diodes (LEDs) and some kind of III–V concentrator solar cells (same semiconductor materials and similar manufacturing processes), an analysis to determine to what extent III–V high-concentration solar cells are less stressed than LEDs has been carried out in a previous work [77]. As a result, it has been concluded that solar cells handle lower current densities than LEDs and that solar cells are also expected to operate at their maximum power point for most of their useful life (i.e. the point at which the maximum fraction of the power is delivered to a load and thus the minimum fraction of the power is internally available for degradation). In that paper, it was shown that in this situation, there is a higher stress on series resistance and a lower one on the junction. Therefore, a significant effort must be made to reduce the series resistance in order to decrease the degradation of the solar cells. The effect of temperature is less severe in solar cells than in high-power LEDs. It was also concluded that encapsulating and protecting the solar cells against moisture is a must for obtaining highly reliable devices. Finally, a theoretical prediction showed that an MTTF as high as 10^5 h (this is equivalent to 34 years assuming 8 h of average operation per day) could be reached for these solar cells [77].

After this prediction, the first reliability analysis derived from a stress-temperature ageing test performed on a III–V high concentrator solar cells was carried out in [78]. Unprotected single-junction GaAs solar cells were subjected to temperature step-stress ageing test in climatic chambers. Working conditions were simulated by forward biasing the 1 mm^2 solar cells at the same current level (250 mA, i.e. 25 A cm^{-2}) they would handle at the operating concentration (i.e. 1,000 suns). Dark I–V curves were routinely registered during the test at the temperature of every step (90, 110, 130 and 150°C, respectively). By using the method described in [79], the experimental data recorded for the dark I–V curve were used to reproduce the illumination curve.

Data were analyzed according to the Weibull reliability function. This analysis yielded a lower value of the MTTF of $2.02 \cdot 10^5$ h (i.e. about 69.2 years assuming 8 h of average operation per day in a year) for a confidence interval of 90%. The activation energy was not determined in this test, but the selected value for the calculation of the acceleration factor (0.9 eV) can be considered to be a low value if compared to the literature. The reliability at 25 years was higher than 65% (see Fig. 2.22). This is a low value for a solar cell, but that work dealt with devices manufactured in a research laboratory, therefore could be considered as a promising result. Besides, the failure rate in the initial stage decreased very steeply.

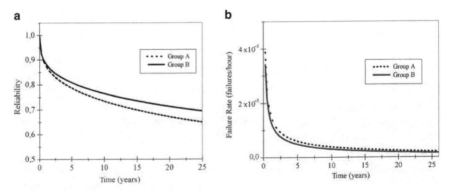

Fig. 2.22 (a) Reliability and (b) failure rate estimations for a period of 25 years of operation of the GaAs single-junction solar cells operating at 1,000 suns, obtained in [78] (Reprinted with permission from [78]. Copyright 2008. Elsevier B.V.)

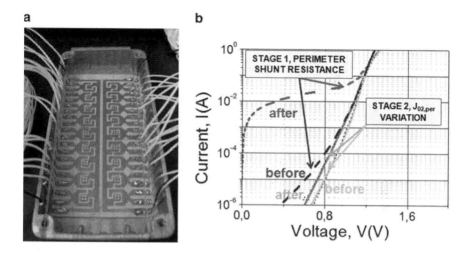

Fig. 2.23 (a) GaAs solar cells encapsulated on a DBC substrate and covered with silicone [80]. The caption shows the wiring inside the climatic chamber before the beginning of the test. (b) *Dark* I–V curves of two different cells before and after the degradation process. Solar cell of stage 1 corresponds to the experiment of [78] while solar cell of stage 2 corresponds to the experiment of [80]

Then the failure rate trends asymptotically to a saturation value as can be seen in Fig. 2.22. The failure rate estimation at 25 years is $1.67 \cdot 10^{-6}$ failures/year, which is a promising result, mostly bearing in mind that the experiment was performed on uncoated and bare devices. In this sense, the origin of degradation was shown to be related to the appearance of short circuits at the perimeter of the devices [78].

A recent work has tried to circumvent the failure originated at the perimeter by protecting the cells with a transparent cover [80]. Again small size (about 1 mm^2) GaAs single-junction solar cells were tested in order to compare them with the study in [78] but now the cells were covered with silicone to protect their perimeter which, in addition, reproduces the real situation of a cell inside a secondary optics in a concentrator (Fig. 2.23).

The main conclusions of that work show that perimeter protection seems to be a good way of enhancing the reliability of the solar cells because no degradation is found related to short-circuits at the perimeter. Besides, the low degradation suggests an MTTF value one order the magnitude larger than the one obtained in the thermal stress step test [78] where the perimeter was uncoated. In that work, a simple method based on the aforementioned 3-D distributed model of the solar cells has been used for the failure analysis [81], which was also employed in [78].

A step forward has been carried out in [82] where again small size GaAs solar cells protected with silicone were degraded at different temperatures and injection currents. At an injected current equivalent to 1,050 suns and a temperature of 150°C, a power degradation of 4% is observed after around 4,000 h of accelerated test.

Considering a nominal operation temperature of the cells of 65°C, the test at 150°C implies an acceleration factor of about 800. Therefore, those solar cells operating 8 h/day at 1,050 suns would experiment a power degradation of 4% in 550 years.

Consequently, III–V UHCPV solar cells seem to be very robust devices with much longer MTTFs than LEDs because operating conditions of solar cells are less demanding than those of LEDs. However, the results presented in [78, 80, 82] are based on tests in which the injected current does not traverse all the parts of the device such as in photogeration conditions happens. Besides, there is no stress caused by illumination. Therefore, a wider type of test set must be developed [75]. In spite of the fact that the aforementioned works have established a proper methodology, failure criteria, etc. for determining the reliability of cells, the experiments have been carried out on GaAs single junction cells. Accordingly, it is necessary to develop similar tests with MJSCs in order to detect anomalous origins of failure such as those detected when germanium is the substrate of the MJSC [83]. Degradation tests developed on parts surrounding the MJSC such as those described in [84, 85] are also required.

Finally, these accelerated ageing tests must be complemented with real-time tests in which the different parts of module and specially the MJSCs must be analysed. To our knowledge, the first attempt has been carried out in [86] where a statistical model for assessing the reliability of a CPV module based on degradation data was presented. The model was applied to a CPV module manufactured specifically with the purpose of assessing its reliability. Independent electrical access was provided to each receiver in the module to evaluate its performance over time without the masking effects that a series or parallel connection configuration may introduce.

In a more general sense, the valuable work developed at ISFOC (Puerto llano, Spain) [87] on the global performance and reliability of complete systems seems to be key in order to get confidence in the UHCPV technology.

Acknowledgements The authors would like to express their gratitude to the current and former members of the Group of III–V Semiconductors of the Solar Energy Institute of the Technical University of Madrid (IES-UPM) for their close collaboration and involvement in the subjects covered in this chapter.

This work has been supported by the Spanish Ministerio de Educación y Ciencia with the CONSOLIDER-INGENIO 2010 program by means of the GENESIS FV project (CSD2006-0004). The Spanish Ministerio de Ciencia e Innovación has also contributed with the SIGMA-SOLES project (PSS-440000-2009-30) and with the projects with references TEC2008-01226 and TEC 2009-11143 as well as the Comunidad de Madrid under the NUMANCIA II programme (S2009/ENE1477).

References

1. J.F. Geisz, D.J. Friedman, J.S. Ward, A. Duda, W.J. Olavarria, T.E. Moriarty, J.T. Kiehl, M.J. Romero, A.G. Norman, K.M. Jones, Appl. Phys. Lett. **93**(12), 123505 (2008)
2. W. Guter, J. Schone, S.P. Philipps, M. Steiner, G. Siefer, A. Wekkeli, E. Welser, E. Oliva, A.W. Bett, F. Dimroth, Appl. Phys. Lett. **94**(22), 223505 (2009)

3. R.R. King, A. Boca, W. Hong, X.Q. Liu, D. Bhusari, D. Larrabee, K.M. Edmondson, D.C. Law, C.M. Fetzer, S. Mesropian, N.H. Karam, in *Conference Record of the 24th European Photovoltaic Solar Energy Conference and Exhibition*, Hamburg, Germany, 21–25 Sep 2009, pp. 55–61
4. Http://www.spectrolab.com/concentrator.htm. Accessed: 2011-02-11. Archived by WebCite© at http://www.webcitation.org/5wPp5yOS0)
5. http://www.emcore.com/assets/photovoltaics/CTJ_Cell_10mm.pdf. Accessed: 2011-11-03. "http://www.webcitation.org/62vETsvHn"(Archived by WebCite© at http://www.webcitation.org/62vETsvHn)
6. http://www.azurspace.com. 2010-10-27. (Archived by WebCite© at http://www.webcitation.org/5tnCeUIs1)
7. C. Algora, *Concentrator Photovoltaics*, ed. by A. Luque, V. Andreev. Chap. Very High Concentration Challenges of III-V Multijunction solar Cells (Springer, Heidelberg, 2007)
8. A. Feltrin, A. Freundlich, in *Conference Record of the 2006 IEEE 4th World Conference on Photovoltaic Energy Conversion*, Waikoloa, HI, 07–12 May 2006, pp. 2469–2472
9. G. Sala, A. Luque, *Concentrator Photovoltaics*, ed. by A. Luque, V. Andreev. Chap. Past Experiences and New Challenges of PV Concentrators (Springer, Heidelberg, 2007)
10. LED market explosion hampered by materials shortage Semiconductor Today, 25th May (2010) http://www.semiconductor-today.com/news_items/2010/MAY/STRATEGY_250510.htm
11. W. Wilming, Sun Wind Energ. **12**, 98 (2009)
12. C. Algora, *Next Generation Photovoltaics: High Efficiency Through Full Spectrum Utilization*, ed. by A. Martí, A. Luque. Chap. The Importance of the Very High Concentration in 3rd Generation Solar Cells (Institute of Physics, Bristol, 2004)
13. A. Luque, Progr. Photovoltaics **9**(4), 303 (2001)
14. C. Algora, J.C. Minano, P. Benítez, I. Rey-Stolle, J.L. Álvarez, V. Díaz, M. Hernández, E. Ortíz, F. Munoz, R. Pena, R. Mohedano, A. Luque, G. Smekens, T. de Villers, H. Nather, K. Viehmann, S. Saveliev, in *Conference Record of the 16th European Photovoltaic Solar Energy Conference*, Glasgow, UK, 01–05 May 2000, pp. 2241–2242
15. G.L. Araújo, in *Physical Limitations to Photovoltaic Energy Conversion*, ed. by G.L. Araújo, A. Luque (Adam Hilger, Bristol, 1990)
16. European Photovoltaic Industry Association, EPIA. Global Market Outlook until 2013 (2009), http://www.epia.org/publications/epia-publications.html. Accessed 18 Oct 2010 (Archived by WebCite© at http://www.webcitation.org/5tZKPq1zc)
17. I. García, P. Espinet, C. Algora, I. Rey-Stolle, M. Baudrit (2010). European Patent Applications EP10382076.7. 5 Apr 2010
18. C. Algora, I. Rey-Stolle, D. Martín, R. Pena, B. Galiana, J.R. González, in *Conference Record of the 19th European Photovoltaic Solar Energy Conference*, Paris, France, 7–11 Jun 2004, pp. 34–37
19. H.C. Hamaker, J. Appl. Phys. **58**(6), 2344 (1985)
20. B. Galiana, C. Algora, I. Rey-Stolle, I.G. Vara, IEEE Trans. Electron Dev. **52**(12), 2552 (2005)
21. I. Rey-Stolle, C. Algora, I. García, M. Baudrit, P. Espinet, B. Galiana, E. Barrigón, in *Conference Record of the 34th IEEE Photovoltaic Specialists Conference*, Philadelphia, USA, 7–12 Jun 2009, pp. 1622–1627
22. B. Galiana, C. Algora, I. Rey-Stolle, Progr. Photovoltaics **16**(4), 331 (2008)
23. B. Galiana, C. Algora, I. Rey-Stolle, Sol. Energ. Mater. Sol. Cell. **90**(16), 2589 (2006)
24. I. García, C. Algora, I. Rey-Stolle, B. Galiana, in *Conference Record of the PVSC: 2008 33rd IEEE Photovoltaic Specialist Conference*, San Diego, USA, 11–16 May 2008, pp. 92–97
25. I. García Vara, Desarrollo de células solares de doble unión de GaInP/GaAs para concentraciones luminosas elevadas. Ph.D. thesis, Telecomunication School. Polytechnic University of Madrid (2010)
26. I. García, I. Rey-Stolle, B. Galiana, C. Algora, Appl. Phys. Lett. **94**(5), 053509 (2009)
27. C. Algora, V. Díaz, Progr. Photovoltaics **8**(2), 211 (2000)

28. J.C. Zolper, J.F. Klem, T.A. Plut, C.P. Tigges, in *Conference Record of the 1994 IEEE 1st World Conference on Photovoltaic Energy Conversion*, Waikoloa, USA, 5–9 Dec 1994, pp. 1843–1846
29. K.A. Bertness, D.J. Friedman, J.M. Olson, in *Conference Record of the 1994 IEEE 1st World Conference on Photovoltaic Energy Conversion*, Waikoloa, USA, 5–9 Dec 1994, pp. 1859–1862
30. A.W. Bett, B. Burger, F. Dimroth, G. Siefer, H. Lerchenmüller, in *Conference Record of the 2006 IEEE 4th World Conference on Photovoltaic Energy Conversion*, Waikoloa, USA, 07–12 May 2006, pp. 615–620
31. K. Nishioka, T. Takamoto, W. Nakajima, T. Agui, M. Kaneiwa, Y. Uraoka, T. Fuyuki, in *Conference Record of the 3rd World Conference on Photovoltaic Energy Conversion*, Osaka, Japan, 11–18 May 2003, pp. 869–872
32. P. Espinet, I. García, I. Rey-Stolle, C. Algora, M. Baudrit, in *Conference Record of the 6th International Conference on Concentrating Photovoltaic Systems*, Freiburg, Germany, 7–9 Apr 2010, pp. 24–27
33. W. Guter, A.W. Bett, IEEE Trans. Electron Dev. **53**(9), 2216 (2006)
34. A. Braun, B. Hirsch, E.A. Katz, J.M. Gordon, W. Guter, A.W. Bett, Sol. Energ. Mater. Sol. Cell. **93**(9), 1692 (2009)
35. J.M. Olson, in *Conference Record of the 35th Photovoltaic Specialists Conference (PVSC)*, HI, USA, 20–25 Jun 2010, pp. 201–204
36. M. Baudrit, C. Algora, Phys. Status Solidi a-Appl. Mater. Sci. **207**(2), 474 (2010)
37. S.P. Philipps, M. Hermle, G. Létay, F. Dimroth, B.M. George, A.W. Bett, Phys. Status Solidi-Rapid Res. Lett. **2**(4), 166 (2008)
38. B. Galiana, I. Rey-Stolle, M. Baudrit, I. García, C. Algora, Semicond. Sci. Tech. **21**(10), 1387 (2006)
39. L.J. Brillson, *Contacts to Semiconductors: Fundamentals and Technology* (Noyes Publications, Park Ridge, 1993)
40. J.M. Olson, W.E. McMahon, S. Kurtz, in *Conference Record of the 2006 IEEE 4th World Conference on Photovoltaic Energy Conversion*, Waikoloa, HI, 07–12 May 2006, pp. 787–790
41. J. Yoon, S. Jo, I.S. Chun, I. Jung, H.S. Kim, M. Meitl, E. Menard, X.L. Li, J.J. Coleman, U. Paik, J.A. Rogers, Nature **465**(7296), 329 (2010)
42. C. Algora, I. Rey-Stolle, I. García, B. Galiana, M. Baudrit, P. Espinet, E. Barrigón, J.R. González, in *Conference Record of the 34th IEEE Photovoltaic Specialists Conference*, Philadelphia, USA, 07–12 Jun 2009, pp. 1571–1575
43. L. Lazzarini, L. Nasi, G. Salviati, C.Z. Fregonara, Y. Li, L.J. Giling, C. Hardingham, D.B. Holt, Micron **31**(3), 217 (2000)
44. B. Galiana, E. Barrigón, I. Rey-Stolle, V. Corregidor, P. Espinet, C. Algora, E. Alves, Superlattice. Microst. **45**(4–5), 277 (2009)
45. I. Beinik, B. Galiana, M. Kratzer, C. Teichert, I. Rey-Stolle, C. Algora, P. Tejedor, J. Vac. Sci. Tech. B **28**(4), C5G5 (2010)
46. W.E. McMahon, J.M. Olson, J. Cryst. Growth **225**(2–4), 410 (2001)
47. B. Galiana, K. Volz, I. Rey-Stolle, W. Stolz, C. Algora, in *Conference Record of the 2006 IEEE 4th World Conference on Photovoltaic Energy Conversion*, Waikoloa, HI, 07–12 May 2006, pp. 807–810
48. J.M. Olson, D.J. Friedman, S. Kurtz, in *Handbook of Photovoltaic Science and Engineering*, ed. by A. Luque, S. Hegedus. Chap. 9: High Efficiency III-V Multijunction Solar Cells (Wiley, NY, 2003)
49. C. Domínguez, I. Antón, G. Sala, Optic. Express **16**(19), 14894 (2008)
50. ASTM Standard E1021-95: Standard Test Methods for Measuring Spectral Response of Photovoltaic Cells. American Society for Testing and Materials (1995)
51. IEC Standard 60904-8. Measurement of Spectral Response of Photovoltaic (PV) Device. International Electrochemical Commission (1998)

52. G. Stryi-Hipp, A. Schoenecker, K. Schitterer, K. Bücher, K. Heidler, in *Conference Record of the 23rd IEEE Photovoltaic Specialists Conference*, Kentucky, USA, 10–14 May 1993, pp. 303–308
53. J.M. Gee, in *Conference Record of the 19th IEEE Photovoltaic Specialists Conference*, New Orleans, USA, 4–8 May 1987, pp. 1390–1395
54. I. García, I. Rey-Stolle, B. Galiana, C. Algora, in *Conference Record of the 2006 IEEE 4th World Conference on Photovoltaic Energy Conversion*, Waikoloa, HI, 07–12 May 2006, pp. 830–833
55. P. Bhattacharya, *Semiconductor Optoelectronic Devices*, 2nd edn. (Prentice Hall, NJ, 1998)
56. M. Zazoui, M. Mbarki, A.Z. Aldin, J.C. Bourgoin, O. Gilard, G. Strobl, J. Appl. Phys. **93**(9), 5080 (2003)
57. R.F. Reyna, A. Martí, J.C. Maroto, Solid-State Electron. **42**(4), 567 (1998)
58. V.D. Rumyantsev, J.A. Rodriguez, Sol. Energ. Mater. Sol. Cell. **31**(3), 357 (1993)
59. G. Smestad, H. Ries, Sol. Energ. Mater. Sol. Cell. **25**(1–2), 51 (1992)
60. M.I. Gercenshtain, M.M. Koltun, Sol. Energ. Mater. Sol. Cell. **28**(1), 1 (1992)
61. T. Kirchartz, U. Rau, M. Hermle, A.W. Bett, A. Helbig, J.H. Werner, Appl. Phys. Lett. **92**(12) (2008)
62. U. Rau, Phys. Rev. B **76**(8) (2007)
63. P. Espinet, C. Algora, I. Rey-Stolle, M. Baudrit, I. García, in *Conference Record of the PVSC: 2008 33rd IEEE Photovoltaic Specialist Conference*, San Diego, USA, 11–16 May 2008, p. 147
64. C. Honsberg, A.M. Barnett, in *Conference Record of the 21st IEEE Photovoltaic Specialists Conference*, FL, USA, 21–25 May 1990, pp. 772–776
65. L.C. Kilmer, C. Honsberg, J.E. Phillips, A.M. Barnett, in *Conference Record of the 20th IEEE Photovoltaic Specialists Conference*, NV, USA, 26–30 Sep 1988, pp. 785–787
66. G.L. Araújo, J.C. Maroto, O. Ciorda, A. Martí, C. Algora, G.C. Filgueira, J.C. Minano, in *Conference Record of the 11th European Photovoltaic Solar Energy Conference*, Montreux, Switzerland, 12–16 Oct 1992, pp. 229–232
67. J.R. González, I. Rey-Stolle, C. Algora, B. Galiana, IEEE Electron Dev. Lett. **26**(12), 867 (2005)
68. IEC International Electrochemical Commision Concentrator photovoltaic (CPV) modules and assemblies – Design qualification and type approval (2007)
69. A. Skoczek, T. Sample, E.D. Dunlop, Progr. Photovoltaics **17**(4), 227 (2008)
70. J.H. Wohlgemuth, D.W. Cunningham, A.M. Nguyen, J. Miller, in *Conference Record of the 20th European Photovoltaic Solar Energy Conference*, Barcelona, Spain, 6–10 Jun 2005, pp. 1942–1946
71. I. Rey-Stolle, C. Algora, Progr. Photovoltaics **11**(4), 249 (2003)
72. I. Rey-Stolle, C. Algora, in *Conference Record of the 17th European Photovoltaic Solar Energy Conference*, Munich, Germany, 21–26 Oct 2001, pp. 2223–2226.
73. K. Araki, M. Kondo, H. Uomizi, Y. Kemmoku, T. Egami, M. Hiramatsu, Y. Miyazaki, N.J. Ekins-Daukes, M. Yamaguchi, G. Siefert, A.W. Bett, in *Conference Record of the 19th European Photovoltaic Solar Energy Conference*, Paris, France, 7–11 Jun 2004, pp. 2495–2498
74. R.A. Sherif, in *4th International Conference on Solar Concentrators for the Generation of Electricity or Hydrogen*, El Escorial, Spain, 12–16 Mar 2007.
75. J.R. González, C. Algora, Vázquez, N. Núnez, I. Rey-Stolle, in *Conference Record of the 5th International Conference on Solar Concentrators*, CA, USA, 16–19 Nov 2008
76. McGraw-Hill Dictionary of Scientific and Technical Terms. The McGraw-Hill Companies, Inc, NY. ISBN: 007042313X (2002)
77. M. Vázquez, C. Algora, I. Rey-Stolle, J.R. Gónzalez, Progr. Photovoltaics **15**(6), 477 (2007)
78. J.R. Gónzalez, M. Vázquez, N. Núnez, C. Algora, I. Rey-Stolle, B. Galiana, Microelectron. Reliab. **49**(7), 673 (2009)
79. I. Antón, G. Sala, K. Heasman, R. Kern, T.M. Bruton, Progr. Photovoltaics **11**(3), 165 (2003)
80. J.R. González, Vázquez, N. Núnez, C. Algora, P. Espinet, in *Conference Record of the 6th International Conference on Concentrating Photovoltaic Systems*, Freiburg, Germany, 7–9 Apr 2010, pp. 221–224

81. P. Espinet, C. Algora, J.R. Gónzalez, N. Nunez, M. Vázquez, Microelectron. Reliab. **50**, 1875 (2010)
82. N. Núnez, M. Vázquez, J.R. Gónzalez, C. Algora, P. Espinet, Microelectron. Reliab. **50**, 1880 (2010)
83. J. Schöne, G. Peharz, R. Hoheysel, G. Siefer, W. Guter, F. Dimroth, A.W. Bett, in *Conference Record of the 23rd European Photovoltaic Solar Energy Conference and Exhibition*, Valencia, Spain, 1–5 Sep 2008, pp. 118–122
84. N. Bosco, C. Sweet, S. Kurtz, in *Conference Record of the 34th IEEE Photovoltaic Specialists Conference*, Philadelphia, USA, 7–12 Jun 2009, pp. 917–922
85. I. Aeby, in *Conference Record of the Photovoltaic Module Reliability Workshop. Archived by WebCite© at http://www.webcitation.org/5wPqCvMjI*, Denver, USA. Accessed 18–19 Feb 2010
86. J.R. Gónzalez, M. Vázquez, C. Algora, N. Nunez, Progr. Photovoltaics **19**, 113 (2010)
87. www.isfoc.com

Chapter 3
The Sun Tracker in Concentrator Photovoltaics

Ignacio Luque-Heredia, Goulven Quémére, Rafael Cervantes, Olivier Laurent, Emmanuele Chiappori, and Jing Ying Chong

Abstract This chapter provides an updated insight into the specifications and design issues associated with the sun tracker in photovoltaic concentrators, regarding both the mechanical structure and the electronic control unit, along with the description of a set of representative examples. It continues with the description of specific quality assessment instrumentation, developed for the measurement of sun tracking accuracy. It ends by presenting and analyzing the results of a monitoring campaign carried out with these instruments in a demo CPV system for a range of different control strategies.

3.1 Introduction

Most PV concentrators use only direct solar radiation, and therefore must permanently track the sun's apparent daytime motion. They are hence to integrate an automatic sun tracking structure, able to mount and position the concentrator optics in such a way that direct sunlight is always focused on the cells. This sun tracker is basically composed of a structure presenting a sunlight collecting surface in which to attach concentrator modules or systems, which is somehow coupled to a one or two-axis mechanical drive, and also of some sun tracking control system which operates the drive axes, and maintains an optimum aiming of the collecting surface or aperture toward the sun.

Static mounts are only feasible today for low concentration factors (below 5×). However in the long term, static concentrators with higher ratios may appear making use of luminescence and photonic crystals. All these issues are, nevertheless, beyond the scope of this chapter.

I. Luque-Heredia (✉)
BSQ Solar. C/ del Vivero 5, 28040 Madrid, Spain
e-mail: iluque@bsqsolar.com

Line focus reflective concentrators, such as troughs, only require one axis tracking to maintain the PV receiver along the focus line. However, due to the daily variations in the sun elevation, sunlight incidence on the tracker's aperture is usually somewhat oblique, thus reducing the intercepted energy and causing the sun's image to move up and down within the focus axis, and producing further losses whenever it surpasses the receiver's ends. Line focus refractive concentrators, such as those based on linear Fresnel lenses, experience severe optical aberrations whenever light incidence is not normal, thus requiring two-axis sun tracking. The same happens to most point focus concentration concepts developed, excepting some low concentration factor devices with enough acceptance angle to admit sun's altitude variations.

Nearly all PV concentrators already commercial or currently under development use two-axis tracking being the so-called pedestal tracker, as depicted in Fig. 3.1, with its azimuth-elevation axes the most common configuration, followed by the tilt-roll tracker operating on the declination-hour angle axes. With regard to sun tracking control, most of the early systems consisted on analog sun pointing sensors based on the shadowing or illumination differences of a couple or quad of PV cells, integrated in an automatic closed loop with the tracker's driving motors. The advent of cheap microcontrollers motivated the appearance of sun tracking control systems requiring no sun sensing, and based only on the digital computation of precise analytic sun ephemeris equations. To date the need for an efficient and reliable sun tracking control in CPV applications has driven the state of the art toward a blend of these two initial approaches, producing hybrid strategies that integrate both sun alignment error feedback and ephemeris-based positioning.

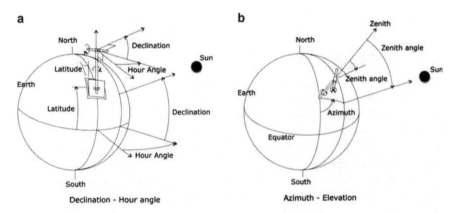

Fig. 3.1 (a) The two most common sun tracking axes geometries used in solar trackers, declination-hour angle or also called tilt-roll trackers; and (b) azimuth-elevation, usually implemented with pedestal trackers

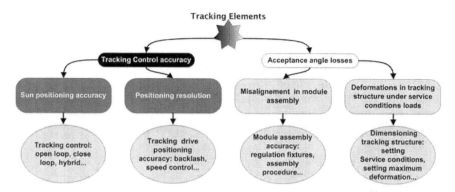

Fig. 3.2 Factors conditioning sun tracking performance related either to the tracking control accuracy or to the acceptance angle losses

3.2 Requirements and Specifications

Strictly speaking, the main commitment to be fulfilled by a CPV sun tracker is to permanently align the pointing axis of the supported concentration system with the local sun vector, in this way producing maximum power output. As we will see throughout this chapter, there are several error sources to take into account and therefore, some off-tracking tolerance is required. Usually, this tolerance, or minimum tracking accuracy required, is characterized by means of the acceptance angle of the concentration system, usually defined as the off-tracking angle at which power output drops below 90%.

As presented in Fig. 3.2, reasons for the decrease of sun tracking performance can be classified into two main types: (1) those purely related to the precise pointing of the tracker to the sun and (2) those which provoke shrinkage of the overall acceptance angle of the concentrator system, thus indirectly increasing the tracking accuracy required. Among those related to acceptance angle losses caused by the tracking system, these are, on the one hand, due to the accuracy which can be attained in the mounting and alignment of the concentrator system atop of the tracker. This is in first instance a design problem having to do with the mechanical fixtures provided for this purpose, their accurate assembly, and the regulation means provided for module levelling, but also with the mounting protocols devised to carry out these tasks. Also resulting in acceptance angle cuts is the stiffness conferred to the tracker, which is to say the bending allowed in the different elements of its structure under service conditions.

Regarding those having to do with tracking accuracy, these are basically, on the one hand, the exactness of the sun position coordinates generated by the control system, expressed in terms of rotation angles of the tracking axes. These will be produced either by means of sun ephemeris-based computations, or result from the feedback of sun pointing sensor readings. Also a combination of both these techniques is used by some but, whatever the method employed to generate sun position coordinates, it will be affected by several error sources. On the other hand, we will have to count with the precision with which the tracker can be positioned at these dictated orientations, i.e., the positioning resolution of the tracking drive and its control system, which essentially depends on the performance of tracking speed control and on the mechanical backlash in the drive's gearings.

Characterization of service conditions for a CPV tracker deserves some discussion. Basically, these consist in fixing a value for the maximum wind speed, i.e. wind load, to be withstood during sun tracking operation. The bigger this value, the heavier and more expensive tracking structure required to maintain bending under the threshold required for accurate tracking. A cost-effective approach is to determine this so-called maximum service wind speed (MSWS) value from the cross correlation between wind speed and direct radiation, in the location or set of locations in which the trackers are planned to be marketed and installed. When above this wind speed value, stiffness specifications do not have to be met and the tracker can switch to some low wind profile stow position.

These correlations have been estimated in a systematic way to aid the design of solar collectors employed in solar thermal plants by Sandia National Laboratories [1], by using records of insolation and surface meteorological conditions obtained at 26 weather stations of the US National Climatic Center, distributed over the contiguous United States and available through the so-called SOLMET data tapes. These tapes provide hourly observations of wind speed and direction in addition to the normal direct irradiance spanning the 1952–1964 time frame. By computing cumulated direct irradiation occurring below a certain wind speed for double and single axis trackers, it was found that for 22 of the 26 SOLMET stations, at least 95% of the direct irradiation is available at wind speeds not higher than $11\,\mathrm{m\,s^{-1}}$. The windiest station, the one at Great Falls, MT, collects 95% of the direct insolation at wind speeds up to $13\,\mathrm{m\,s^{-1}}$. This wind speed value below which this 95% cumulated direct irradiation is comprised seems in first instance a reasonable choice to be taken as the maximum service wind speed, over which the tracking control can order a stow position. A case example of this type of analysis is presented in Fig. 3.3, worked out with 1 year of continuous wind speed and direct irradiation (considering a two-axis tracker) hourly data, for the Spanish city of Granada.

Further fine-tuning of this wind speed threshold can be achieved if we are able to obtain the function of tracker cost vs. MSWS, for the particular tracking design chosen for our project. Considering we can also estimate the energy produced by the concentrator system for, say, its assumed operative lifetime, also as a function of the maximum service wind speed, we will be able to obtain an electricity cost and determine its optimum value as that MSWS in which an electricity cost minimum occurs. This exercise was done by the Spanish tracker specialist Inspira [2], using a $9\,\mathrm{m^2}$ pedestal tracker, and in which service stiffness was specified in such a way that aperture bending was to remain always below $0.1°$. This minimum, again using the 1 year data collected for Granada, was found to occur at a maximum service wind speed of $22\,\mathrm{Km\,h^{-1}}$, which as seen in Fig. 3.3, allows for a direct irradiance cumulated collection in the 95% range which, to some extent, corroborates the preliminary choice of this 95% threshold as a reasonable one.

3 The Sun Tracker in Concentrator Photovoltaics

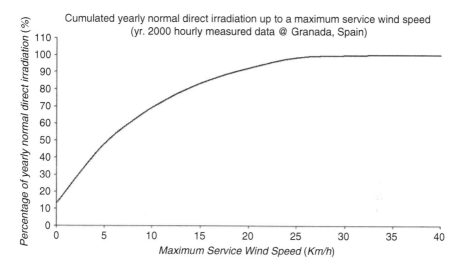

Fig. 3.3 Yearly cumulated DNI (Direct Normal Irradiance) vs. wind speed for the determination of maximum service wind speed, here computed for Granada, Spain, and showing availability of 95% DNI below 22 Km h^{-1} (6.1 m s^{-1}) wind speed

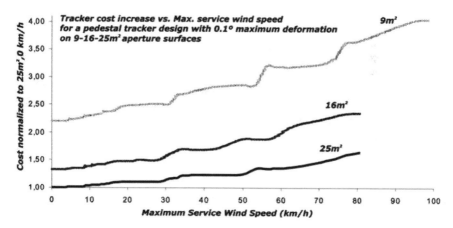

Fig. 3.4 Normalized tracker cost vs. maximum service wind speed for a specific tracker design dimensioned for three different aperture surfaces and costed for 1,000 units/yr. productions

In Fig. 3.4, we see how tracker cost increased with the maximum service wind speed for the referred 9 m^2 design, and also for its upscaling to 16 and 25 m^2 apertures considering 10,000 units/yr. production volumes. Design re-dimensioning at constant wind speed interval, seeking for compliancy with the 0.1° maximum bending criterion, implies an almost linear smooth cost increase in what respects metal structural elements, mainly driven by weight increase. However, some leaps can be observed in Fig. 3.4, which are due to the introduction of new part numbers in the drive components (bearings, gearing sets, etc.) whose price is partly driven upward by their nominal load, but also eventually downward by their bigger market and demand, and therefore the higher volumes in which their manufacturers produce them.

We therefore infer how structural dimensioning of a tracking design can be cost optimized for a given location, an option which may be worthwhile when building large-scale CPV plants provided there are some tuning possibilities in the tracking structure design and manufacturing, and also if it can be done without compromising the cost effectiveness of the supply chain.

The other variable involved in this characterization of service conditions, the maximum allowed structural bending measured in the aperture surface, which for this example has been set to 0.1°, intends to place a bound on the losses caused by the tracker's flexure on the acceptance angle of its CPV modules. In the following section, we will present a procedure developed to provide an estimate of these losses, for a design carried out in observance of this maximum bending criterion. Determination of the value for this maximum possible bending is to depend on the acceptance angle of the particular CPV technology object of the design, bearing in mind that the finally obtained tracking accuracy is to be comprised within the overall CPV array acceptance angle.

Measuring this tracking accuracy is still an unsettled issue, provided there is to date no standard instrumentation and measurement procedures, capable of providing enough sensitivity to gauge the sub-degree accuracy ranges usually involved. CPV developers frequently overlook this critical issue, and loosely talk about the usually very high tracking accuracies they can achieve, without any explanation on how they measure these. Precise tracking accuracy measurement basically entails the continuous monitoring of difference angles with respect to the position producing maximum power output. In the last section of this chapter, we will present a system based in solid state image sensors, which we propose as an efficient tool to determine the tracking accuracy statistics of a given CPV system.

Apart from sheer tracking performance, downtime or the availability ratio is the other main concern with a CPV tracker. Here, the mechanical tracker is usually free of suspect, provided the pertinent structural codes are respected in its design, and taking into account that when recurring to off-the-shelf drive gearings, these are subject to very mild operation conditions – one axis turn per day – when compared with their usual market applications in machine tools, cranes, etc. Instead, most of the reported problems arise in the electrical and electronic parts which, first of all, are to be designed to operate reliably in outdoor conditions, but also comply with a suitably chosen set of electromagnetic compatibility (EMC) and electrical safety standards, thus anticipating common field problems such as power spikes or surges. When considerable amounts of software are involved, as happens with todays tracking control systems integrating microprocessors, it is not just a matter of a reliable and well-protected electronic design but also of a redundant code, immune to hangs and able to gracefully recover from power outages or sags.

3.3 The Tracker Structure

In this section, an overview of how the stiffness issue is tackled in one of the first tracker designs that in Spain specifically targeted CPV volume deployment is presented [3]. It was a two-axis pedestal tracker designed by Inspira, with a 30 m^2 aperture surface, customized for the CPV modules with 1,000× concentration factor being developed by Spanish PV manufacturer Isofoton (Fig. 3.5). The modules had a nominal acceptance angle of ±0.6°, determined through indoor lab measurements using collimated laser light. Subtracting the approximately 0.26° subtended half angle of the sun, this means we require a 0.34° minimum tracking accuracy. As we will see below, a feasible value for the minimum tracking accuracy is 0.1° (i.e., 95% probability that the off-track angle remains below 0.1°) so acceptance angle overall

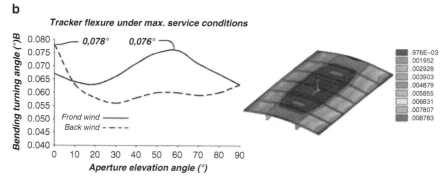

Fig. 3.5 (**a**) CPV pedestal tracker designed and produced by Inspira was subject to flexure analysis. (**b**-*left*) Maximum bending turning angle in aperture surface when subject to maximum service conditions (43 km h^{-1} wind speed) as a function of aperture elevation function. (**b**-*right*) Sample finite elements analysis of flexure bending

loss on the array must not surpass 0.24°. Introducing some overestimation, to allow for extra acceptance angle losses due to some degree of error in module levelling atop of the tracker's aperture surface, a maximum 0.1° bending was the starting point set for the tracking structure design. This means that this is to be the maximum allowed turn induced by structural flexure for any vector normal to the aperture surface when subject to maximum service conditions (maximum operating wind speed set in this case to 43 km h^{-1}, both blowing from the front or the back of the tracker) at any aperture elevation angle. First in the design of the metal structure forming the tracker's aperture is choosing what we can loosely call its topology. Here, only the lengths of dimensionless metal beams and the connections among them and with the drive block are decided, seeking here the optimization of different aspects such as transportation, in field installation, mounting of CPV modules, etc. Once the tracker frame is settled, this is to be sized playing with the precise form of the structural beams. For example using I-beams, angles and channels, tubes, etc. if directly opting for off-the-shelf construction standards, or using instead others requiring more processing such as trussed parts, but in all cases taking into consideration their stiffness to weight ratios and their manufacturing costs.

It is when getting to this point that the stiffness constraints start to rule over the design, and precise finite elements (FE) analysis are to be carried out over the complete tracker structure, when subject to the specified maximum service loads (CPV modules payload and maximum operative wind loads). When this was done for the referred tracker, a solution based on standard structural beams was obtained, which resulted in the least tracker's self weight, and according to FE simulations, did not surpass the 0.1° bending at any aperture elevation. In the case of this pedestal tracker, the design was separately considered in three segments, (1) aperture frame (2) pedestal and drive block and (3) foundation. First, a certain percentage of that total maximum 0.1° maximum flexure was allocated to each segment, taking into account that, while bending in the aperture will usually result in overall acceptance angle shrinkage, bending in the pedestal or the foundation works as an overall pointing vector turn, which as will be seen in the sun tracking control section can eventually be characterized and handled by a tracking controller. In the case of the tracker's foundation, meeting its flexure quota requires a standard geotechnical analysis of the ground where it will be installed, in order to choose the best suitable solution. Quite obviously, in a pedestal tracker with a rectangular aperture surface, maximum bending at whichever elevation will occur in its corners. Final results for this design can be seen in Fig. 3.5b, where maximum bending when maximum service wind load comes frontally is 0.076° and occurs at 57° aperture elevation, while when this same wind speed is received in the aperture's rear face this maximum bending is slightly bigger, 0.078° and happens at 0° elevation. In any case, maximum structural bending remains under the 0.1° threshold.

Once the structure sizing is optimally below this maximum bending threshold, next step is to estimate the acceptance angle losses induced by structural flexure, using the bending rotation values of the set of vectors normal to the aperture, obtained in the FE simulation. An approximation to this problem was attempted through a geometrical model in which, assuming that each CPV module mounted

3 The Sun Tracker in Concentrator Photovoltaics

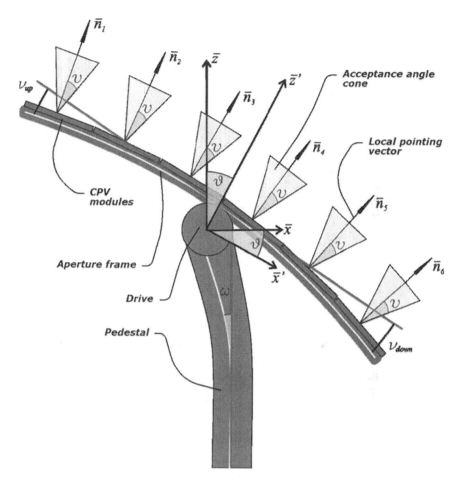

Fig. 3.6 Cross section of pedestal tracker subject to flexure. At a certain aperture elevation (zenith angle θ). The local pointing vector to each module $\bar{n}_i$ and its acceptance angle (v) cone within the apertures local reference system (**x, y, z**)

on the aperture could be considered to remain undeformed under service loads, a single normal vector is considered per CPV module, as seen in Fig. 3.6. This normal vector is taken as the pointing vector of the module, i.e. that vector that when aligned with the local sun vector produces the modules maximum power output. Acceptance angle for each module is characterized by the cone drawn by the vectors at this angle from the pointing vector which is then the cones axis. Simplifying, power is assumed to drop down to zero outside the acceptance angle cone, and also a worst case scenario is assumed in terms of acceptance angle losses, in which all modules are supposed to be connected in series. Thus, the set of tracker orientations producing nominal power output for a certain aperture elevation angle is taken, as the set of vectors pertaining to the acceptance angle cones of all the modules, i.e.

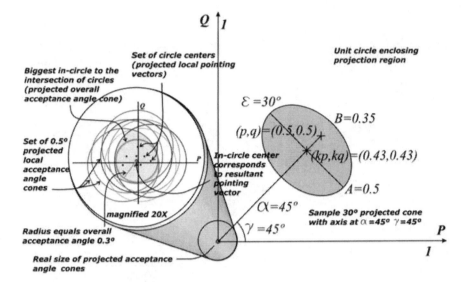

Fig. 3.7 Illustration of the plane projection of the modules local pointing vectors and associated acceptance angle cones, for the determination of worst case complete array pointing vector and acceptance angle

their intersection. The acceptance angle at this aperture elevation can then be defined as the maximum cone contained in this intersection of cones, and the axis of this cone is taken as the overall concentrator pointing vector.

The problem of determining this overall acceptance angle cone can be better viewed and solved if the pointing vectors and their respective acceptance angle cones are projected in the plane. This more precisely meaning the projection of the intersection of pointing vectors and cones with a unit radius sphere whose center coincides with the origin of all the pointing vectors. In this way, as can be seen in Fig. 3.7, every module pointing vector is transformed into a point in the plane, having as Cartesian coordinates its direction cosines with respect to plane reference axes and cones are transformed into ellipses. The flexure turning angle of a certain pointing vector will be small and therefore its projected coordinates appear very close to the reference system origin, which represents the pointing vector of the concentrator if the tracker was ideally rigid and undeformable, and the distance of each pointing vector to the origin is its bending rotation angle. For these pointing vector points located close to the origin, their corresponding acceptance angle ellipses can be approximated by a circle, centered in the pointing vector coordinates. On the other hand, high concentration CPV modules usually have small acceptance angles (in the sub-degree range) and, in this case, the radius of the projected circle representing the acceptance angle cone equals the acceptance angle itself.

So, after this projection, we can reformulate the problem of obtaining the maximum cone contained in the intersection of module acceptance angle cones, as that of determining the maximum incircle to the intersection of acceptance angle

3 The Sun Tracker in Concentrator Photovoltaics

circles in the plane. The center of this incircle – the incenter – will then represent the projection of the overall concentrator pointing vector. It can be proven that finding this maximum incircle is equivalent to determining the minimum enclosing circle (MEC) containing all the pointing vector points, where the center of this MEC coincides with the incenter of the maximum incircle. Also the radius of the maximum incircle, i.e. the overall acceptance angle, equals the acceptance angle of a single module minus the radius of the obtained MEC, which in this way represents the acceptance angle loss due to flexure. MEC determination for a set of points in the plane is a classical computational geometry problem first stated by Sylvester in 1857, for whose solution we implemented the most efficient algorithm to date due to Welzl, able to achieve $O(n)$ linear running time [4].

Applying this model to the FE simulations obtained from the pedestal tracker of our case example produced the plot of acceptance angle loss as a function of aperture elevation for both front and back maximum service wind speeds shown in Fig. 3.8. Also in this figure, the different MECs for aperture elevation angles taken every 10° from 0 to 90° are shown along with the centers of each MEC, which shows how the overall pointing vector also moves due to flexure. In this case, the local pointing vectors used at each elevation are only those corresponding to the modules mounted along the aperture perimeter, which are the ones that suffer the biggest bending. As can be seen from the acceptance angle loss graph, the maximum acceptance angle loss is 0.063°, and it occurs with maximum service front wind at 70° of aperture elevation. So having started with 0.1° maximum bending threshold under maximum service conditions, and then obtaining an optimum design with 0.076° maximum flexure, has finally resulted in a maximum acceptance angle loss of 0.063°. Getting back, at this point, to our modules nominal acceptance angle,

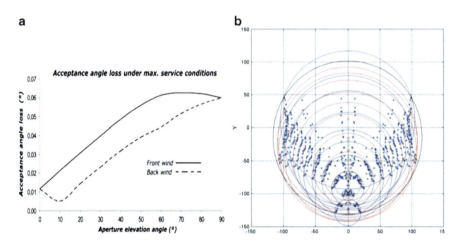

Fig. 3.8 (a) Estimation of a worst case acceptance angle loss in the 30 m^2 pedestal tracker as a function of aperture elevation and subject to maximum service conditions. (b) Pointing vectors and MECs at different elevations with maximum service wind speed windward and leeward to modules active surface

and the expected tracking accuracy, suggests we could try to further relax the bending threshold and go for a second optimization iteration to further lighten the structure and reduce its cost. It is to be pointed out that the aperture elevation angles, producing on the one hand maximum bending of local pointing vectors, and on the other maximum acceptance angle loss, do not necessarily coincide because as said, the turning angle computed is also affected by the pedestal and other global components which equally affect all aperture pointing vectors, and therefore do not contribute to acceptance angle losses.

Several other trackers were designed and manufactured by Inspira following this flexure constrained approach, being representative samples the ones built for Concentrix and Solfocus CPV modules, with maximum flexure 0.2° and 0.3°, respectively (Fig. 3.9).

Fig. 3.9 Examples of trackers designed and manufactured by Inspira for different CPV module technologies subject to flexure constraints. (**a**) 48 m^2 aperture surface tracker with 0.3° maximum flexure for Solfocus CPV (**b**) 36 m^2 aperture surface tracker with 0.2° maximum flexure for Concentrix CPV

3 The Sun Tracker in Concentrator Photovoltaics

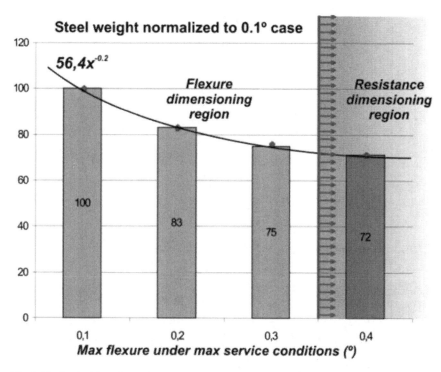

Fig. 3.10 Steel weight for a CPV tracker design when varying its maximum flexure constraint normalized to the 0.1° most restrictive case

The cost reduction achievable in the tracking structure through the increase of the flexure threshold and hence, weight reduction, has been plotted for a particular CPV tracker design. This was resized for each different threshold [3] and and its overall weight decreased following a potential law (Fig. 3.10). It must be said that this behavior is to some extent design specific, and other design approaches to the tracker structure, that for example do not rely exclusively in standard construction parts, may behave differently. However, this analysis does provide a good example of how the reduction of the flexure constraint reaches a limit, that we can call the flexure floor, beyond which it is ultimate structural strength under the maximum loads specified by standard building codes, the more stringent condition determining the sizing of the tracking structure, limit which in this analysis occurs in between 0.3 and 0.4°. Obviously, this flexure floor will also depend on the maximum service wind speed and the set of maximum loads, due to snow or wind, specified by the ruling construction codes in the region for which the tracker design is intended.

This design and sizing process for the structural part of the tracker is presented here in the sequential and iterative way in which it happened in the Inspira early designs, which on the other hand, appears as the most intelligible when compared to present design methods which have evolved afterward, more automated and parallel. For example, present design method at BSQ Solar, sketched in Fig. 3.11, has further

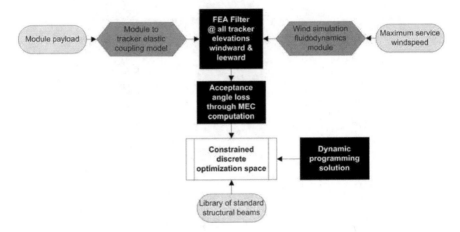

Fig. 3.11 Tracker structural dimensioning process carried out by at BSQ solar

gained accuracy by introducing additional modelling to simulate the elastic coupling of quasi-rigid modules and the aperture frame as well as fluidodynamics modelling to simulate wind loads. This method also merges FEA structural bending simulation and acceptance angle loss estimation through the MEC method, in a single process that enables obtaining the optimum tracker topology and sizing, in a constrained discrete optimization space built from a library of standard structural beams, through the use of dynamic programming methods devised for the so-called Knapsack kind problems [5].

However, further refinements of this design methods are still being developed such as the very significant of not considering overall tracker weight as the only merit function to minimize, but also take into account materials, manufacturing, and installation costs, which even if being much more case dependant, will provide when the data is available, the last fine tuning to the optimization. On the other hand, the MEC method that has proven to be computationally simple in most practical cases requires further linkage with the electric behavior of the CPV array so that the interconnection of modules and cells within them can also be taken into account and optimized.

3.4 Sun Tracking Control

3.4.1 Background

Early sun tracking controllers were developed following the classical control system closed-loop approach by integrating a sun sensor able to provide pointing error signals, one per tracking axes, which generate motor correction movements [6, 7]. This sun sensor is essentially constituted by a pair of photodiodes and some sort

3 The Sun Tracker in Concentrator Photovoltaics

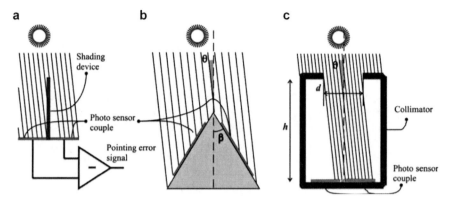

Fig. 3.12 (a) Shade balancing principle of sun pointing sensors. (b) Tilted mount of photos sensors to increase sensitivity. (c) Precise sun pointing by means of a collimator

of shading device, which casts a different shade on these photodiodes, therefore generating different photocurrents, whenever it is not aligned with the local sun vector (Fig. 3.12a). Added to this, the photodiodes can be mounted on tilted planes in order to increase the photocurrent sensitivity through cosine effect (Fig. 3.12b) and, very commonly in CPV applications, the shading device is presented as a collimating tube which prevents diffuse irradiation from entering the sensor and masking a precise measurement of the sun alignment position (Fig. 3.12c).

Even if this closed-loop approach can be very cheap and simple to implement, it has already gathered significant field experience to unveil some recurrent problems affecting its reliability [8], mostly caused by drifts in the analog electronics involved, and the requirement of cleanliness. This imposes the requirement of frequent maintenance which may possibly be affordable in research centres, where cared by attentive technical personnel, but is not feasible for the control of large scale industrial tracker fields. Furthermore, closed loop controllers have proved not to perform well under less than ideal illumination conditions. For example, due to the fact that the irradiance within a sensors acceptance angle is averaged, there is a funny phenomenon by which the bright reflection of a nearby cloud can cause tracking errors in the one degree range even when the sun is visible. This to the extent that, when the sun is hidden, closed-loop controllers have been reported to track bright clouds drifting away from the sun. This simple closed loop controller is also by itself unfit to manage nontracking and stowing situations. Due to their limited acceptance angle, the reacquisition of the sun after overcast periods is usually time consuming and inefficient, if not complemented with auxiliary control electronics. Finally, in high accuracy applications, a fundamental handicap arises provided they are to be aligned with the peak power output of the CPV array under control which, being a difficult operator requirement in itself, may even not suffice in big aperture trackers where, as seen in the past section, structural flexure varying with the tracker orientation impede a stable alignment. Nevertheless, these sun pointing sensors remain a fast pathway to CPV compliant sun tracking control. In recent times, highly integrated versions of these devices have been developed [9], and they remain an auxiliary constituent of the tracking control unit in several CPV systems.

Also brought up in the early days of sun tracking, as an alternative to sun pointing sensor controllers, the possibility of digital computing of sun ephemeris and converting this output into tracking axes turning angles, gave way to, again using the control theory term, open-loop controllers, which required no feedback of sun position measurements. These were able, in principle, to keep on tracking no matter the degree of clearness of the sky, and easily programmed the management of nontracking situations such as night or emergency stowage, e.g., when subject to high winds. However, a precise timing source is to be provided to feed the computation of the ephemeris equations, and also, in implementations seeking subdegree accuracy, some sensing device able to measure axes turning angles. Heliostat fields in solar thermal, such as in DOE's precursory Solar One plant (10 MW, yr. 1981) were the first to implement open-loop tracking, soon followed by also a grand CPV forerunner such as ARCO's Carissa Plains plant (6 MW, yr. 1985). Being digital computers still expensive, this first open-loop demonstrations were carried out in a centralized way, where a single computer continuously calculated turning angles for all trackers in the plant and transmitted them using a field data network. The advent of cheap microprocessors and embedded electronic systems enabled the development of specific open-loop tracking controllers, at feasible unit prices, which enabled the autonomous control of every tracker in a plant. Autonomous tracking control is not only inherently more reliable due to its distributed approach, but also because it gets rid of a complex and expensive field communication system, which due to its broad coverage, was frequently reported vulnerable to, for example, ground loop currents. Even if first patents and publications proposing these specific open-loop controllers can be traced back to the 1980s [10], it is the $SolarTrak^{TM}$ controller developed in the first nineties by Sandia Labs' Alexander Maish the first serious and well documented effort done in this direction.

However, an open-loop controller, even if operating on the very precise sun ephemeris equations available to date, is affected, once connected in the field to its corresponding concentrator, by a set of error sources which can highly degrade its final tracking accuracy well below its ephemeris' nominal value to the point of even missing concentrator's specifications. Among these error sources, the most significant ones have a deterministic nature. They result from a defective characterization of the concentrator by the controller, and operate over the transform employed to convert sun ephemeris coordinates, usually in the Az.-El horizontal topocentric format used in solar applications, into tracking axis turns. Tolerances of the manufacturing, assembly and installation processes of a concentrator will produce some deviations with respect to specifications and therefore, also to the assumptions made in the sun coordinates to axes turning angles transform. Drifts in the internal timing required for the computation of the sun ephemeris is the other major error source which is to be restrained. Second-order error sources, and also to some extent predictable, such as gravitational bending in wide aperture trackers, the effect of mismatch in multi-secondary axis trackers, or even ephemeris inaccuracies, due to the effect of local atmospheric refraction, might have to be considered as well. Feedback of the tracking errors caused by the referred sources is to be somehow integrated, in the control strategy, in order to suppress them. This open-loop core strategy blended with a feeding backloop is sometimes referred as the hybrid approach.

We can talk of basically two types of hybrid sun tracking controllers, whether we follow (1) the model-based calibrated approach, or the (2) the model free predictive approach. The calibrated approach relies on a mathematical error model, able to characterize the set of systematic error sources responsible of degrading tracking accuracy below that provided by the core sun ephemeris equations. After a full clear day session obtaining tracking error measurements, these are used to fit the model parameters. Provided the acquisition of error measurements is a time-consuming task, some degree of automation in this process is required when used in large tracker fields, in order to permit the simultaneous set up of them all and avoid the need of personnel to carry out this task. After the calibration session, the error model tuned with these best fit parameters will be used as the transform converting the sun coordinates supplied by the sun ephemeris to trackers axes turning angles, and thus will, in principle, operate from then on, on a purely open-loop basis with no further requirement of tracking error feedback.

Automatic calibration routines are commonly featured in electronic instrumentation products. Very similar approaches to this type of hybrid sun tracking control are the ones commonly found for the calibration of the pointing control of many telescopes in scientific observatories worldwide, such as happens with the commonly used TPoint software [11]. Among the early developers of this technique is Nobel Laureate Arno Penzias, the discoverer of the background radiation. When Penzias first joined Bell Laboratories, he was put on the pointing committee of an antenna built to communicate with the Telstar satellite. Aiming errors occurred because the steel antenna bents under gravity, wind load, and temperature changes, nor were the antenna's gears perfect, and its foundation was not perfectly horizontal either. Penzias solution was to calibrate it using an error model fitted by pointing to a known and precisely located radio galaxy [12].

On the other hand, the predictive approach to hybrid sun tracking [13] avoids getting into any error modelling and its subsequent fitting. It intends, instead, to avoid initial assumptions regarding the tracking errors that will be encountered, thus seeking a general purpose conception able to cope with any sort of tracking errors at whichever tracker design. However, to achieve this it requires the integration of permanent tracking error surveillance implying, once again, as in the case of a calibration session, some scanning scheme to determine correct sun positions. So, in this case, corrections to sun positions provided by computed ephemeris will result from an estimation based on some set of past tracking error measurements and estimations, and for this purpose, the wide mathematical toolbox for time series forecasting is at hand. The more general form of this approach is presented in Fig. 3.13 in which it all begins with the computation of the sun ephemeris to provide a first set of sun coordinates. As represented these ephemeris have to be corrected, due to whichever error sources or simply because the type of sun orientation

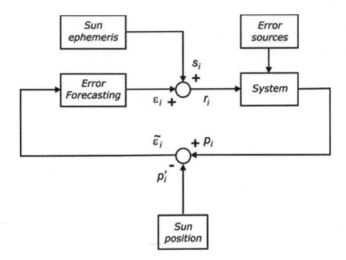

Fig. 3.13 Error model free hybrid sun tracking control relying on error scanning and iterative forecasting

Fig. 3.14 View of the string of 14 *EPS-Tenerife* tracking controllers designed and produced by Inspira for the EUCLIDES 480 kWp CPV plant in Tenerife

coordinates employed are not matched with the real tracking axes employed, such as for example would happen if providing horizontal Az.-El. coordinates to a two axis Tilt-Roll tracker. Some scanning scheme is used by the tracking axes to obtain some precise sun pointing and the correct axes turning angles from which to obtain tracking error measurements, which are then to enter the box labelled error forecasting, which produces corrections to be added to the next "raw" sun ephemeris coordinates.

First implementation of this tracking control approach was that developed by Inspira for the EUCLIDES[TM] CPV technology called *EPS–Tenerife* (Fig. 3.14), and in which its error correction estimates were computed using one of the most simple and widely used time series forecasting methods, as is exponential smoothing, however, in this case with an adaptive scheme for the variation of its parameter. The required gathering of tracking error measurements to feed the estimator is highly simplified whenever EUCLIDES was a single axis linear trough and accurate sun pointing measurements at each time could be acquired by exploring with back and forth turns [14, 15]. Other model free hybrid approaches have been proposed such as the one using a discrete version of a classical proportional-integral (PI) controller as the error forecasting method [13]. Correction estimation makes sense when precise sun pointing is a costly task, such as can happen when this is obtained through power output maximization, so that in this case, prediction will to some extent reduce scanning time and increase mean tracking accuracy. However, as said, some present concentrator tracking controllers work on a two-stage basis, first coarsely aiming based on sun ephemeris computed coordinates followed then by fine pointing using a sun sensor. Leaving aside the discussed reliability of a sun pointing sensor, and provided it is always kept well aligned with maximum power output, this is a feasible method moreover when pointing a sensor is simpler than seeking for maximum power orientation, and can be classified with the hybrid model-free approaches as the simplest case involving no forecasting at all.

3.4.2 The Autocalibrated Sun Tracking Control Unit

Inspira's *EPS-Tenerife* tracking control unit made the correction estimates dependent on the tracking angle of the EUCLIDES single axis tracker, and these estimates were kept in a memory stored look up table, one per each one degree tracking sector, being permanently updated based on the referred adaptive forecasting process. Even if these forecast estimates, together with the tracking errors measured for their generation happen to vary continuously during the year, this variation is basically seasonal because it is mostly caused by the above referred systematic characterization errors. We can attempt to model these systematic errors and correct them from a start, in this way making the scheme of permanent scanning movements unnecessary, and thus reducing motor fatigue and increasing tracking accuracy. This will be even more advantageous in the case of two axis trackers which require more complex scanning routines, which will further subtract from the accurate tracking operating time.

Later on, an error model was also developed by Inspira, which was termed calibration model, that assumed the tracker's axes and their reference orientations have the same reference system as the horizontal Az.-El. coordinate system used by the ephemeris. This means that the axis connected to the foundation, defined as the primary axis, is pointing to the local zenith with its reference orientation pointing south, and the secondary axis, the one which is linked to the primary, always remains at right angles with it, and has its reference orientation pointing the horizon, in other words, the ideal Az.-El. pedestal tracker.

The error model is based on a six-parameter kernel characterizing the departure of the real tracker to be controlled from that which ideally assumed:

- *Primary axis azimuth (φ) and zenith angle (θ):* These two parameters are the azimuth and zenith angle coordinates determining the real orientation of the primary axis, which no matter the axes configuration is always defined as the axis, which is fixed to the ground. This is mainly an installation error due to the imprecise foundation of the tracker.
- *Primary axis offset (β):* This parameter determines the location of the reference orientation of the primary axis. Reference orientation is usually determined by a specific sensor, or the index mark when working with incremental optical encoders directly installed in the primary axis. Misplacement of this sensor during manufacturing or its incorrect alignment at installation may cause this error. When $\varphi = \theta = 0$, β simply becomes the angular offset to the south. These first three parameters are in the referred order, the nutation, precession, and spin Euler angles, which relate any two reference systems with a common origin, and only these will be required if our tracker has only one axis, the primary axis, such as in present polar, azimuthal, or EW or NS horizontal axis trackers. When a secondary axis is attached to the primary, three more parameters are required and the pointing vector is defined as that one which is oriented by the joint action of the two tracking axis, and if aligned with the sun vector produces the maximum power output of the concentrator array.
- *Nonorthogonality of axes (λ):* This parameter takes the value of the difference to the right angle between the primary and secondary axes. This mainly being a manufacturing error source, a nonzero value for this angle implies the two-axis tracker is no longer ideal, and a cone of orientations around the primary axis will remain out of reach.
- *Pointing vector axial tilt (δ):* The pointing vector is assumed normal to the secondary axis and contained in the horizontal plane when this axis rotation is zero. The axial tilt of the pointing vector is the difference angle to a plane normal to the secondary axis. This error can have its origin not only in the defective assembly of the trackers aperture frame, but also in the misalignment of the concentrator optics.
- *Secondary axis offset (η):* The secondary axis offset accounts not only for the difference angle between the plane normal to the primary axis and the reference orientation of the secondary axis, but also for the difference angle between this reference orientation and the plane containing the pointing vector and the secondary axis, i.e. a radial tilt which is the second value characterizing pointing vector departure from assumptions. This error therefore derives from both the misplacement of the secondary axis reference sensor or, again, the improper assembly of aperture frame or optics.

These six parameters appear in a $\mathbb{R}^2 \to \mathbb{R}^2$ function, consisting in the composition of five partial transforms, which convert the ephemeris horizontal coordinates into pairs of angular rotations for both tracking axes. For single axis trackers, only the first three parameters enter into play and it is the primary axis turning angle

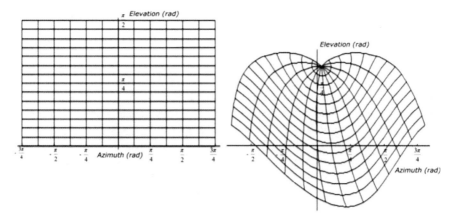

Fig. 3.15 Calibration model transform of a rectangular grid in the Az.-El. coordinates plane, into the two tracking axes turning angles plane for parameter values $\theta = 0$, $\lambda = 20°$ $\delta =, -20°$

the valid output. Behavior of this calibration function can be visualized through the usual grid transform representation in complex variable analysis (Fig. 3.15). The fact that the assumed reference system for the tracker under control is that of an ideal Az.-El. pedestal tracker is just a convention, and the model is able to correct horizontal ephemeris coordinates to any one or two axes configuration, including others frequently used such as, for example, the tilt-roll assembly (ideally $\varphi = \theta = \pi/2$). In order to maintain the generality of the model, no simplifying assumptions are made regarding the transform parameters, which otherwise would restrain its application range.

As said, the parameters characterizing a specific tracker and its in-field installation are to be fitted to a set of tracking error observations and, due to the nonlinear nature of the model, it is to be by means of numerical optimization techniques. Inspira's target was to integrate this numerical procedure in a low cost embedded system, and therefore programming efficiency was a must, as well as the good conditioning of the maximum likelihood estimation (MLE) function. Least squares (LS), is the MLE chosen which, even if there are other more robust estimators less sensitive to outlier measurements and fat tailed distributions, this one is by far the one presenting the most effective nonlinear minimization techniques. The existence of local minima obliges to resort to the global optimization toolbox, and finally a clustered multi-start algorithm with Levenberg–Marquardt (LM) [16, 17] based local searching is implemented. These local minima sometimes depend on the day of the year on which the tracking error measurements are made. For example, in NS oriented single axis trackers particularly strong local minima appear when calibrating on the eve of the equinoxes.

Even if Inspira's first development prototypes relied on manual collection of tracking errors, this proved to be a tiresome and error prone task, which had to be automated in order to prevent outliers from creeping in and thus increase the accuracy of the corrected ephemeris. An automatic error collection scheme was developed, in which direct search of the alignment of the CPV array pointing vector and the local sun vector, in order to obtain each tracking error measurement, required the maximization of the concentrator's power output. However, in first instance and in order to avoid interaction with the inverter's MPPT, or to be obliged to dissipate a high power, an approximately equivalent variable such as the CPV array's short circuit current was employed as feedback signal. Sun precise alignment proceeds in three stages. First a coarse approach by maximizing the irradiance in a PV cell mounted parallel to the concentrator's aperture. Beyond this point search proceeds blindly scanning by means of spiral search of the Koopman kind [18] until the sun enters the concentrator's acceptance angle and then, a two dimensional short circuit current maximization is to be carried out [19] whose complexity largely depends on whether the power output vs. off-track angle function exhibits rotational symmetry or not [20].

The above described capacities were implemented by Inspira in an electronic embedded system, based on an 8-bit microprocessor, along with the required chipset and sensors to carry out the described algorithms and perform the analog measurements, and also to provide motor driving capacities [21, 22] (Fig. 3.16). Remarkable hardware elements further enhancing performance are the temperature compensation circuit devised to restrain drifts in the quartz oscillator responsible for internal timing, as well as its encoder decoding and interpolation subsystem which increases the axis turn measurement accuracy. Named *SunDog*® STCU – always follows his master the Sun – it was supplied with *SunDog STCU Monitor*, a Windows application to run in a locally or remotely connected PC as a virtual user interface. It also integrated an interchangeable modem for PSTN, RF, Ethernet or GSM/GPRS Internet connectivity, which enabled e-mail reporting and

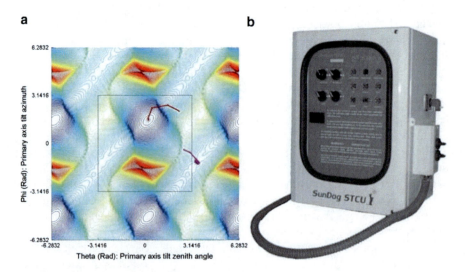

Fig. 3.16 (a) Levenberg–Marquardt local searches in the least squares function used to fit a simulated error model transform using for visualization purposes only its two first parameters (φ, θ). (b) *SunDog* STCU

web-based control and monitoring. Intended for operation in harsh environments, it was tested in electronic certification labs attaining CE labelling covering the corresponding EMC and electrical safety standards, and also successfully passed climatic tests (temperature cycling, humid and high salinity ambient, water and dust tightness, etc.)

This tracking controller designed by Inspira should rightly be considered a landmark, in the development of sun tracking control systems specifically designed for CPV Systems, as it is the first that fully automated the referred model-based calibrated approach, being in this way able to attain, as we will see below, very high tracking accuracies. However, further refinements can still be introduced in this type of controllers, such as, for example, extensions in the calibration model that account for second-order error sources such as flexure effects on the pointing vector position, which are usually specific to the tracker concept employed. Also in this line of possible improvements, regarding the very important adjustment of internal clock drifts. In this respect, even if Internet connectivity or GPS might provide atomic time synchronicity and this may be cost efficiently implemented in networked CPV tracker fields, the availability of high accuracy ephemeredes provides an immediate and autonomous alternative to precise time-keeping. A time drift parameter can be included in the calibration model in such a way that it can be fitted with a tracking error set either, jointly with the rest of the parameters or individually within periodic time adjustment procedures. Finally, the tracking error set in calibrated controllers is not necessarily to be obtained from Sun position measurements, as in principle any other light source with precise analytic kinematics, and with enough emitting power to extract a measurable output from the concentrator, will suffice. In this respect, the full Moon proves to be a good candidate, as far as it will enable night calibrations not interfering with concentrators daily production and even more, it will also permit these to be done with a maximum power point bias but at much lower power levels than nominal thus highly easing its handling in terms of switching and biasing hardware. As is well known, the moon's 0.49° apparent diameter is very similar to that of the suns, while its irradiance is six orders of magnitude smaller than the sun, so its photo-generated current is still within reach of cheap current sensing devices. On the other hand, the full moon irradiance is three orders of magnitude above that of the most brilliant planets and stars so it will be easily distinguishable by the concentrator when searching the night sky.

3.5 Sun Tracking Accuracy Monitoring

Assessment of sun tracking accuracy should not be overlooked during the development of CPV technologies, and even more, by those players raising very high concentration concepts over the 100× frontier. Some analyses point out that the acceptance angle of present designs in concentration optics may be overestimated even from a theoretical point of view, which added to the still uncertain acceptance angle losses inflicted on the overall system by mass assembly processes, may finally shrink the allowable tolerance and divert the entire burden to the tracking accuracy. Instrumentation for the monitoring of sun tracking operative performance, providing enough sensitivity to gauge the sub-degree accuracy ranges required by high concentration systems is therefore needed, and in this direction we can again refer the Tracking Accuracy Sensor (TAS) developed by Inspira for this kind of application [23]. This TAS, commercially named *SunSpear*® TAS, was essentially based on a so-called Position Sensitive Device (PSD), a monolithic optoelectronic sensor, which was housed along with the signal conditioning, digitizing, and transmission electronics, inside a watertight enclosure which in addition integrated a sunlight collimating tube placed right over the PSD's surface. The TAS was then to be installed on the aperture of the tracker to be monitored, and whenever the direct sunlight is received within its acceptance angle, the collimated sunbeam impinges on the PSD surface with the sensor then producing, as a voltage, the Cartesian coordinates of this sunspot, which can be further converted to an off-track angle with respect to the TAS' axis. The TAS was then linked by means of a serial connection to a PC which was to process the in-streaming sampling of sunspot coordinates, both displaying time series for significant variables and also producing its statistics for a specified time frame.

3.5.1 The Tracking Accuracy Sensor

In the PV field, there are few past experiences on which to base the development of a sensor, able to measure the incidence angle of direct sun radiation with respect to some built-in axis [24]. However, fairly accurate devices of this kind can be found in the aerospace sector, which based on CCD and CMOS arrays contribute to satellite attitude control [25]. These usually feature hemispheric acceptance angles, which preclude the extraction of higher accuracies from their very high resolution image sensors, nonetheless attaining the $0.05°–0.01°$ range. These are produced at very high costs due to their required compliance with spacecraft specifications, and usually on a custom-made basis without an open commercial intention, so even if they could serve our means, some CPV-related companies have recently found it worthwhile to develop a specific sensor.

This was the case of Inspira, which to our knowledge produced the first TAS for CPV applications, which they refer as the *SunSpear* TAS. The PSD sensor that was chosen for this TAS has no discrete elements such as in CCDs, and provides continuous data of a light spot on its surface by making use of the surface resistance of a planar PIN photodiode. Due to its analog nature, these sensors feature excellent position resolution in the micron range and very high speed; moreover, they detect the "center of gravity" position of the light spot and have proven a very high reliability. As usual, placing a light collimator on top of this sensor will produce the required light spot, and enable the measurement of the angle of the incoming light beam with respect to sensor's axis, where the acceptance angle and also the angular resolution of this measurement is basically determined by the height of

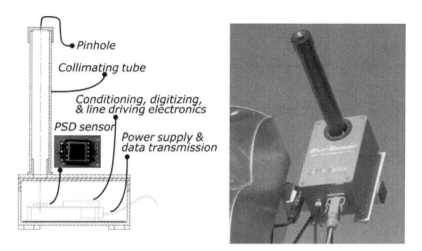

Fig. 3.17 The *SunSpear* tracking accuracy sensor

the collimator's pinhole over the sensors surface. With this *SunSpear* TAS design, resolutions in the 1/1,000th° range at a ±1° acceptance angle can be achieved (Fig. 3.17).

Along with the off-track angle measurement system, Inspira's TAS was completed by a custom-made electronic system providing signal conditioning to the PSD's output, AD conversion and driving a RS-232 serial output. It is this electronic system that hosts the PSD sensor in its PCB and is contained in a watertight enclosure also providing a fixture point for the machined collimator tube.

Confirming the requirement for a rigorous control of tracking accuracy for the development and manufacturing of CPV products, once this industry is growing in maturity, other tracking accuracy measurement sensors have later been developed following the trail of the *SunSpear* TAS that based on similar design concepts, are at last truly commercial products such as that of the Silicon Valley company GreenMountain [26].

3.5.2 The Monitoring System

In the case of the Inspira *SunSpear* TAS, the sampled position data generated is sent through a serial link to a PC, which is to run a monitoring software application able to store it but also to display it in real time, convert sunspot Cartesian coordinates to off-tracking angles, generate plots and statistics of selected tracking periods, and to periodically e-mail tracking data and reports (Fig. 3.18).

The first application given by Inspira to this tracking error monitor was in the assessment of the tracking accuracy of whatever hybrid tracking routines, which no matter if based on an error model or on error forecasting schemes, all have

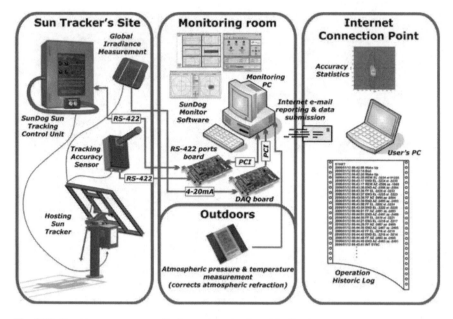

Fig. 3.18 Tracking accuracy monitoring system developed by Inspira

in common the requirement of obtaining power output feedback, or some other equivalent of this variable, from the CPV array, in order to get precise sun position measurements. The method used to test these strategies without having to mount a CPV array and its power output measurement electronics relies on using the TAS as a virtual power output, i.e., perfect alignment of the TAS built in axis with the solar vector is assumed as peak power output. Aside from the simplicity, an added advantage of this setup is that, in principle, it allows for very precise measurements at a faster rate than real power output maximization requiring the involvement of the MPPT stage. In this way, this method exposes the specific weaknesses of a certain tracking strategy, when almost not affected by errors in the sun position measurements whether these are used to feed the calibration model fitting or in error forecasting schemes.

Inspira gets into the specific case of testing an autocalibrated *SunDog* STCU controller, in such a way that once these calibration measurements are completed, some selection of them are fed fed into the error model Levenberg-Marquardt LS fitting routine, whose basic parameters regarding e.g., the clustered multi-start optimization routine or the stopping criteria, can also be fixed within the software running in the monitoring PC. Depending on the selection of error observations, some variation in the best fit parameters obtained can appear. In this respect, this software application also offers the possibility to periodically change the parameters being used by a connected *SunDog* STCU unit for sun ephemeris correction, in order to jointly obtain the tracking accuracy statistics of an assortment of varying sets of parameters and thus help to estimate those measurement schemes obtaining the best performance. In addition, the monitoring software application version implementing this tracking accuracy assessment tools when using the TAS as virtual power output for calibration, features additional resources to further restrict tracking error margins, by integrating the connection to an outdoors thermometer and a barometer. Ambient temperature and atmospheric pressure readings are employed in the computation of atmospheric refraction corrections to the sun ephemeris elevation coordinate, based on Bennet's model [27]. Measured values of these corrections have been reported to amount up to an average 0.6° for near the horizon elevations [28], and as will be seen in next section, they do have a measurable impact in a tracking accuracy monitoring campaign. These corrections are applied to computed sun ephemeris both in the error model fitting stage and also afterward during real-time tracking, in which the corrections are periodically transmitted to the *SunDog* STCU controller. Finally, and also contributing to tracking accuracy enhancement, the monitoring software can use top quality arcsecond precise ephemeris during the error model fitting data, thanks to a built-in direct connection to the USNO MICA software [29].

The second and more general application of the tracking error monitoring system consists in directly measuring the real tracking accuracy of a concentrator. This system will have to be initially calibrated against the peak power output of the CPV array, i.e. recording the sunspot coordinates on the PSD surface when maximum power is delivered and taking these as the TAS' reference system origin when converting its readings to off-track angle. Provided the effect of the tracker's self-weight and its CPV array payload upon the bending of the concentrator's structure varies with its orientation, the calibration is to be carried out at a set of different positions. Then the reference points to use in off-track angle conversion at each orientation will be interpolated from them [30]. This second function is independent of the tracking control means employed. However, during calibration it requires combined automatic readings both from the TAS and array's power in order to precisely locate the sunspot coordinates on the PSD when power output is maximum.

3.5.3 Accuracy Assessment: Example of the Autocalibrated Tracking Strategy

Following the procedure advanced above, we present here an example of long-term monitoring with the tracking error monitor applied to the assessment of the tracking accuracy performance of a *SunDog* STCU, when controlling a small aperture (4 m^2) laboratory sun tracker. The calibration was based on an error measurement session carried out on 31/01/06 that included 368 points. Four different model-based hybrid routines were set to compete:

- *Case no 1:* The six core parameters of the autocalibrated *SunDog* STCU error model were fitted directly in this controller's processor using all the error measurements. This case represents the normal *SunDog* STCU performance.

- *Case no 2:* Same as no 1 but the best fit values for the six parameters were calculated by the monitoring SW application running in a PC, with its added float point accuracy, and also using the superior performance of the MICA ephemeris when compared to the analytic ones computed by *SunDog* STCU (0.03° mean accuracy taking MICA as reference). This case works with more precise fit than that being attained by the LM embedded implementation in *SunDog* STCU. However as in case no 1, once fitting is completed, tracking control solely relies on *SunDog* STCU and its less accurate built-in ephemeris. It represents an operative alternative in which *SunDog* STCU units operating a CPV plant are networked and send their tracking error measurement sets for fitting to a more powerful central computer.
- *Case no 3:* Same as no 2 but temperature and atmospheric pressure measurements are activated, and atmospheric refraction correction in the ephemeris' elevation is integrated both during the tracking error acquisition session and also later providing the *SunDog* STCU real-time measurement pairs every 30s for it to internally calculate the corrections. This case is in principle the one which should present the highest performance and is useful to assess the possibility of integrating a thermometer and barometer in the *SunDog* STCU hardware.
- *Case no 4:* A simple two-parameter linear model is used to fit the error measurement set essentially obtaining the mean offsets in both tracking axes. This is the most straightforward tracking error model, which requires no numerical fitting algorithm, and in the same way that the nominal *SunDog* STCU performance of case no 1, it should rank below the enhanced no 3. This case is to serve as a low performance benchmark to rate the benefit of using the *SunDog* STCU nonlinear model along with its fitting procedures.

The monitoring ran uninterruptedly till 30/04/06. Every day tracking control of the lab tracker was assumed by a different case, following a fixed sequence. Every TAS sample includes along with the Cartesian coordinates of the sunspot on the PSD surface, the incidence light level. This incidence light level measurement is to be above a certain threshold to ensure the minimum required resolution, and in order to accept a monitoring daily session, 90% of its samples were to have its light level above this threshold. This basically means that only full clear sky days were considered for the analysis in order to compare the performance of the different cases on identical grounds. This means that, at the end of the monitoring period, some cases have had more valid days than others, but nevertheless, all had enough to draw some interesting conclusions. For every day the tracking accuracy statistics were calculated: mean, standard deviation, and the daily probabilities of accuracies below 0.1 and 0.05°. Probability density and distribution functions were plotted, and also the probability density of the sunspot point over the PSD surface was obtained.

In Fig. 3.19, daily probability density plots of the collimated sunspot over the PSD surface are included for all four cases, both for the first valid day and the last one comprised in the monitoring period. The 0.1 and 0.2° tracking accuracy rings are displayed in these plots. The purpose of presenting plots for the two ends of the monitoring period is to see how tracking accuracy may drift with time. In Fig. 3.20b,

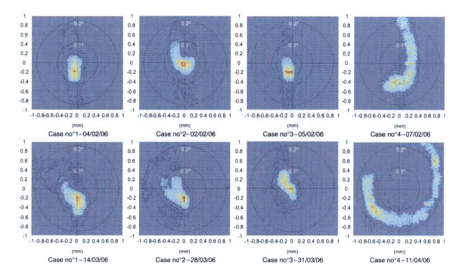

Fig. 3.19 Daily probability density of the sunspot over the PSD surface for the first and last monitoring day in each of the four calibration cases. The 0.1 and 0.2° off-tracking circles are represented. Mean value approaches zero with increasing precision of calibrated ephemeris, and standard deviation relates to positioning resolution

a tracking accuracy statistics sample is shown. This is representative of the best ratings obtained with the Case no 3 setup in this monitoring period, showing a mean daily tracking error of 0.05°, with Std. Dev. 0.02°, and having accuracy better than 0.1° with 98% probability and better than 0.05° with 50%. Also shown is the sunspot trace over the PSD's surface from which all statistics are obtained. Also, Fig. 3.20a, the daily plot of the tracking error is presented and it can be seen this may also prove to be useful to detect positioning resolution defects in a certain tracking drive, such as occur in the two error peaks appearing symmetrically with respect to solar noon, and which have to do with a momentary unleashing of the drives backlash at that precise elevation in which push and pull loads equate in the tracker's aperture.

Figure 3.21 plots the evolution of daily mean tracking error for the four cases during the monitoring period. Most obvious result, as can be also inferred from density plots in Fig. 3.19, is the superiority of the complete six-parameter error model (Cases 1–3) over the simplified mean offsets model benchmark (Case 4), moreover when the former drifts further apart during monitoring period and, as can be seen in its last recorded density plot of 11/04/06, it is finally incapable of entering the 0.1° ring with its mean accuracy rising over 0.2°. Besides, a more subtle drift is also found in the tracking errors of Cases no 1 and no 2 which disregard the atmospheric refraction effect. Even if Case no 3 accuracy was already slightly better than Cases no 1 or no 2 from a start, just after the calibration day, Case no 3 maintains the same ratings all over the reported monitoring period while Cases no 1 and no 2 suffer a slight decrease in accuracy clearly noticeable when entering the 2nd monitoring month. Main reason for this effect is to be found

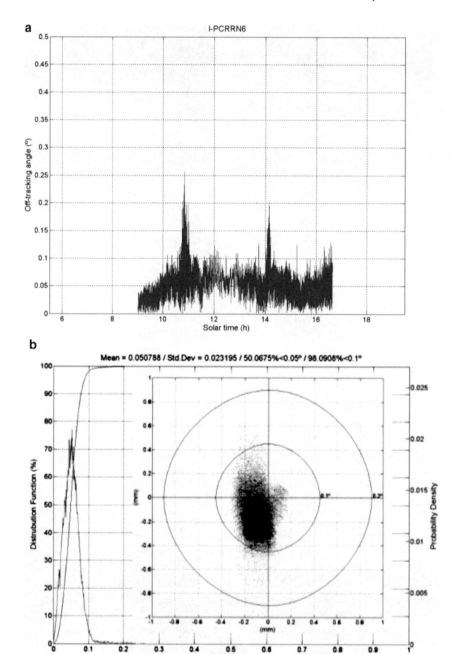

Fig. 3.20 (**a**) off-tracking angle during a day remains below 0.1 excepting two symmetric points with respect to noon at which aperture elevation has equal push and pull forces thus releasing backlash, (**b**) a typical daily off-track angle probability density and distribution functions with 0.05° mean and 0.02° std. dev., and superimposed the sunspot trace over the PSD

3 The Sun Tracker in Concentrator Photovoltaics

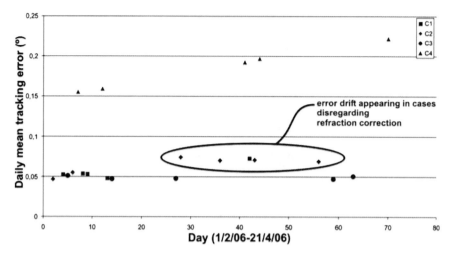

Fig. 3.21 Daily mean tracking error for the fully clear days during the 3 months monitoring period

in the fact that calibration is done in winter when sun elevations are lower and therefore, the atmospheric refraction correction is relatively more important during the day. In Cases no 1 or no 2, the atmospheric refraction effect is erroneously taken for an elevation axis offset effect and absorbed by the corresponding error model parameter (η secondary axis offset above). This defective identification is exposed when approaching summer and low elevations requiring atmospheric refraction correction become relatively less important. Mean tracking error slightly increases as seen encircled in Fig. 3.21 and regarding the probability distribution first 93–95% probabilities for errors below 0.1° drop down to 83–85%. Also remarkable is the fact that Cases no 1 and no 2 have a similar behavior and no significant advantage seems to derive from using the PC's higher accuracy float point arithmetic for the model parameter fitting, or the very accurate MICA ephemeris which based on the interpolation of tabulated numerically integrated solutions of the equations of celestial motion are not feasible for a low cost embedded integration.

References

1. D.E. Randall, N.R. Grandjean, Correlation of insolation and wind data for solmet stations (1982). SAND82-0094, Sandia National Laboratories
2. I. Luque-Heredia, C. Martín, T. Mañanes, J.M. Moreno, J.L. Auger, V. Bodin, in *Conference Record of the 3rd World Conference on Photovoltaic Energy Conversion*, Osaka, Japan, 11–18 May 2003, pp. 857–860
3. I. Luque-Heredia, R. Cervantes, G. Quéméré, in *Conference Record of the 21st European Photovoltaic Solar Energy Conference and Exhibition*, Dresden, Germany, 4–8 Sep 2006, pp. 2179–2183

4. E. Welzl, in *New results and New Trends in Computer Science, Lecture Notes in Computer Science*, Ganz, Austria, 20–21 Jun 1991, pp. 359–370
5. I. Luque-Heredia, E. Chiappori, O. Laurent, in *Conference Record of the 2nd Concentrated Photovoltaic Summit*, Toledo, Spain, 28–30 Apr 2009, p.
6. T.E. McCay, Sun tracking control apparatus (1992). US Patent no. 3,917,942
7. A. Luque, G. Sala, A. Alonso, J. Ruiz, J. Fraile, G. Araujo, J. Sangrador, J. Gomez Agost, J. Eguren, J. Sanz, E. Lorenzo, in *Conference Record of the 13th IEEE Photovoltaic Specialist Conference*, Washingtown DC, USA, 5–8 Jun 1978, pp. 1139–1146
8. A.B. Maish, The solartrack solar array tracking controller (1991). SAND90-1471, Sandia National Laboratories
9. J.M. Quero, L. García, A. Guerrero, in *Conference Record of the 2001 IEEE International Symposium on Circuits and Systems*, Sydney, Australia, 6–9 May 2001, pp. 648–651
10. J.H. Weslow, J.A. Rodrian, Solar tracker (1980). US Patent no. 4,215,410
11. Software bisque. t-point telescope pointing analysis software (1980). 2011-01-10. URL:http://www.bisque.com/sc/shops/store/All+Products/Hot+Products/tpoint-add-on-upgrade-windows.aspx. Archived by WebCite© at http://www.webcitation.org/5vd0M2Vhn)
12. J. Bernstein, *Three Degrees Above Zero: Bell Labs in the Information Age* (Charles Scribner's Sons, New York, 1984)
13. I. Luque-Heredia, F. Gordillo, F. Rodríguez, in *Conference Record of the 19th European Photovoltaic Solar Energy Conference and Exhibition*, Paris, France, 7–11 Jun 2004, pp. 2383–2387
14. J.C. Arboiro, G. Sala, Progr. Photovoltaics **5**(3), 213 (1997)
15. I. Luque-Heredia, J.M. Moreno, Quéméré, R. Cervantes, M. P.H., in *Conference Record of the 3rd International Conference on Solar Concentrators for the Generation of Electricity or Hydrogen*, Scottsdale, AZ, USA, 1–5 May 2005
16. D.W. Marquardt, J. Soc. Ind. Appl. Math. **11**(2), 431 (1963)
17. A. Törn, S. Viitanen, in *Recent Advanced in Global Optimization*, ed. by C.F. Floudas, P.M. Pardalos. Chap. Topographical Global Optimization, pp. 431–441 (Princenton University Press, Pricenton, 1963)
18. B. Koopman, *Search and Screening: General Principles with Historical Applications* (Pergamon Press, NY, 1980)
19. R.P. Brent, Algorithms for finding zeros and extrema of functions without calculating derivatives. PhD. Discussion, Standford University, 1971
20. G. Quéméré, R. Cervantes, J.M. Moreno, I. Luque-Heredia, in *Conference Record of the 23rd European Photovoltaic Solar Energy Conference and Exhibition*, Valencia, Spain, 1–5 Sep 2008, pp. 887–890
21. I. Luque-Heredia, J.M. Moreno, G. Quéméré, R. Cervantes, P.H. Magalhães, in *Conference Record of the 20th European Photovoltaic Solar Energy Conference and Exhibition*, Barcelona, Spain, 6–10 Jun 2005, pp. 2047–2050
22. I. Luque-Heredia, J.M. Moreno, G. Quéméré, R. Cervantes, P.H. Magalhães. Equipo y procedimiento de control de seguimiento solar con autocalibración para concentradores fotovolaticos (2008). Spanish Patent no. 2273576
23. I. Luque-Heredia, R. Cervantes, G. Quéméré, in *Conference Record of the 2006 IEEE 4th World Conference on Photovoltaic Energy Conversion*, Waikoloa, HI, 07–15 May 2006, pp. 706–709
24. G. Galbraith, Development and evaluation of a tracking error monitor for solar trackers (1988). SAND88-7025, Sandia National Laboratories
25. A.S. Zabiyakin, V.O. Prasolov, A.I. Baklanov, A.V. Eltsov, O.V. Shalnev, in *Conference Record of the SPIE the International Society for Optical Engineering*, Prague, Czech Republic, vol. 3901, 10 Mar 1999, pp. 106–111
26. M. Davis, J. Lawler, J. Coyle, A. Reich, T. Williams, in *Conference Record of the PVSC: 2008 33rd IEEE Photovoltaic Specialist Conference*, San Diego, USA, 11–16 May 2008, pp. 1–6
27. G.G. Bennett, J. Instit. Navigation **35**, 255 (1982)
28. B.E. Schaefer, Publ. Astron. Soc. Pac. **102**, 796 (1990)

29. Multilayer interactive computer almanac 1990-2005 version 1.5. U.S. Naval observatory (Willmann-Bell, Richmond, 1998)
30. R. Cervantes, G. Quéméré, I. Luque-Heredia, in *Conference Record of the 23rd European Photovoltaic Solar Energy Conference and Exhibition*, Valencia, Spain, 1–5 Sep 2008, pp. 891–894

Chapter 4
Light Management in Thin-Film Solar Cell

Janez Krč, Benjamin Lipovšek, and Marko Topič

Abstract Employment of advanced light management techniques presents an important aspect in the design of thin-film solar cells. In this chapter, we highlight a number of light management approaches leading towards higher cell performances. Efficient light scattering within the cell, which can boost the photocurrent generation, can be achieved by optimised surface textures. Random textures and periodic textures for efficient light scattering are addressed. Besides surface texturing, the role of metal nanoparticles in thin-film solar cell structures is investigated in the scope of improved light trapping. To minimise optical losses, antireflective coatings and advanced back reflectors are employed. As examples, photonic crystal structures and diffusive dielectric materials are presented. And finally, better utilisation of the solar spectrum can be achieved by multi-bandgap multi-junction cells. For efficient spectrum harvesting, a concept of wavelength-selective intermediate reflectors is investigated. Further on, a concept of spectrum splitting and dislocated cells connected in a multi-terminal configuration is presented.

4.1 Introduction

Thin-film photovoltaic technologies have gained an important position in the photovoltaic industry and market. Small material consumption and low-temperature deposition processes applicable to large areas in either batch or roll-to-roll production lines make thin-film solar cells and modules competitive to other solar cells, regarding total costs and state-of-the-art performances [1]. In order to boost the conversion efficiency of thin-film solar cells and/or to further reduce the thickness of the absorber layers (meaning shorter deposition times and smaller material and

---

M. Topič (✉)
Faculty of Electrical Engineering, University of Ljubljana, Ljubljana 1000, Slovenia
e-mail: marko.topic@fe.uni-lj.si

energy consumption), advanced light trapping techniques are of great importance [2] . Their aim is to efficiently trap or distribute the light within the cell structure in order to increase the absorption in the thin active (absorber) layers. At the same time, optical losses in the supporting layers (contacts, doped layers) and the reflectance from the cell should be minimised.

To prolong the optical paths inside the absorber layers and, thus, boost the absorption and consequently the photocurrent of the solar cell, light scattering at textured interfaces is most commonly employed. For the solar cells in the so-called *superstrate* configuration (light entering through the transparent front superstrate), textured transparent conductive oxides (TCO) are used to introduce surface roughness and assure the desired light scattering. Thus, superstrates for thin-film Si solar cells typically consist of a glass carrier covered with a layer of either (1) SnO_2:F with pyramidal [3] or, lately, double surface texture [4–6], (2) magnetron-sputtered ZnO:Al [7, 8], or (3) LPCVD-deposited ZnO:B [9, 10] TCO. In the case of substrate configuration, on the other hand, surface roughness is introduced by textured substrates (textured metal sheets, textured plastic foils) [11–13]. Usually, surface textures exhibit a random nature, although periodic-like texturisation is also being employed [11,12,14–20]. Furthermore, as an alternative to textured interfaces, light scattering can also be achieved by metal nanoparticles incorporated inside the solar cell structure [21–27].

To minimise the optical losses in the supporting layers of the solar cells, novel back reflectors besides the conventional metal reflectors (silver, aluminium) have been investigated. Among them, alternatives based on dielectric reflectors, such as white paint [28–30] and photonic crystal structures [31–38], show much potential. The reflectance losses at the critical interfaces in the solar cells can be minimised by specific antireflective coatings and structures [39–42], while a more efficient utilisation of the solar spectrum can be achieved by multi-junction (double-junction tandem, triple-junction) devices. In the case of tandem devices, the role of an intermediate reflector between the top and the bottom cell becomes important to properly balance the light absorption between the cells, which are absorbing different parts of the solar spectrum [43–50]. Besides the approach of stacking the cells in a multi-junction two-terminal device, dislocated single-junction cells of different energy gaps in combination with an advanced spectrum splitting technique have been proposed [51,52]. And finally, nano-wires for efficient light in-coupling [53, 54], intermediate-bands [55, 56], quantum dots [55], and up- and down-converters [57–59] as an advanced approach to manipulate the light wavelength transitions have also been investigated recently.

In this chapter, a selection of advanced light management approaches in thin-film solar cells will be highlighted. First, the optical situation in solar cell structures will be described in detail in the second section. The potential of light trapping and the origin of optical losses in thin-film solar cell structures will be presented by means of optical simulations. Then in the third section, a number of such techniques will be described, covering a broad range of different approaches, such as advanced surface textures, novel back reflectors, photonic crystal structures and enhanced spectrum utilisation concepts. The potential of each of the techniques will be presented by means of the latest numerical and experimental data, and possibilities for further improvements will also be discussed.

4.2 Optical Properties of Thin-Film Photovoltaic Devices

From the optical point of view, thin-film photovoltaic devices (solar cells and photovoltaic modules) are multilayer structures consisting of semi-transparent thin and thick layers. The thickness of the thin layers is in the range of nanometres and micrometres, which is also in the range of light wavelength. Therefore, a coherent propagation of light inside these layers has to be considered. The thickness of the thick layers, on the other hand, is much larger, in the range of millimetres. Consequently, the coherency properties in these layers are lost, and an incoherent analysis of light propagation is required [60]. Further on, at each interface in the multilayer structure, reflection and transmission takes place. Due to reflections, the light is propagating not only in the forward but also in the backward direction. Thus, interactions between the forward- and the backward-travelling waves have to be considered in the case of coherent analysis. This leads to interference fringes, which can be observed in the output spectral characteristics of thin-film solar cells, such as the quantum efficiency and the reflectance of the cell. Besides reflection and transmission at the interfaces, the process of optical absorption within the layers takes place.

Propagation and absorption of light within a layer is determined by the optical properties of the layer material, which are commonly characterised by the wavelength-dependent complex refractive index, $N(\lambda) = n(\lambda) - j \cdot k(\lambda)$. The complex refractive index $N(\lambda)$ consists of the real part, $n(\lambda)$, which is referred to as the *refractive index*, and the imaginary part, $k(\lambda)$, called the *extinction coefficient*. The refractive index determines the effective wavelength of the light inside the material, $\lambda_{\text{eff}} = \lambda_{\text{air}}/n(\lambda)$, whereas the extinction coefficient determines the amount of absorption via the absorption coefficient, $\alpha(\lambda)$; $\alpha(\lambda) = 4\pi \cdot k(\lambda)/\lambda$. Both $n(\lambda)$ and $k(\lambda)$ determine the reflectance and transmittance properties at the bordering interfaces of the layer [61].

In thin-film solar cells, nano-textured interfaces are introduced either by natural poly-crystalline growth of layers (e.g. TCOs or chalcopyrite-based absorbers) or by using additional texturisation processes. Typically, the nature of such interface morphologies is random, though with specific processing steps regular (periodic-like) texturisations can also be achieved. Optically, textured interfaces are favourable, especially in the case of indirect semiconductors (Si), since they enable efficient light scattering inside the solar cell structure. Light scattering can lead to a prolongation of the optical paths inside the thin absorber layer, thus increasing the absorptance in the (long) wavelength region where the optical absorption of the semiconductor material is weak. An increased absorptance in the absorber layer leads to an enhanced photocurrent and thus improved conversion efficiency of the cell.

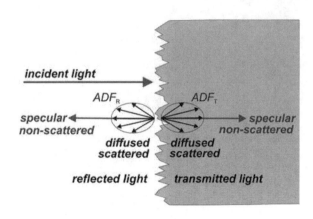

Fig. 4.1 Reflection, transmission and scattering effects taking place at a nano-textured interface between adjacent layers in the solar cell structure

Table 4.1 The primary optical phenomena, parameters, and theoretical background needed for describing the optical behaviour of light propagating through the solar cell structure (the absorptance A here corresponds to a single pass of light rays over a distance l_{opt})

Optical phenomena	Parameters	Theory
Absorption within a layer	Absorptance, A	$A(\lambda) = 1 - \exp(-\alpha(\lambda) \cdot l_{opt})$
Reflection and transmission at the interface	Reflectance, R	$R(\lambda) = \left(\dfrac{N_1(\lambda) - N_2(\lambda)}{N_1(\lambda) + N_2(\lambda)}\right)^2$
	Transmittance, T (total, specular, or diffused)	$T(\lambda) = 1 - R(\lambda) = \dfrac{4N_1 N_2}{(N_1 + N_2)^2}$
Scatterring at the interface	Haze, H	$H_R(\lambda) = \dfrac{R_{dif}}{R_{tot}} = \dfrac{R_{dif}}{R_{spec} + R_{dif}}$
		$H_T(\lambda) = \dfrac{T_{dif}}{T_{tot}} = \dfrac{T_{dif}}{T_{spec} + T_{dif}}$
	Angular distribution function, ADF	$ADF_R(\lambda) = I_{Rdif}(\varphi)$
		$ADF_T(\lambda) = I_{Tdif}(\varphi)$

In general, light scattering can be described by two types of descriptive scattering parameters: *haze*, H, and the *angular distribution function, ADF* [62]. Haze is defined as the ratio of the scattered light to the total reflected or transmitted light, whereas the ADF describes the angular dependency of the scattered light. Both types of scattering parameters are defined for reflected and transmitted light (H_R, H_T, ADF_R, ADF_T). Experimental and theoretical approaches are used to link the textured interface morphology to light scattering. One of the most commonly used statistical parameters to characterise the random texturisation is the vertical root mean square roughness, σ_{rms}.

The optical phenomena taking place at an interface between two layers and the descriptive scattering parameters are presented schematically in Fig. 4.1. Additionally, the optical effects described above, together with the primary parameters and main theoretical background, are summarised in Table 4.1.

4 Light Management in Thin-Film Solar Cell

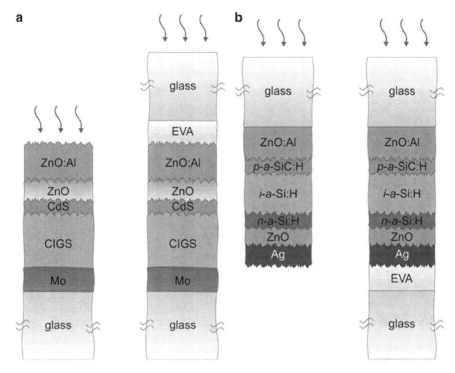

Fig. 4.2 Substrate (**a**) and superstrate (**b**) configuration of thin-film solar cells, on the cell (a and b, *left* figures) and PV module (a and b, *right* figures) level

Solar cell structures appear in two configurations, called *substrate* and *superstrate* configuration, as shown in Fig. 4.2 for the case of chalcopyrite-based Cu(In,Ga)Se$_2$ (CIGS) and amorphous silicon (a–Si:H) p–i–n solar cells. In the case of substrate configuration (Fig. 4.2a, left), the solar cell structure is deposited on top of the substrate which is located at the back side of the cell, whereas in the case of superstrate configuration (Fig. 4.2b, left), the superstrate (i.e. front substrate) is located at the front of the device and thus has to be transparent. For the use in PV modules, cells are encapsulated by a film of ethyl-vinyl-acetate (EVA) foil and an additional protective layer (glass or metal sheet). From the optical point of view, this encapsulation changes the situation in the case of substrate solar cells (Fig. 4.2a, right), since the optical properties of the additional two layers at front have to be considered when predicting the performance of the final encapsulated device. In the case of superstrate solar cells (Fig. 4.2b, right), on the other hand, no additional layers need to be considered, since the light does not enter the EVA film and the protective layer at the back side.

To summarise, a thin-film solar cell as a multilayer structure with thin (coherent) and thick (incoherent) layers with flat and textured interfaces (multiple reflections/transmissions, light scattering) presents a relatively complex optical system.

To thoroughly analyse the optical behaviour of such a system, advanced numerical modelling approaches hand in hand with different characterisation techniques are required.

4.2.1 Optical Modelling and Simulation

Over the years, several tools have been developed to simulate the optical behaviour of thin-film photovoltaic devices [63–70]. Most of them are based on optical models derived from the theory of optics and electromagnetic waves, which are then implemented in numerical simulators. An important part of optical modelling consists of the determination of realistic input parameters for the simulations. Here, employment of different characterisation methods is necessary [62, 68]. After the initial verification of the model, i.e. comparing the simulation results with the experimental ones, the simulation tool can then be employed effectively to study and optimise the device with respect to potential further improvements in any or all of the device components. Further on, novel device structures can be designed and their performance predicted. As this can be done quickly, inexpensively, and can cover a broad range of influencing factors, numerical simulation has become a powerful and invaluable tool for thin-film solar cell design and optimisation.

Different ratios in lateral (length, L) over vertical (thickness, d) dimensions of conventional and thin-film PV devices determine the number of space dimensions included in the model that are used for simulations. This, along with the method of discretisation, presents one of the critical characteristics of the simulation tool. On the one hand, increasing the number of space dimensions (up to three) and the number of discrete elements of the domain (e.g. in a finite element method) adds to the overal accuracy of the model, while on the other hand, it leads to an increased complexity of the system and is much more demanding on the computational power and resources. Thus, a careful balance between the two extrema must be achieved.

In thin-film photovoltaic devices, however, the lateral dimensions of the device are generally much larger than the vertical dimensions (typically, $L/d > 2{,}000$). Therefore, the lateral size of the device can be assumed as infinite, and simulation in just one (vertical) dimension (1D) is justified in many cases. Nevertheless, for the analysis of specific structures, for example periodic textures (diffractive gratings), the model must still be extended to the second (2D) or even third dimension (3D), depending on the nature of the periodicity of the texture [71].

As an example, the general principle of the 1D optical simulator Sun*Shine* [66, 67] is explained by means of Fig. 4.3. Each layer in the multilayer structure is described by its position in the stack, the thickness, and the wavelength-dependent complex refractive index. The realistic (measured) complex refractive indices of individual layers and the layer thicknesses are taken as the input parameters. Additionally, the measured roughness and scattering parameters of the interfaces also have to be included. In the model, the light is divided into two parts: the direct (specular) light with its coherent nature inside thin layers, and the scattered

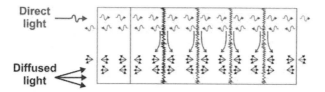

Fig. 4.3 The principle of operation of the 1D simulator SunShine. The propagation of the direct and the diffused light is analysed separately, while all the optical effects taking place at the interfaces and within the individual layers are taken into account

(diffused) light, which is considered to be incoherent everywhere (assuming random morphology of textured interfaces). For the evaluation of the coherent light propagation, the equations from the electromagnetic wave theory are employed, whereas the incoherent light is treated by the method of ray-tracing projected to one dimension. The scattering at each interface is also taken into account, based on the input parameters. As the output data of the simulations, we can determine the charge-carrier generation rate profile as the main input for further electrical simulations, the optical absorptances in the individual layers, the reflectance from the cell, etc.

The 1D optical simulator Sun*Shine* can be employed for analysis and optimisation of numerous advanced thin-film solar cell concepts, such as single and tandem structures with flat and textured interfaces [62, 72], intermediate reflector concepts [45], photonic crystal or white paint back reflectors [36, 73] (some of them will be discussed in detail in the later subsections).

As mentioned, for analysis of special structures, such as thin-film solar cells with periodic texturisation at the interfaces, a 2D or 3D approach is more convenient. As an example, results obtained with the 2D optical numerical simulator FEMOS [70, 71] are presented in Figure 4.4. The figure shows the electric field of a monochromatic gaussian light beam diffracting at a two-dimensional rectangular diffracting grating implemented at the TCO/p–a–SiC:H interface. The FEMOS simulator is based on the finite element method (FEM) and numerical solving of the Maxwell equations. The 2D domain discretisation into equisize triangular elements is shown in the inset of the figure.

4.2.2 Importance of Light Management in Thin-Film Solar Cells

Thin-film solar cells are based on different semiconductor materials with a wide range of energy bandgaps (E_G) and other material parameters. The wavelength-dependent absorption coefficients of the three most common thin-film semiconductors (a–Si:H, μc–Si:H, CIGS) are plotted in Fig. 4.5. Since silicon is an indirect semiconductor, a–Si:H and μc–Si:H do not exhibit a sharp absorption

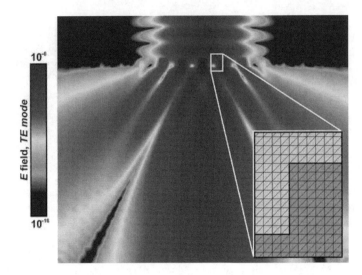

Fig. 4.4 The electric field intensity of a light beam diffracting at a rectangular grating, calculated by the 2D simulator FEMOS

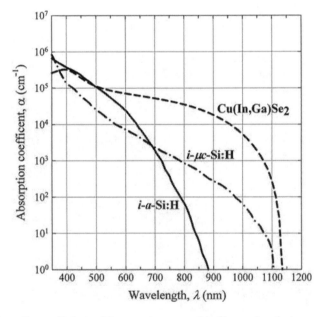

Fig. 4.5 Absorption coefficients of three most common thin-film semiconductors

edge at the bandgap wavelength as in the case of CIGS. While amorphous silicon absorbs heavily in the short-wavelength region, the absorption range is limited to $\lambda \approx 850$ nm by the high optical gap of 1.74 eV. Microcrystalline silicon, on the other hand, exhibits lower absorption coefficient in the short-wavelength region, but can absorb up to $\lambda \approx 1,100$ nm due to a lower optical gap of about 1.12 eV. Furthermore, if the energy carried by the photons is much larger than the optical gap of the material, the excess is manifested in thermalisation losses, which result in heating up the cell.

The challenge in the optical design of thin-film solar cells is to increase the absorption within the thin absorber layers and thus boost the photocurrent and consequently the conversion efficiency of the device. In this respect, a number of light-trapping approaches have been developed, which can be generally divided into three categories. First, techniques of efficient *light scattering* aim toward prolonging the effective optical path of the light passing through the cell. Second, the cell design should strive to *minimise the optical losses* within the supporting layers of the device (contacts, p- and n-layers in p–i–n cells) and due to the light reflected from the front interfaces. And finally, approaches towards an *enhanced utilisation of the solar spectrum* enable a broader part of the spectrum to be exploited and at the same time a reduction of the thermalisation losses caused by the photons carrying an excess of energy. In this concept light absorption is distributed between two or more absorber layers with different energy gaps. These, along with the supporting components of the cell, are usually stacked in a single two-terminal device with sub-cells electrically connected in series (double-junction tandem or triple-junction solar cell). Thus, the gain here is related to an increased open-circuit voltage, V_{OC}, and not to the photocurrent of the device, which is in general lower than in single-junction devices.

To demonstrate the importance of light trapping, we determined the potential increase in the solar spectrum absorption in a 300 nm thick i–a–Si:H and 1 μm thick i–μc–Si:H absorber layer if the effective optical path is increased by a factor of 10 and 50 (multiple passing is possible due to light scattering and trapping). In the simulations, this prolongation of the optical path was realised simply by increasing the thickness of the layers. In order to study only the effects related to the optical path length inside the layer, the reflections at the front and rear interfaces were excluded from the simulation. The short circuit current densities, J_{SC}, were calculated by assuming that each absorbed photon generates an electron-hole pair and that all the generated electrons and holes contribute to the electrical current (ideal extraction). The results for an optical path increase by a factor of 10 and 50 compared to a single pass are presented in Fig. 4.6. For one pass (1×) through the 300 nm thick i–a–Si:H layer, $J_{SC} = 13.11$ mA cm^{-2} was determined (Fig. 4.6a). If the effective optical path is 10× longer, the absorption is extended towards longer wavelengths, and J_{SC} is increased by 52%. Finally, if the optical path is prolonged by 50 times, an almost 80% increase in J_{SC} is achieved. These effects are even more apparent in the i–μc–Si:H absorber due to the extended absorption coefficient of the material towards longer wavelengths. In Fig. 4.6b, we can see that the initial J_{SC} generated in the 1 μm thick i–μc–Si:H layer (14.87 mA cm^{-2}) can be increased by 90% and 137% if the optical path is prolonged by a factor of 10 and 50, respectively. These results show that the prolongation of optical paths, which can be realised by means of efficient light scattering and trapping inside thin-film Si solar cell structures, is essential. A number of possible realisations will be presented in Sect. 4.3.1.

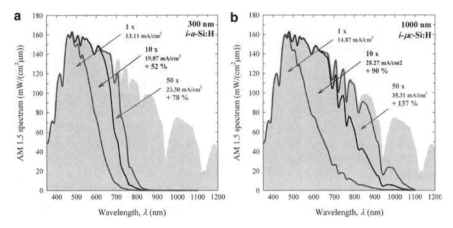

Fig. 4.6 The influence of optical path prolongation in thin-film i–a–Si:H (**a**) and i–μc–Si:H (**b**) layers on the utilisation of the AM1.5 solar spectrum

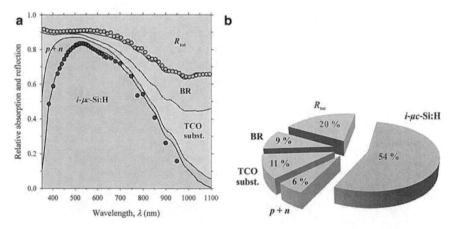

Fig. 4.7 The break-down of optical losses in a state-of-the-art μc–Si:H solar cell, presented as the light absorption distribution (**a**) and relative share of the solar spectrum in terms of potential J_{SC} (**b**) between the components of the device. Circles correspond to measurements, lines are simulation

In the solar cell structure, however, not all of the light spectrum is available for charge-carrier generation, since optical losses due to the optical absorption in the supporting layers and due to reflection occur. In Fig. 4.7a, the break-down of optical losses based on optical simulations is presented for the case of state-of-the-art μc–Si:H single-junction solar cell in the superstrate configuration with a 1.2 μm thick i–μc–Si:H absorber. Also shown are the measured external quantum efficiency, and total reflectance, which match the simulated results very well. The plot presents a relative distribution of the absorption of the solar spectrum between different layers, as well as the part corresponding to the reflected light. We can see that, generally, there are four sources of optical losses: the absorption in the p- and n-layers, the absorption in the TCO substrate (glass/textured ZnO:Al), the absorption and the transparency losses attributed to the back reflector (ZnO/Ag BR) and the reflection from the solar cell. For further evaluation and comparison, we calculate the distribution of the solar spectrum absorption between each of the components of the cell from the point of view of potential J_{SC} contributions (Fig. 4.7b). Whereas the J_{SC} in the absorber layer can be related to the actual J_{SC} of the cell, J_{SC} values in the other layers are attributed to the optical losses. The simulations show that from the maximum J_{SC} available from the full energy of the AM1.5 solar spectrum (40.9 mA cm^{-2}, $\lambda = 350$–$1,100$ nm) only 54% is absorbed in the i–μc–Si:H absorber layer. The rest are optical losses due to the reflection (20%), the absorption in the TCO substrate (11%), BR (9%), and p- and n-layers (6%). Therefore, approaches towards the reduction of these limiting effects are required and will be the topic of Sect. 4.3.2 .

A better utilisation of the solar spectrum can be in general achieved by multi-bandgap multi-junction solar cells. In Fig. 4.8a, the structure of a triple-junction a–Si:H/a–SiGe:H/μc–Si:H solar cell is presented schematically. The structure consists of the top a–Si:H cell with $E_{G,i-a-\text{Si:H}} \approx 1.74$ eV, which efficiently absorbs the shorter wavelengths, the middle a–SiGe:H cell with $E_{G,i-a-\text{SiGe:H}} \approx 1.5$ eV, absorbing the middle wavelengths, and the bottom μc–Si:H cell with $E_{G,i-\mu c-\text{Si:H}} \approx 1.12$ eV, which absorbs the longer wavelengths. As the individual cells are connected in series, a careful distribution of the solar spectrum between them must be assured since the total J_{SC} is limited by the lowest J_{SC} generated in a cell. In the optimal case, all three should generate approximately the same

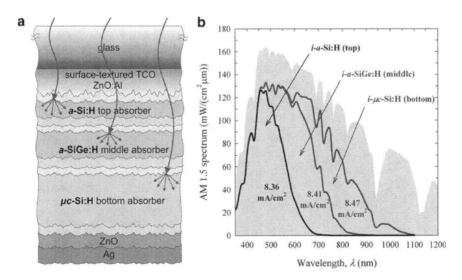

Fig. 4.8 The structure of a triple-junction a–Si:H/a–SiGe:H/μc–Si:H solar cell (**a**) and the corresponding utilisation of the solar spectrum (**b**)

amount of current (current matching). Here, optical modelling and simulations play an important role in the design of the device. The regions of the solar spectrum harvested by each of the three absorbers in a current-matched triple-junction cell are presented in Fig. 4.8). Compared to a single-junction μc–Si:H cell, higher cell performance can now be achieved due to the gain in V_{OC}, which is proportional to the sum of the three energy gaps. Approaches towards an efficient current matching and multi-junction cell design will be further discussed in Sect. 4.3.3.

4.3 Advanced Light Management in Thin-Film Solar Cells

In this section, we highlight a number of different light management approaches for thin-film solar cells which have been developed over the past years. Their common goal is to increase the efficiency of thin-film solar cells by enhancing the light absorption within the thin absorber layers. As mentioned in the previous section, the concepts are generally divided under three categories:

1. Improvements of light scattering
2. Reduction of optical losses, and
3. Enhanced utilisation of the solar spectrum.

All three categories can be interconnected.

4.3.1 Efficient Light Scattering

Light scattering in thin-film solar cells presents one of the most important techniques for substantially increasing the short-circuit current density and quantum efficiency of the device. This can mainly be attributed to two specific optical effects, both of which are schematically presented in Fig. 4.9. First, light scattering prolongs the effective optical paths of the light passing through the thin absorber layer. Large scattering angles are especially desired (broad ADF of scattered light). And second, since the refractive index of the absorber layer is usually larger than the surrounding media, such as TCO, the light travelling through the absorber at high angles

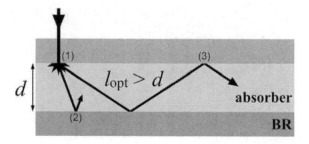

Fig. 4.9 Light-trapping in the absorber layer, as a result of light scattering (1), reflection at the back reflector (2), and total reflection for high incident angles at the front (3)

4 Light Management in Thin-Film Solar Cell

may experience total reflection from the bordering interfaces. In the case of TCO ($n \approx 2$)/$p-a$-SiC:H ($n \approx 3.5$) interface, for example, this happens for incident angles larger than 35°, according to Snell's law [61]. Thus, in effect, the light can be *trapped* within the absorber between the front TCO layer and the back reflector, which contributes to a further boost in charge-carrier generation. In thin-film solar cells, light scattering is most commonly realised by random or periodic surface-textured interfaces. However, novel techniques such as metal nanoparticles and diffusive back reflectors have also emerged and show much potential.

4.3.1.1 Random Textured Interfaces

Textured interfaces in thin-film silicon-based solar cells are introduced by textured glass/TCO superstrates or textured back contact substrates. By successive deposition of layers, the texture is transferred to the internal interfaces throughout the solar cell structure. In Fig. 4.10, SEM pictures of four types of textured TCOs are presented: (a) surface-textured SnO_2:F (Asahi U type), (b) magnetron-sputtered and chemically etched ZnO:Al, (c) LPCVD-deposited ZnO:B and (d) double (W-) textured SnO_2:F (Asahi). By tuning the deposition process parameters (such as temperature, pressure, dopant concentration) and thickness (in case of LPCVD), different texturisations of the same type of TCO can be achieved [10, 74]. In thin-film poly-crystalline solar

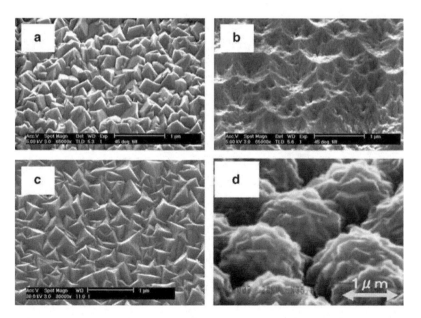

Fig. 4.10 SEM pictures of four types of textured TCOs: (**a**) surface-textured SnO_2:F (Asahi U type), (**b**) magnetron-sputtered and etched ZnO:Al, (**c**) LPCVD-deposited ZnO:B, and (**d**) double (W-) textured SnO_2:F (Asahi)

cells (CIGS, CdTe), on the other hand, surface texture is achieved by native growth of poly-crystalline grains.

Optical simulations can be employed to determine directions towards the optimal texturisation. For this reason, approaches of three-dimensional optical simulations based on finite difference time domain (FDTD) and other methods have recently been utilised [17, 75, 76]. In principle, the simulations are based on rigorously solving the wave equations throughout the structure, while taking into account the realistic morphology of the interfaces. The near field of light and the local absorption inside the thin absorbers are also considered in the calculations. These simulations show, as indicated in [75], that the sharp edges (spikes) and the rims of the holes (craters) in the texture are the most beneficial for high levels of absorption (effective scattering), whereas the valleys themselves do not lead to an increased absorption. Moreover, the simulations showed that the light absorption is in fact improved by partially filling up the valleys, even though the root-mean-square roughness of the interface is thus lowered [75].

TCOs with regular types of roughness exhibit randomly distributed surface features such as crater-like holes or pyramid grains. Additionally, TCOs with double textures have also been developed (Fig. 4.10d) in which two types of surface features are superimposed: small pyramids on top of larger hemi-spheres. The haze parameters for the transmitted light of the pyramidal-type SnO_2:F and (double) W-textured SnO_2:F are plotted in Fig. 4.11a, b, respectively. Measurements in air are shown by circles, whereas lines represent the simulations in which the equations from the scalar scattering theory were used to approximate the wavelength-dependent haze characteristics [77, 78]. Compared to the pyramidal-type TCO superstrate, a significantly higher haze in the short- and especially in the long-wavelength region is measured for the W-textured TCO. Furthermore, a shallow peak at $\lambda \approx 500$ nm

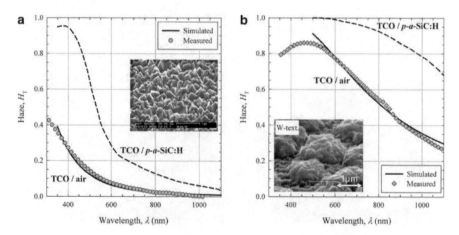

Fig. 4.11 The measured (*circles*) and the calculated (*lines*) haze of pyramidal-textured (**a**) and (double) W-textured SnO_2:F (**b**), observed in air. The predicted haze values at the internal SnO_2:F/p–a–SiC:H interface are also presented (Reproduced from [6] with permission. Copyright 2010. Elsevier)

is also observed in this case and can be attributed to the presence of two types of texturing. The haze measurements at the TCO/air interface were used to calibrate the equations of the scalar scattering theory. In simulations, these were then further employed to determine the haze parameter at the internal interfaces (e.g. TCO/p-a-SiC:H, as shown in the graphs) of a solar cell structure [62]. Simulations reveal that the simulated haze at the TCO/p-a-SiC:H interface has significantly higher values than the one determined at the TCO/air interface. Following the scalar scattering theory, this can be attributed to the higher difference in the refractive indices between the incident medium (TCO) and the medium in transmission (p-a-SiC:H), which leads to a higher scattering level of the transmitted light [78]. Thus, at the internal interfaces formed by layers with a high difference in the refractive indices (e.g. TCO with $n \approx 1.9$ and p-a-SiC:H with $n \approx 3.3$ at $\lambda = 600$ nm), a much higher haze is expected than the one determined in air surrounding for all types of TCO superstrates.

Tandem a-Si:H/μc-Si:H solar cells were deposited by PECVD on both types of glass/SnO$_2$:F superstrates [79]. The thicknesses of the top i-a-Si:H and the bottom i-μc-Si:H absorber layer were 0.35 μm and 1.5 μm, respectively. For the back reflector, ZnO/Ag was used. No intermediate reflector was applied between the top and the bottom cell. The results of the measured external quantum efficiency, QE, are shown in Fig. 4.12 (symbols). Almost no changes are observed in the quantum efficiencies corresponding to the top cell, QE_{top}, when using pyramidal-type or W-textured SnO$_2$:F superstrate. However, in the quantum efficiency of the bottom cell, QE_{bot}, a noticeable increase in the long-wavelength region can be observed for the W-textured superstrate (due to the enhanced long-wavelength scattering). The corresponding J_{SC} values are $J_{SCtop} \approx 11.8$ mA cm^{-2} for the top cells on both types of glass/SnO$_2$:F superstrates and $J_{SCbot} \approx 10.4$ mA cm^{-2} and 11.9 mA cm^{-2} for the

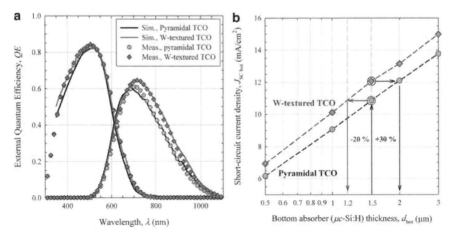

Fig. 4.12 The measured and the simulated external quantum efficiency of the top and the bottom cells on both types of SnO$_2$:F superstrates (**a**), and the possibilities of bottom cell thickness reduction related to the use of high-haze W-textured SnO$_2$:F superstrate (**b**)

bottom cells on pyramidal- and W-textured SnO_2:F superstrates, respectively. This means that the W-textured TCO superstrate leads to around 12% increase in J_{SCbot}.

Assuming good extraction of the generated charge carriers in the absorber layers (i–a–Si:H and i–μc–Si:H) and neglecting the small contributions from the p- and n-doped layers, the simulated absorptances in the top and the bottom absorber (full lines in Fig. 4.12a) can be directly compared with the measured QEs of the analysed solar cell, showing good agreement for both superstrates. Based on this verification of the model, simulations were then used further to analyse the optical situation in the cells with W-textured SnO_2:F superstrates and to examine the potential of possible further improvements in solar cell performance, related to the enhanced light scattering properties.

First, potential thickness reductions of the bottom i–μc–Si:H absorber, d_{bot}, are determined. In Fig. 4.12b, the J_{SCbot} is shown as a function of d_{bot} for the solar cells on both types of TCO superstrates. Linear dependency between J_{SCbot} and d_{bot} is observed in the semi-logarithmic plot. From the plot, one can see, for example, that a 20% thinner i–μc–Si:H absorber (1.2 μm) can be used in the case of the W-textured SnO_2:F superstrate compared to the same J_{SCbot} value achieved in the case of the pyramidal-type SnO_2:F superstrate (reference $d_{bot} = 1.5$μm). Further on, about 30% thicker i–μc–Si:H absorber (1.95 μm) should be used in the cell with the pyramidal-type SnO_2:F superstrate to achieve the same J_{SCbot} as can be achieved in the cell with the W-textured superstrate ($d_{bot} = 1.5$μm). These results indicate the potential of high-haze W-textured SnO_2:F to enable noticeable thickness reductions (and thus shortening of the deposition times) of the i–μc–Si:H absorber layer in tandem a–Si:H/μc–Si:H solar cells.

4.3.1.2 Periodically Textured Interfaces

However, random-like textures are not the only option for achieving high scattering angles of the light transmitted into the absorber layers of the cell. Periodic surface textures such as diffractive gratings present another promising approach. As the light of the wavelength λ is transmitted through a diffractive grating with the groove period P, it is scattered into discrete beams travelling at specific angles φ_{scatt}. These angles can be calculated by the grating equation,

$$n_{inc} \cdot \sin \varphi_{inc} + n_{out} \cdot \sin \varphi_{scatt} = \frac{m \cdot \lambda_0}{P}, \quad (4.1)$$

where n_{inc} and n_{out} are the refractive indices of the incident medium and the medium in transmission, respectively, φ_{inc} is the incident angle, m is the scattering order, λ_0 is the wavelength of light in the air and P is the grating (groove) period. The same effect is also observed when the light is reflected from the grating (Fig. 4.13), in which case n_{inc} and n_{out} represent the same medium and are therefore the same. The height of the groove and the shape of the grating feature do not affect the direction of scattering, but the intensity of the beam travelling at a specific angle.

4 Light Management in Thin-Film Solar Cell

Fig. 4.13 Light impinging on a grating reflector. Besides the specular reflection, the reflected light is diffracted into discrete angles (scattering orders)

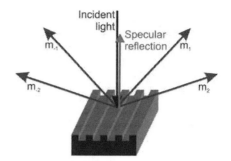

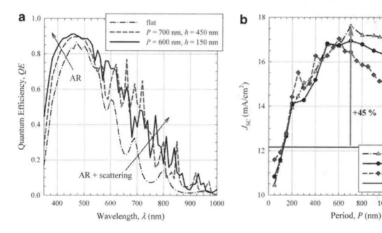

Fig. 4.14 The simulated quantum efficiency (**a**) and short-circuit current density (**b**) of a μc–Si:H solar cell with periodically textured interfaces with different P/h combinations (Reproduced from [20] with permission. Copyright 2009. American Institute of Physics)

In thin-film solar cells periodic textures have a potential to scatter the light into large discrete angles, according to the grating equation, which can lead to a prolongation of the optical paths and thus improved light trapping. In Fig. 4.14, the simulation results of a single-junction microcrystalline Si solar cell with periodically textured interfaces (1D grating) are shown [20]. The configuration of the cell was: ITO (70 nm)/p–μc–Si:H (15 nm)/i–μc–Si:H (500 nm)/n–μc–Si:H (20 nm)/ZnO (100 nm)/Ag. The shape of the texture was sine-like and was applied to all the interfaces. The two-dimensional optical simulator FEMOS [20, 71] was used to calculate the QE and J_{SC} of the structure (although an analytical solution has also been predicted in [80]). A simplified electrical analysis as described in the previous section was applied. In Fig. 4.14a, QE of the solar cells with different P and h of the periodical texture are plotted. Improvements in the short- and the long-wavelength QE are observed compared to the QE of the cell with flat interfaces. The analysis showed that, in the short-wavelength part, the antireflective effect due to the surface nano-texturing of the air/ITO and ITO/p-layer interfaces improves. In the long-wavelength part, on the other hand, the antireflective effects are supported by light scattering at the front and back interfaces, which leads to the observed enhancements. In Fig. 4.14b, the J_{SC} values of the cells are presented as a function of P for three different h. For the optimal case ($P \approx 700$ nm, $h \approx 450$ nm), an almost 45% improvement in J_{SC} is determined, compared to the J_{SC} of the flat cell.

Practically, periodically textured structures are especially convenient for cells deposited on plastic foil substrates. The texturisation of the foil can be obtained by hot embossing techniques, for example, where periodic structures with relatively small (a few 100 nm) or larger (micrometre range) features can be realised. By optimising the deposition parameters of the layers, comparable electrical characteristics to those of the layers deposited on randomly textured substrates can be achieved. However, the state-of-the-art cells with randomly textured interfaces still exhibit better performances than the cells with periodically textured interfaces. Further optimisation and novel designs of periodically textured interfaces are thus required. An option is even to combine periodic and random interfaces to gain the beneficial features of both [81].

4.3.1.3 Nanoparticles and Nanocomposites

Scattering at nanoparticles can be introduced as an alternative to scattering at textured interfaces to prolong the optical paths and improve light trapping in thin-film solar cell structures. Especially metal nanoparticles (Ag, Au) show much potential because the incident light in the wavelength range of interest can excite the surface plasmon polaritons to oscillate, leading to re-irradiation of the light in all directions around the particles [25]. This can be considered as an efficient scattering mechanism with a broad ADF (Fig. 4.15). The most intensive irradiation can be achieved if the wavelength of the incident light is close to the resonant wavelength (frequency) of the metal nanoparticle embedded in the surrounding medium. However, at the resonant frequency, the parasitic absorption losses are enhanced as well. By optimising the particle size, shape, and the surrounding medium of the particle, the ratio between the irradiated light and the absorbed light can be tailored [23, 25].

Figure 4.16a shows the SEM picture of nanoparticles obtained after thermal annealing of 15 nm thick Ag layer on glass. Their haze and absorption characteristics are shown in Fig. 4.16b. The sample was illuminated from the glass side, and the

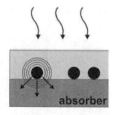

Fig. 4.15 The incident light is exciting the surface plasmon polaritons in nano-particles to oscillate, which leads to re-irradiation of the light

4 Light Management in Thin-Film Solar Cell

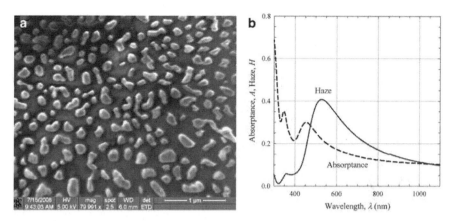

Fig. 4.16 SEM picture of silver nanoparticles obtained after thermal annealing of 15 nm thick silver layer on glass (**a**) and the corresponding haze and absorption characteristics (**b**)

haze of transmitted light (at the particle side) was determined. The absorptance was calculated as $A = 1 - R - T$. The selective peak in haze as well as in the absorptance characteristics can be observed due to oscillations of the surface plasmon polaritons in the metal nanoparticles.

In recent years nanoparticles have been studied extensively in the field of solar cells [21–27]. Improvements due to the integration of metal nanoparticles in thin-film Si solar cells have been indicated [21]. However, challenges still remain with respect to the proper position of the particles in the cell, in order to decrease the absorption losses and charge carrier recombination losses.

4.3.2 Minimisation of Optical Losses

The goal in thin-film solar cells is to draw as much light as possible into the thin absorber layer(s), while at the same time keeping the optical losses due to the reflected light and due to the absorption in the other layers at minimum. However, in state-of-the-art thin-film solar cells, a significant portion of light still escapes out of the structure or is absorbed in the supporting layers of the device. Advanced light management approaches are thus required to mitigate these losses. Besides efficient light scattering, which helps to trap the light inside the solar cell structure, one has to pay attention to minimise the undesired reflectance losses at the front interfaces of the structure, as well as to improve the optical properties of the supporting layers, such as the front and the back contact, in order to minimise the light absorption therein. In the following subsections, we will focus on antireflective coatings and on the realisation of highly reflective back reflectors based on photonic crystals and diffusive dielectric materials.

4.3.2.1 Antireflective Coatings and Structures

Light reflection at an interface between two different materials appears due to the abrupt change in the refractive index. The larger is the difference, the more pronounced is the effect and consequently more of the incident light is reflected (see the Fresnel equation in Table 4.1). In thin-film solar cells, all the reflection appearing before the incident light finally reaches the absorber layer contributes to optical losses. In this respect, two of the interfaces at the front are especially critical. The first is the interface between the surrounding air and the first layer. This can be either glass in the case of cells in superstrate configurations and PV modules covered with glass ($R \approx 4.5\%$), or the front TCO layer in the case of uncovered solar cells in substrate configurations ($R \approx 10\%$). The reflectance of the latter is more pronounced, and thus more critical, since the refractive indices of TCO materials are higher ($n \approx 2$) than those of glasses ($n \approx 1.5$). The second interface in the front part of the solar cell with a large change in the refractive index is the internal interface between the front TCO layer and the doped p-layer. The refractive indices in this case are about 2 and 3.5, respectively. In order to improve the performance of the solar cell, techniques for reducing the reflection of light at these critical interfaces need to be employed.

In principle, the solution is to make the transition of the refractive index at the interface less abrupt. This can be realised by refractive index grading, i.e. introducing a layer or a stack of layers where the values of the refractive indices are gradually changed from those of the surrounding medium towards those of the first layer in the structure. Another option for index grading is texturing of the interface. If texturing is on the nanometre scale, the exchange of two materials in lateral direction in the volume occupied by the corrugations can be considered as an effective layer, which has a refractive index between the two layers (media) forming the interface [42]. If applying texturisation with larger features (in micro or millimetre scale), multiple reflections at the geometrical features can lead to improved light in-coupling (the reflected rays have another chance to enter the structure) [82]. In the following, we focus on the first approach – introduction of additional layers in the role of antireflective coatings (ARC). For example, the reflection at the front air / glass interface can be reduced by depositing a single layer with the refractive index between 1 and 1.5. This can be, for instance, MgF_2 with $n = 1.38$. Similarly, the reflection at the front TCO/p-type absorber interface can be reduced by introducing for example a very thin layer of TiO_2, whose refractive index is about 2.3, i.e. between the value of refractive index of TCO ($n \approx 2$) and p-layer ($n \approx 3.5$) [83].

In theory, to minimize reflection, the refractive index of an optimal single-layer ARC inserted between two media should be equal to the geometric mean of the two refractive indices of the surrounding media; $n_{ARC} = \sqrt{n_1 \cdot n_2}$. For the case of air/glass interface, the optimum n_{ARC} is thus about 1.24. The thickness of the ARC layer is also important and should be optimised in order to produce destructive interference effects in reflection. For a single-layer ARC, this condition can be achieved when $d_{ARC} = \lambda/4n_{ARC}$, where d_{ARC} is the thickness of the ARC layer and λ is the target wavelength in air. Antireflective coatings can also be comprised

4 Light Management in Thin-Film Solar Cell

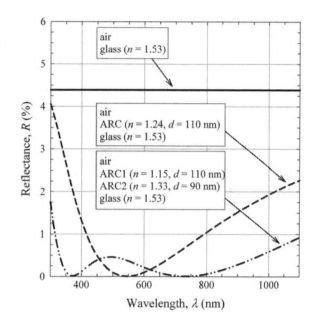

Fig. 4.17 Calculated reflectance of bare air/glass interface (*full line*) compared to the reflectance of glass coated with optimal single- and double-layer ARC

of more than just one layer, in which case the wavelength region of low reflectance can be extended to a broader range. An optimal double-layer ARC deposited on top of the front glass sheet, for example, should consist of two layers with refractive indices of 1.15 and 1.33, according to the theory. Such layers with refractive indices between 1 and 1.5 can be realized by using porous Si-oxide materials [39, 42].

The influence of the refractive index and the thickness of the ARC layer on the reflectance characteristics are presented by the simulation results plotted in Fig. 4.17. The flat full line shows the reflectance from the air/glass interface, which is about 4.5%. By introducing a single-layer ARC with $n_{ARC} = 1.24$ and $d_{ARC} = 110$ nm (values optimised according to the analytical equations mentioned above for $\lambda = 550$ nm), a significant reduction of the reflectance can already be achieved (dashed line). Furthermore, a destructive interference can be observed at $\lambda = 550$ nm, where the reflectance drops to zero. For the case of a double-layer ARC ($n_{ARC1} = 1.15$, $d_{ARC1} = 110$ nm; $n_{ARC2} = 1.33$, $d_{ARC2} = 90$ nm), on the other hand, an even further decrease in the reflectance can be achieved in a much wider spectral region (dash-dotted line).

To indicate how the antireflective coatings on glass improve the performance of a solar cell, we employed the optical simulator Sun*Shine* to simulate a μc–Si:H cell in a standard superstrate configuration ((ARC)/glass/ZnO:Al/p/i–μc–Si:H/n/ZnO/Ag), with the thickness of the i–μc–Si:H absorber layer of 1.2 μm. The J_{SC} value of the cell without any ARC on glass was 26.00 mA cm^{-2}. By introducing a single-layer ARC ($n_{ARC} = 1.24$, $d_{ARC} = 110$ nm) and a double-layer ARC ($n_{ARC1} = 1.15$, $d_{ARC1} = 110$ nm; $n_{ARC2} = 1.33$, $d_{ARC2} = 90$ nm), the J_{SC} values are improved by 3.6% (26.94 mA cm^{-2}) and 4.1% (27.07 mA cm^{-2}), respectively.

4.3.2.2 Advanced Back Reflector Concepts

The back reflector presents another important component of a thin-film solar cell. Its primary function is to reflect the light, which is not fully absorbed during the first passing through the cell back towards the absorber layer. This is especially important for the low-energy, long-wavelength part of the spectrum near the bandgap of the absorber material, where the chance of a successful absorption diminishes rapidly due to the low absorption coefficient. This behaviour is noticed above all in thin-film cells with indirect semiconductor absorber (like Si). The back reflector should exhibit a high reflectance, especially in this near-bandgap spectral region of the absorber.

Conventional thin-film solar cell technology relies on metal-based back reflectors, which commonly consist of a layer of silver (Ag) or aluminium (Al) in combination with a thin (∼80 nm) undoped TCO (ZnO) layer. Such reflectors exhibit excellent electrical conductivity and relatively high reflectance values in a broad wavelength region. However, they also show some disadvantages. Textured Ag reflectors suffer from the effect of surface plasmon absorption, which reduces the reflectivity properties of the reflector and thus presents one of the primary sources of optical losses [84]. Further on, high costs related to the metal deposition in the cell manufacturing process and the pronounced sensitivity of the material to the moisture in the environment are also an issue. Therefore, to circumvent these drawbacks and further increase the performance of thin-film solar cells, advanced back reflector concepts are being investigated. Among them, photonic crystals and white diffusive dielectrics have been indicated as two of the promising novel approaches [28, 31].

A one-dimensional (1D) photonic crystal (PC) is essentially a periodic stack comprised of two interchanging materials with different complex refractive indices and layer thicknesses, schematically presented in Fig. 4.18a. A 1D PC can act as a *distributed Bragg reflector* (DBR). As the light propagates through such a stack, it experiences multiple reflections at the internal interfaces. However, since the thicknesses of the layers are in the order of tens to hundreds of nm, the light coherency conditions are maintained. Consequently, the forward and the backward travelling electromagnetic waves within a layer are subject to constructive and destructive interference, depending on the wavelength. Thus, a specific reflectance characteristic is formed reminiscent of a band-stop filter. By carefully trimming the layer thicknesses, the number of layers in the stack, and the complex refractive indices of the two materials, it is possible to achieve a reflectance characteristic with an extremely high (∼ 100%) reflectance in the desired wavelength region. Theoretically, to obtain high reflectance values around the central wavelength λ_0, the following condition regarding the thicknesses and the refractive indices of the two interchanging materials has to be fulfilled: $d_1 \approx \lambda_0/4n_1$ and $d_2 \approx \lambda_0/4n_2$.

As an example, the measured reflectances of three 1D photonic crystals comprised of 12 layers of interchanging a–Si:H and a–SiN$_x$:H layers are presented in Fig. 4.18b [85]. The graph confirms that close to 100% reflectance can be achieved in the band-stop wavelength region. This is highly beneficial in thin-film solar cells where an appropriately optimised photonic crystal can be employed as an efficient

4 Light Management in Thin-Film Solar Cell

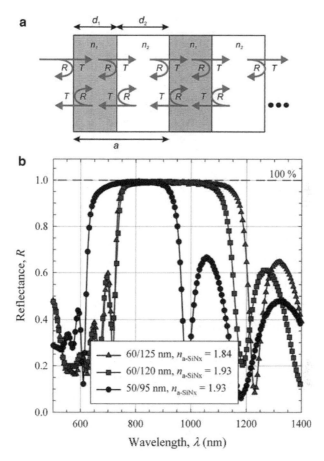

Fig. 4.18 Schematic representation of a 1D photonic crystal (**a**). Measured (*symbols*) and simulated (*lines*) reflectance of different PCs comprised of 12-layer of a–Si:H and a–SiN$_x$:H (**b**)

back reflector. Furthermore, it is also evident that both the refractive indices and the combination of the individual layer thicknesses significantly affect the nature of the reflectance characteristics (i.e. the width of the band-stop region, its location on the wavelength axis, etc.). Since there is a relatively large number of interconnecting parameters present in the design of PCs, optical simulations again prove to be a valuable tool for profound analysis and optimisation.

The concept of *modulated photonic crystals* was introduced to extend the wavelength region of high reflectance [36]. The modulation refers to a proper change of thicknesses or optical properties of the two exchanging layers. As an example of thickness modulation, we present a PC which consists of two integrated PC parts – in the first part the thicknesses of the two interchanging layers are smaller, whereas in the second part the thicknesses are larger (see inset in Fig. 4.19a). In this way, the first part assures high reflectance in the short-wavelength region

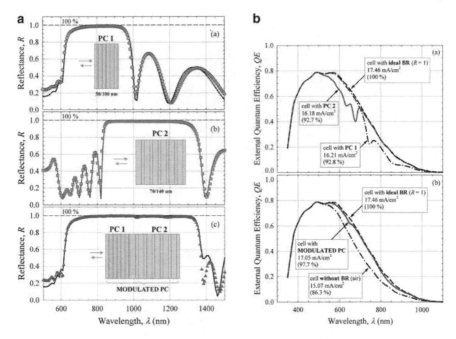

Fig. 4.19 Modulated photonic crystal vs. regular photonic crystals based on a–Si:H and a–SiN$_x$:H layers from the point of view of reflectance (**a**). Integration of a modulated PC based on conductive a–Si:H and ZnO:Al layers in the role of back reflector in a μc–Si:H solar cell (**b**) (Reproduced from [36] with permission. Copyright 2009. American Institute of Physics)

and the second in the long-wavelength region. To test and confirm the approach, formation and characterisation of the modulated PC is presented in Fig. 4.19a. The upper two plots in the figure show the measured and the simulated reflectance of two individual PCs, each comprised of 12 interchanging layers of a–Si:H and a–SiN$_x$:H ($n_{a-\text{Si:H}} = 3.83$ and $n_{a-\text{SiN}_x} = 1.81$ at $\lambda = 800$ nm). The only difference between them are the layer thicknesses (thickness pairs a–Si:H/a–SiN$_x$:H are given in the figure). The third plot in the figure corresponds to a modulated PC created by the two parts. The plot shows that the reflectance characteristic of the modulated PC exhibits a broad band-stop region, which in fact presents the combination of the two narrower band-stop regions observed for the separate PCs.

After demonstrating that the concept works, we have designed a suitable modulated PC for a broadband back reflector in thin-film μc–Si:H solar cells. In this case, the reflector must also serve as the back electrical contact. Therefore, conductive n-doped amorphous silicon (n–a–Si:H) and Al-doped magnetron sputtered ZnO (ZnO:Al) were used. The refractive indices of these layers are $n_{n-a-Si:H} = 3.32$ and $n_{\text{ZnO:Al}} = 1.77$ ($\lambda = 800$ nm). For μc–Si:H solar cells, it is important that the back reflector has a high reflectivity over a much broader wavelength region ($\lambda \approx 500$–$1{,}100$ nm), something that cannot be achieved by a single PC comprised of the two conductive materials. Therefore, to extend the region of high reflectance, we designed a modulated structure which consists of three periods (layer pairs) with the thicknesses $d_{n-a-Si:H} = 40$ nm and $d_{ZnO:Al} = 80$ nm (PC 1 part) and four periods with the layer thicknesses $d_{n-a-Si:H} = 60$ nm and $d_{ZnO:Al} = 120$ nm (PC 2I part).

In Fig. 4.19b (top), the simulated EQE is plotted for the μc–Si:H solar cells ($d_{absorber} = 2\,\mu$m) with regular PC back reflectors (PC 1 and PC 2). While the EQE of the cell with the PC 1 back reflector is lowered in the long-wavelength region ($\lambda > 700$ nm), the cell with the PC 2 exhibits reduced EQE in the middle-wavelength region ($\lambda = 500$–700 nm). The corresponding J_{SC} values under the reference AM1.5 solar irradiance reach 92.8 and 92.7% of the J_{SC} calculated for the cell with an ideal back reflector, respectively. In Fig. 4.19b) (bottom), the simulated EQE of the cell with the modulated PC is shown. In this case, the EQE over the entire wavelength region resembles closely that of the cell with an ideal back reflector. The observed small deviations are related to decreases in the internal reflectance. The cell with the modulated PC back reflector achieves 97.7% of the J_{SC} of the cell with an ideal back reflector. This is an additional 5% increase with respect to the cells with regular PC's.

Unlike the photonic crystals, however, the second advanced back reflector concept, which has emerged recently and is showing much potential is based on only one material – a layer of highly reflective (i.e. white) diffusive dielectric material, such as white paint or white EVA foil. Diffusive dielectrics, in particular white paint, can be described as heterogeneous composites of two materials – white (pigment) particles, acting as scattering centres, and the surrounding medium (binder). As light enters such a material (Fig. 4.20a), it passes through the binder and eventually hits a pigment particle. At the interface, it scatters both in transmission and reflection. The transmitted light then passes through the pigment material and scatters again at the next pigment/binder interface. Multiple reflections at the pigment/binder interfaces and light scattering due to the dimensions of the particles (200–300 nm) leads to high reflectance and excellent scattering characteristics of such films. In typical commercial white paints, for example, high reflectances above 0.9 can be determined in a broad part of the spectrum (500–1,000 nm). In Fig. 4.20b, the haze parameter and the ADF of a commercial white paint are presented in comparison with the scattering properties of a conventional Ag reflector deposited on textured SnO_2:F Asahi U type TCO ($\sigma_{rms} \approx 40$ nm, pyramidal texturisation features). The results show that white paint is a significantly better light scatterer, especially in the long-wavelength region. The haze parameter is almost ideal, and the angular distribution function is broad, close to the Lambertian (cosine) function.

Although the reflective properties of inexpensive white diffusive dielectric materials are not as high as those of typical Ag, the approach nevertheless offers a number of advantages over the conventional metal-based reflectors. First, the scattering properties of white diffusive dielectrics are far superior and assure high levels of light trapping within the cells. Second, the dielectric material itself is much cheaper and often easy to deploy (in case of white paint, simple spraying or printing techniques). And finally, dielectrics such as white paint and foils are also much more durable, they can resist severe weather conditions, and in fact they can even serve

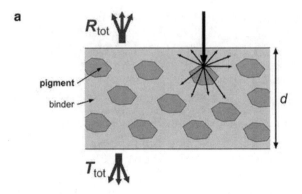
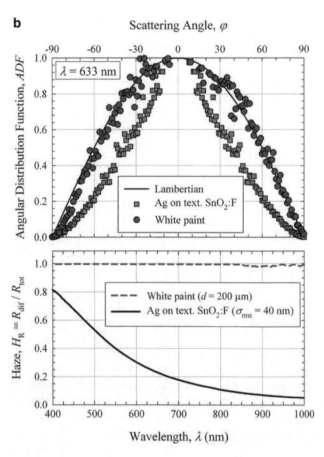

Fig. 4.20 A schematic of light scattering in white paint (**a**). The measured scattering parameters (**b**) are much higher than for the case of textured Ag reflector

for the back-encapsulation of the modules. The drawback of the diffusive dielectric back reflectors, on the other hand, is their lack of electrical conductivity. Therefore, they can only be used in combination with a layer of conductive TCO, such as LPCVD-deposited ZnO:B or magnetron-sputtered ZnO:Al. For high performance of the cells, a sufficiently thick (~1,000 nm) layer of high-quality TCO with good vertical and lateral conductivity and low levels of light absorption must be employed.

4.3.3 Enhanced Utilisation of the Solar Spectrum

In single-junction solar cells, a part of the energy of the photons, which carry more energy than is the energy gap of the absorber layer, is thermally lost. These thermalisation losses, however, can be minimised by using multi-bandgap multi-junction devices, which allow us to tune the bandgaps of the individual absorber layers. In the case of thin-film solar cells, double- and triple-junction devices are present in the market additionally to the single-junction ones. Examples are double-junction (tandem) a–Si:H/μc–Si:H *micromorph* cells and triple-junction a–Si:H/a–SiGe:H/μc–Si:H cells. Chalcopyrite-based $CuGaSe_2$/$Cu(In,Ga)Se_2$ (CGS/CIGS) tandems are also under investigation [86] and show a potential since it is easy to alter the bandgap of the bottom (CIGS) absorber and thus optimise the performance of the device. The challenge in the design of two-terminal multi-junction cells, however, is to assure an equal distribution of the solar spectrum between the cells connected in series and thus achieve the required current matching. Furthermore, the parasitic sub-bandgap absorption in the top cell which does not contribute to J_{SCtop} presents here a significant source of losses, especially in the proposed CGS top cell. While the top CGS cell absorbs in the short-wavelength spectrum, which leads to photocurrent generation, it also absorbs substantially in the long-wavelength region. This, however, does not contribute to J_{SC} and instead hinders the current generation in the bottom CIGS cell. Therefore, solutions in the multi-junction cell design that would lead to very thin top cells, or even realisations based on spectrum splitting (two separate cells) are of interest. Both concepts will be addressed in the following two subsections.

4.3.3.1 Intermediate Reflectors in Multi-Junction Solar Cells

To circumvent these issues and also to further boost the performance of the multi-junction cells, the concept of a *wavelength-selective intermediate reflector* (WSIR) has been studied to be introduced between the top CGS and the bottom CIGS cell of the chalcopyrite-based tandem device. The WSIR should perform the function of an optical high-pass filter – it should efficiently reflect the short-wavelength part of the spectrum back into the preceding (top) cell, and at the same time fully transmit the long-wavelength part absorbed in the following (bottom) cell. The desired consequence of this approach is an enhanced light absorption in both cells, which leads to higher generated currents and potentially lower absorber thicknesses.

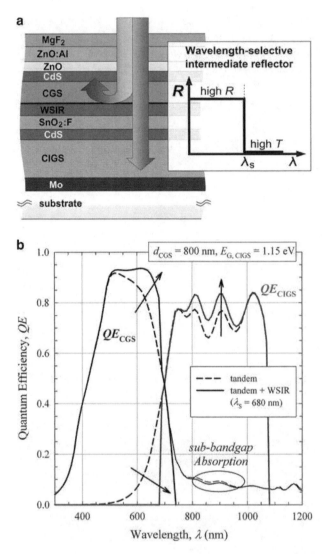

Fig. 4.21 The structure of a CGS/CIGS tandem with the WSIR (**a**) and the simulated QE_{CGS} and QE_{CIGS} of the tandem ($d_{CGS} = 800$ nm, $d_{CIGS} = 2{,}500$ nm) with and without the WSIR ($\lambda_0 = 680$ nm) (**b**)

Naturally, the WSIR itself must be low absorbent, and it must also exhibit a sufficient vertical conductivity in order to electrically connect the individual cells of the two-terminal multi-junction device.

The potential of WSIR in CGS/CIGS tandems can be investigated by means of optical simulations. The structure of the tandem is presented in Fig. 4.21a. The bandgaps of the top CGS and bottom CIGS cells are 1.68 and 1.15 eV, respectively, which is close to the theoretical optimum for tandems [87]. By simulating the photo-generation profile in both cells and calculating the current-voltage characteristics according to the diode equations, the efficiency of the tandem can be determined. For these calculations, realistic electrical parameters of the CGS and CIGS absorbers (i.e. the saturation current density, J_0, and the diode ideality factor, A) are assumed, which are extracted from the latest reported state-of-the-art cells: (1) CGS: $J_0 = 7 \cdot 10^{-7}\,\text{mA}\,\text{cm}^{-2}$, $A = 2.1$ [88]; and (2) CIGS: $J_0 = 2.1 \cdot 10^{-9}\,\text{mA}\,\text{cm}^{-2}$, $A = 1.14$ [89].

The simulated tandem structure (without the WSIR) was as follows: MgF_2 (100 nm)/ZnO:Al (120 nm)/ZnO (50 nm)/CdS (50 nm)/CGS (d_{CGS})/SnO$_2$:F (500 nm)/CdS (50)/CIGS (d_{CIGS})/Mo (300 nm). Simulations show that, in this case, the maximal efficiency of 20.3% can be achieved with the optimised values for the cell thickness, which are 1,050 for the CGS (d_{CGS}) and 2,500 for the CIGS cell (d_{CIGS}). However, if an appropriately tailored WSIR is introduced between the top and the bottom cells (see Fig. 4.21a), the thickness of the top cell can be reduced substantially. For an ideal WSIR exhibiting the optimal reflectance characteristics with the roll-off wavelength $\lambda_S = 680$ nm (as show in the inset of the figure), the maximal efficiency of 20.8% can be achieved for $d_{CGS} = 800$ nm and $d_{CIGS} = 2,500$ nm. This is an almost 25% reduction in the top cell thickness compared to the case with no WSIR. Without the WSIR, on the other hand, the efficiency of this tandem ($d_{CGS} = 800$ nm, $d_{CIGS} = 2,500$ nm) would be 19.8%. The effects of the WSIR on QE_{CGS} and QE_{CIGS} are shown in Fig. 4.21b, where it can be observed that introducing the WSIR boosts the absorption in both CGS and CIGS cells.

In reality, intermediate reflectors may be realised most effectively by photonic-crystal-like structures in the role of DBR (the principles of such structures have already been discussed in Sect. 4.3.2). The materials chosen for the photonic crystals must exhibit low optical absorption in the active wavelength region of the tandem and high vertical electrical conductivity (or local interconnections) for good ohmic connection between the two cells.

4.3.3.2 Spectrum Splitting

To further reduce (eliminate) the effects of sub-bandgap absorption in the top cell, the concept of spectrum splitting, which is schematically presented in Fig. 4.22a, is introduced. The idea is to separate the incident light spectrum by means of dichroic beam splitters into different parts. Each of them is then illuminating a solar cell whose properties are tailored specifically to yield high performance in this spectral region. The separate cells are then connected in a hybrid multi-terminal configuration.

Optical simulations allow us to test this approach in a four-terminal hybrid configuration of CGS and CIGS cells and compare it to the two-terminal concept with WSIR. Since there is no interconnection between the individual cell thicknesses, both cells can now be 2,500 nm thick, while the parameters which need to be optimised are the bandgap of the CIGS cell and the spectrum split point (λ_S) of the dichroic mirror. The optimisation results are presented in Fig. 4.22b. The results

Fig. 4.22 Schematic representation of the concept of spectrum splitting (**a**) and the effects of the spectrum split point and the CIGS bandgap on the total conversion efficiency of the cells connected in a hybrid four-terminal configuration (**b**)

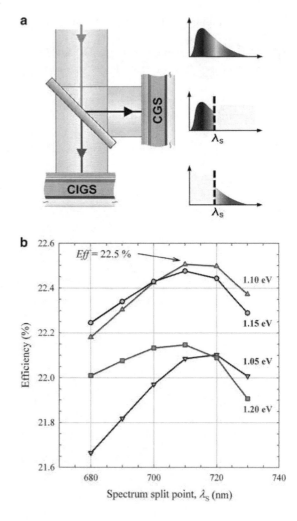

indicate that the carefully selected λ_S is crucial for a high efficiency and should not deviate much from the optimal value (deviations should be less than 20 nm). The $E_{G,CIGS}$, on the other hand, allows for more freedom since similar results are rendered for CIGS bandgaps between 1.10 and 1.15 eV. In our case, the maximal efficiency of 22.5% can be achieved for $E_{G,CIGS} = 1.10$ eV and $\lambda_S = 710$ nm.

4.4 Conclusion

Light management in thin-film solar cells is of great importance with respect to efficient utilisation of the solar spectrum in thin absorber layers of the cells. To analyse and optimise the optical situation in the multilayer structures with flat and textured interfaces, optical modelling and simulations are indispensable tools.

They enable to locate and determine optical losses inside the structure and indicate the potential of further improvements.

Surface-textured interfaces present a crucial aspect in efficient light trapping, especially in thin-film solar cells with indirect semiconductor absorbers (Si) and to some extent also in cells made of direct semiconductors (CIGS, CdTe) with respect to thinning down the absorber layers. Regarding their scattering characteristics inside the solar cells, optimised random-like texturisations still prevail over the regular periodical textures. However, novel concepts to improve both approaches or even to combine them (modulated texturisations) are being investigated extensively.

As an interesting alternative to surface scattering at the interface corrugations, metal nanoparticles can be employed. Excitation of surface plasmon polaritons by incident light leads to re-irradiation of light in different directions around the particles, prolonging optical paths and thus absorption of light in thin absorbers. However, how to decrease the optical absorption inside the particles themselves and how to avoid recombination losses of charge carriers at the surfaces remain challenges yet to be resolved.

Besides efficient light trapping, minimisation of optical losses due to the absorption in the supporting layers of the solar cell and due to the light reflected from the surface is also important. Antireflective coatings on the top (air/glass) as well as internal (TCO/p-layer) interfaces have to be used to minimise the negative reflection of light before entering the absorber layer. However, comparing the benefits of antireflective coatings to the effects of light scattering, the latter have more important role.

Next to the reflectance, the front (TCO) and the back contact (metal or dielectric in combination with TCO) present the dominant source of absorption losses in thin-film solar cells. Thus, optimisation of front TCO materials and investigation of novel back contact (reflector) realisations is important. In this chapter, back reflectors based on white paint and photonic-crystal-like structures were investigated as alternatives to the conventional metal-based reflectors.

For efficient use of the solar spectrum, multi-bandgap multi-junction devices and the concept of spectrum splitting are employed. In the case of thin-film double-junction (tandem) solar cells, a wavelength-selective intermediate reflector located between the top and the bottom cell enables thinner top cell absorbers. This reflects in better stability of the tandem cell efficiency (a–Si:H top cell) or lower sub-bandgap absorption losses (CGS top cell). Wavelength-selectivity of the intermediate reflector is an important characteristic for efficient distribution of the solar spectrum between the top and the bottom cell. Another way to avoid the negative influence of the top cell to the bottom cell is by employing the concept of spectrum splitting. In this case, each cell can be optimised independently for the part of the spectrum it is exposed to.

In this chapter, a number of approaches of efficient light management were highlighted, and the importance of their utilisation in thin-film solar cells was indicated. Nevertheless, further optimisation of the existing solutions, as well as development and investigation of novel light trapping designs remains an important challenge.

Acknowledgements The authors thank Miro Zeman and other members of PVMD laboratory at DIMES Institute, Delft University of Technology, The Netherlands, for support on a–Si:H-based solar cell structures and for useful discussions. Takuji Oyama from Asahi Glass Corp., Japan, is acknowledged for support on double-textured TCO superstrates. Martina Schmid from Helmholz Zentrum Berlin, Germany, is acknowledged for useful discussions on chalcopyrite tandems. Finally, Slovenian Research Agency is acknowledged for funding related to this work (Project No. P2-0197 and J2-0851-1538-08).

References

1. A. Jäger-Waldau, *PV Status Report* 2011 (Office for Official Publications of the European Union, Luxembourg, 2011), DOI 10.2788/87966.
2. A.V. Shah, M. Vanecek, J. Meier, F. Meillaud, J. Guillet, D. Fischer, C. Droz, X. Niquille, S. Faÿ, E. Vallat-Sauvain, V. Terrazzoni-Daudrix, J. Bailat, J. Non Cryst. Solids **338**, 639 (2004)
3. K. Sato, Y. Gotoh, Y. Hayashi, K. Adachi, H. Naishimura, Reports Res. Lab., Asahi glass Co., Ltd, vol. 40, p. 129 (1992)
4. M. Kambe, K. Masumo, N. Taneda, T. Oyama, K. Sato, in *Technical Digest of the 17th International Photovoltaic Science and Engineering Conference*, Fukuoka, Japan, 3–7 Dec 2007, pp. 1161–1162.
5. T. Oyama, M. Kambe, N. Taneda, K. Masumo, *Light Management in Photovoltaic Devices-Theory and Practice*, pp. 1101E–KK02. Mater. Res. Soc. Symp. Proc., vol. 1101E, Warrendale, PA, 2008
6. J. Krč, B. Lipovšek, M. Bokalič, A. Čampa, T. Oyama, M. Kambe, T. Matsui, H. Sai, M. Kondo, M. Topič, Thin Solid Films **518**(11), 3054 (2010)
7. O. Kluth, B. Rech, L. Houben, S. Wieder, G. Schope, C. Beneking, H. Wagner, A. Loffl, H.W. Schock, Thin Solid Films **351**(1–2), 247 (1999)
8. M. Berginski, J. Huepkes, W. Reetz, B. Rech, M. Wuttig, Thin Solid Films **516**(17), 5836 (2008)
9. S. Faÿ, D. S., U. Kroll, J. Meier, Y. Ziegler, A. Shah, in *Proceedings of the 16th European Photovoltaic Solar Energy Conference*, Glasgow, UK, 1–5 May 2000, pp. 361–364
10. S. Faÿ, J. Steinhauser, N. Oliveira, E. Vallat-Sauvain, C. Ballif, Thin Solid Films **515**(24), 8558 (2007)
11. T. Söderström, F.J. Haug, X. Niquille, C. Ballif, Progr. Photovoltaics **17**(3), 165 (2009)
12. A.J.M. Van Erven, R.H. Franken, J. de Ruijter, P. Peeters, W. Vugts, O. Isabella, M. Zeman, C. Haase, U. Rau, H. Borg, in *Proceedings of 23th European Photovoltaic Solar Energy Conference and Exhibition*, Valencia, Spain, 1–5 Sep 2008, pp. 2096–2100
13. A. Banerjee, G. DeMaggio, K. Lord, B. Yan, F. Liu, X. Xu, K. Beernink, G. Pietka, C. Worrel, D. B., J. Yang, G. S., in *Proceedings of the 34th IEEE Photovoltaic Specialists Conference*, Philadelphia, USA, 7–12 Jun 2009, pp. 116–119
14. C. Eisele, C.E. Nebel, M. Stutzmann, J. Appl. Phys. **89**(12), 7222 (2001)
15. H. Stiebig, N. Senoussaoui, C. Zahren, C. Haase, J. Muller, Progr. Photovoltaics **14**(1), 13 (2006)
16. C. Haase, H. Stiebig, Progr. Photovoltaics **14**(7), 629 (2006)
17. C. Haase, H. Stiebig, Appl. Phys. Lett. **91**(6), 061116 (2007)
18. O. Isabella, A. Čampa, M.C.R. Heijna, W. Soppe, R. Van Erven, R.H. Franken, H. Borg, M. Zeman, in *Proceedings of 23th European Photovoltaic Solar Energy Conference and Exhibition*, Valencia, Spain, 1–5 Sep 2008, pp. 2320–2324
19. A. Čampa, J. Krč, F. Smole, M. Topič, Thin Solid Films **516**(20), 6963 (2008)
20. A. Čampa, J. Krč, F. Smole, M. Topič, J. Appl. Phys. **105**(8), 083107 (2009)
21. E. Moulin, J. Sukmanowski, M. Schulte, A. Gordijn, F.X. Royer, H. Stiebig, Thin Solid Films **516**(20), 6813 (2008)

22. S. Pillai, K.R. Catchpole, T. Trupke, G. Zhang, J. Zhao, M.A. Green, Appl. Phys. Lett. **88**(16), 161102 (2006)
23. S. Pillai, K.R. Catchpole, T. Trupke, M.A. Green, J. Appl. Phys. **101**(9), 093105 (2007)
24. C. Rockstuhl, F. Lederer, Appl. Phys. Lett. **94**(21), 213102 (2009)
25. K.R. Catchpole, A. Polman, Appl. Phys. Lett. **93**(19), 191113 (2008)
26. Y.A. Akimov, W.S. Koh, K. Ostrikov, Optic. Express **17**, 10195 (2009)
27. T.L. Temple, D.M. Bagnall, in *Technical Digest of the 17th International Photovoltaic Science and Engineering Conference*, Fukuoka, Japan, 3–7 Dec 2007, pp. 526–527
28. J. Meier, U. Kroll, J. Spitznagel, S. Benagli, T. Roschek, G. Pfaneer, C. Ellert, G. Androutsopoulos, A. Hügli, M. Nagel, C. Bucher, L. Feitknecht, G. Bchel, A. Büchel, in *Proceedings of the 31st IEEE Photovoltaic Specialists Conference*, FL, USA, 3–7 Jan 2005, pp. 1464–1467
29. M. Berginski, J. Hüpkes, A. Gordijn, W. Reetz, T. Wätjen, M. Wutting, in *Technical Digest of the 17th International Photovoltaic Science and Engineering Conference*, Fukuoka, Japan, 3–7 Dec 2007, pp. 341–342
30. O. Berger, D. Inns, A.G. Aberle, Optic. Express **91**(13), 1215 (2007)
31. H.Y. Lee, T. Yao, J. Appl. Phys. **93**(2), 819 (2003)
32. H.Y. Lee, S.J. Cho, G.Y. Nam, W.H. Lee, T. Baba, H. Makino, M.W. Cho, T. Yao, J. Appl. Phys. **10**(2), 103111 (2005)
33. L. Zeng, Y. Yi, C. Hong, X. Duan, L.C. Kimerling, *Amorphous and Nanocrystalline Silicon Science and Technology* p. A12.3. Mater. Res. Soc. Symp. Proc. 862, E, Warrendale, PA, 2005
34. L. Zeng, P. Bermel, Y. Yi, B.A. Alamariu, K.A. Broderick, J. Liu, C. Hong, X. Duan, J. Joannopoulos, L.C. Kimerling, Appl. Phys. Lett. **93**(22), 221105 (2008)
35. I.J. Kuzma-Filipek, F. Duerinckx, E. Van Kerschaver, K. Van Nieuwenhuysen, G. Beaucarne, J. Poortmans, J. Appl. Phys. **104**(7), 073529 (2008)
36. J. Krč, M. Zeman, S.L. Luxembourg, M. Topič, Appl. Phys. Lett. **94**(15), 153501 (2009)
37. O. Isabella, B. Lipovšek, J. Krč, M. Zeman, *Amorphous and Polycrystalline Thin-Film Silicon Science and Technology*, pp. 1153–A03–05. Mater. Res. Soc. Symp. Proc., vol. 1153, Warrendale, PA, 2009
38. O. Isabella, J. Krč, M. Zeman, in *Proceedings of the 24th European Photovoltaic Solar Energy Conference and Exhibition*, Hamburg, Germany, Sep 21–25 2009, pp. 2304–2309
39. J. Meier, J. Spitznagel, U. Kroll, C. Bucher, S. Fay, T. Moriarty, A. Shah, Thin Solid Films **451**, 518 (2004)
40. C. Ballif, J. Dicker, D. Borchert, T. Hofmann, in *Proceedings of the 5th ISES Europe Solar Conference*, Freiburg, Germany, 20–23 Jun 2004, pp. 41–49
41. J. Krč, F. Smole, M. Topič, J. Non-Crystalline Solids **352**(9–20), 1892 (2006)
42. S. Chhajed, M.F. Schubert, J.K. Kim, E.F. Schubert, Appl. Phys. Lett. **93**(25), 251108 (2008)
43. D. Fischer, S. Dubail, J. Anna Selvan, N. Pellaton Vaucher, R. Platz, C. Hof, U. Kroll, J. Meier, P. Torres, H. Keppner, N. Wyrsch, M. Goetz, A. Shah, K.D. Ufert, in *Proceedings of the 25th IEEE Photovoltaic Specialists Conference*, Washington DC, USA, 13–17 May 1996, pp. 1053–1056
44. K. Yamamoto, M. Yoshimi, Y. Tawada, S. Fukuda, T. Sawada, T. Meguro, H. Takata, T. Suezaki, Y. Koi, K. Hayashi, T. Suzuki, M. Ichikawa, A. Nakajima, Sol. Energ. Mater. Sol. Cell. **74**(1–4), 449 (2002)
45. J. Krč, F. Smole, M. Topič, Sol. Energ. Mater. Sol. Cell. **86**(4), 537 (2005)
46. B. Lipovšek, J. Krč, M. Topič, Proceedings of Inorganic and Nanostructured Photovoltaics: E-MRS 2009 Symposium B (Energy procedia vol. 2, no. 1), Strasbourg, France, 8–12 Jun 2010, pp. 143–150
47. J. Krč, M. Zeman, Čampa, F. Smole, M. Topič, in *Amorphous and Polycrystalline Thin-Film Silicon Science and Technology-2006*, Washington DC, USA, 13–17 May 2007, pp. 0910–A25–01
48. C. Rockstuhl, F. Lederer, K. Bittkau, T. Beckers, R. Carius, Appl. Phys. Lett. **94**(21), 211101 (2009)
49. D. Domine, P. Buehlmann, J. Bailat, A. Billet, A. Feltrin, C. Ballif, Phys. Stat. Sol. Rapid Res. Lett. **2**(4), 163 (2008)

50. T. Söderström, F.J. Haug, X. Niquille, V. Terazzoni, C. Ballif, Appl. Phys. Lett. **94**(6), 063501 (2009)
51. A. Gombert, A. Luque, Phys. Stat. Sol. A-Appl. Mater. Sci. **205**(12), 2757 (2008)
52. W.B. Chen, P.F. Gu, Display **26**(2), 65 (2005)
53. R. Tena-Zaera, J. Elias, C. Levy-Clement, Appl. Phys. Lett. **93**(23), 233119 (2008)
54. J. Zhu, Z.F. Yu, G.F. Burkhard, C.M. Hsu, S.T. Connor, Y.Q. Xu, Q. Wang, M. McGehee, S.H. Fan, Y. Cui, Nano Lett. **9**(1), 279 (2009)
55. A. Luque, A. Marti, A.J. Nozik, MRS Bull. **32**(3), 236 (2007)
56. S. Tomic, T.S. Jones, N.M. Harrison, Appl. Phys. Lett. **93**(26), 263105 (2008)
57. D. Ginley, M.A. Green, R. Collins, MRS Bull. **33**(4), 355 (2008)
58. P. Löper, J.C. Goldschmidt, S. Fischer, M. Peters, A. Meijerink, D. Biner, K. Krämer, O. Schultz-Wittmann, S.W. Glunz, J. Luther, in *Proceedings of 23th European Photovoltaic Solar Energy Conference and Exhibition*, Valencia, Spain, 1–5 Sep 2008, pp. 173–180
59. S. Ivanova, F. Pellé, R. Esteban, M. Laroche, J.J. Greffet, S. Colin, J.L. Pelouard, J.F. Guillemoles, in *Proceedings of 23th European Photovoltaic Solar Energy Conference and Exhibition*, Valencia, Spain, 1–5 Sep 2008, pp. 734–737
60. L. Ley, in *The Physics of Hydrogenated Amorphous Silicon and its Application* ed. by J.D. Joannopoulus, G. Lucovsky. Chap. Photoemission and Optical Properties (Springer, Heidelberg, 1984)
61. J.A. Kong, *Electromagnetic Wave Theory*, 3rd edn. (Wiley, New York, 1990)
62. J. Krč, M. Zeman, F. Smole, M. Topič, J. Appl. Phys. **92**(2), 749 (2002)
63. F. Leblanc, J. Perrin, J. Schmitt, J. Appl. Phys. **75**(2), 1074 (1994)
64. G. Tao, M. Zeman, J.W. Metselaar, Sol. Energ. Mater. Sol. Cell. **34**(1–4), 359 (1994)
65. B. Sopori, J. Madjdpour, Y. Zhang, W. Chen, S. Guha, J. Yang, A. Banerjee, S. Hegedus, in *Amorphous and Heterogeneous Silicon Thin Films: Fundamentals to Devices*, pp. 755–760. Mater. Res. Soc. Symp. Proc., vol. 557, Warrendale, PA, 1999
66. J. Krč, F. Smole, M. Topič, Progr. Photovoltaics **11**(1), 15 (2003)
67. J. Krč, F. Smole, M. Topič, Informacije MIDEM J. Microelectron. Electron. Comp. Mater. **32**(1), 6 (2002)
68. J. Springer, A. Poruba, M. Vanecek, J. Appl. Phys. **96**(9), 5329 (2004)
69. J. Daey Ouwens, M. Zeman, J. Löffler, R.E.I. Schropp, in *Proceedings of the 16th European Photovoltaic Solar Energy Conference*, Glasgow, UK, 1–5 May 2000, pp. 405–408
70. A. Čampa, J. Krč, M. Topič, J. Appl. Phys. **105**(8), 083107 (2009)
71. A. Čampa, J. Krč, M. Topič, Informacije MIDEM J. Microelectron. Electron. Comp. Mater. **38**(1), 5 (2008)
72. J. Krč, M. Zeman, O. Kluth, E. Smole, M. Topič, Thin Solid Films **426**(1–2), 296 (2003)
73. B. Lipovšek, J. Krč, O. Isabella, M. Zeman, M. Topič, Phys. Stat. Solid. C Curr. Topics Solid State Phys. **7**, 1041 (2010)
74. M. Berginski, J. Hupkes, M. Schulte, G. Schope, H. Stiebig, B. Rech, M. Wuttig, J. Appl. Phys. **101**(7), 074903 (2007)
75. C. Rockstuhl, S. Fahr, F. Lederer, K. Bittkau, T. Beckers, R. Carius, Appl. Phys. Lett. **93**(6), 061105 (2008)
76. K. Bittkau, T. Beckers, S. Fahr, C. Rockstuhl, F. Lederer, R. Carius, Phys. Stat. Solid. A Appl. Mater. Sci. **205**(12), 2766 (2008)
77. P. Beckman, A. Spizzichino, *The Scattering of Electromagnetic Waves from Rough Surfacey* (Pergamon Press, New York, 1963)
78. K. Carniglia, Optic. Eng. **18**(2), 104 (1979)
79. M. Kambe, T. Matsui, H. Sai, N. Taneda, K. Masumo, A. Takahashi, T. Ikeda, T. Oyama, M. Kondo, S. K., in *Proceedings of the 34th IEEE Photovoltaic Specialists Conference*, Philadelphia, USA 7–12 Jun 2009, pp. 1663–1666
80. F.J. Haug, T. Söderström, O. Cubero, V. Terazzoni-Daudrix, X. Niquille, S. Perregeaux, C. Ballif, Light Management in Photovoltaic Devices – Theory and Practice pp. 1001-KK13-01. Mater. Res. Soc. Symp. Proc., vol. 1101E, Warrendale, PA, 2008

81. O. Isabella, F. Moll, J. Krč, M. Zeman, Phys. Stat. Solid. A Appl. Mater. Sci. **207**(3), 642 (2009)
82. M.A. Green, *Operating Principles, Technology and System Applications* (The University of New South Wales, Kensington, 1992)
83. K. Yamamoto, A. Nakajima, M. Yoshimi, T. Sawada, S. Fukuda, T. Suezaki, M. Ichikawa, Y. Koi, M. Goto, T. Meguro, T. Matsuda, M. Kondo, T. Sasaki, Y. Tawada, in *Technical Digest of the 15th International Photovoltaic Science and Engineering Conference*, Shanghai, China, 10–15 Oct 2005, pp. 529–530
84. J. Springer, A. Poruba, L. Mullerova, M. Vanecek, O. Kluth, B. Rech, J. Appl. Phys. **95**(3), 1427 (2004)
85. J. Krč, A. Čampa, S. Luxembourg, M. Zeman, M. Topivc, Light Management in Photovoltaic Devices – Theory and Practice, pp. 1101–KK08–01. Mater. Res. Soc. Symp. Proc., vol. 1101E, Warrendale, PA, 2008
86. M. Schmid, J. Krč, R. Klenk, M. Topič, M.C. Lux-Steiner, Appl. Phys. Lett. **94**(5), 053507 (2009)
87. T.J. Coutts, J.S. Ward, D.L. Young, K.A. Emery, T.A. Gessert, R. Noufi, Progr. Photovoltaics **11**(6), 359 (2003)
88. D.L. Young, J. Keane, A. Duda, J.A.M. AbuShama, C.L. Perkins, M. Romero, R. Noufi, Prog. Photovoltaics **11**(8), 535 (2003)
89. I. Repins, M.A. Contreras, B. Egaas, C. DeHart, J. Scharf, C.L. Perkins, B. To, R. Noufi, Progr. Photovoltaics **16**(3), 235 (2008)

Chapter 5
Surface Plasmon Polaritons in Metallic Nanostructures: Fundamentals and Their Application to Thin-Film Solar Cells

Carsten Rockstuhl, Stephan Fahr, and Falk Lederer

Abstract A surface plasmon polariton is a hybrid excitation where the electromagnetic field is resonantly coupled to a free carrier oscillation in noble metals. Once excited, a large enhancement of the local electromagnetic field and the amount of scattered light can be observed. Since both properties are beneficial for the purpose of photon management, in the past several years an increasing share of research was devoted to exploit such effects in solar cells. In this contribution, we review the fundamentals of surface plasmon polaritons and outline different approaches how to incorporate metallic nanostructures into solar cells. We detail to which extent they are useful to enhance the solar cell efficiency and describe different schemes for their experimental implementation. Emphasis is put on thin-film solar cells, since in this class of solar cells metallic nanostructures may have the largest impact. This chapter is written with the intention to make researchers from either the field of plasmonics or the field of photovoltaics familiar with their respective counterpart to foster research in this applied domain.

5.1 Introduction

With the desire to slow down and eventually to reverse the climate change, concepts to make energy available which is not based on fossil but on regenerative resources became an urgent matter. Approaches relying on wind, water, or tides are currently exploited. A further promising approach takes advantage of solar energy by transforming it to electrical energy using solar cells. A large variety of solar cells is currently established (see for an overview [1, 2]), and a comprehensive overview is well beyond the scope of this chapter. But all of them share an identical

---

C. Rockstuhl (✉)
Institute of Condensed Matter Theory and Optics, Friedrich-Schiller-Universität Jena,
Max-Wien-Platz 1, 07743 Jena, Germany
e-mail: carsten.rockstuhl@uni-jena.de

research aspect, namely to increase their photovoltaic conversion efficiency. The key task consists in increasing this efficiency at low cost to eventually reach the grid parity, requiring the price of solar energy installed at home to be as small as that of energy from fossil resources for the end user. This would appreciably increase the attractiveness of using such regenerative energies.

In addition to improving the electronic properties of solar cell materials themselves, there is a very promising approach towards the efficiency enhancement of a certain solar cell configuration, imposed by external constraints, frequently termed photon management [3]. It is understood here as the incorporation of assisting photonic structures that may serve two different purposes. On the one hand, it may either spectrally redistribute the impinging sunlight such that more photons contribute to the generated current. On the other hand, it may be of use to reduce the reflection losses or to enhance the optical path of photons in the solar cell such that their probability for absorption is enhanced. The latter scheme is sometimes called light trapping.

Here, we outline how to take advantage of surface plasmon polaritons for this purpose where emphasis is put on the second aspect. This chapter is divided into five sections. In the first section, we aim at familiarizing the reader with the basic issues, provide a concise introduction to the field and indicate major streams of research in the past. In the second section, we concisely outline the basic physics of surface plasmon polaritons where we distinguish between localized and propagating surface plasmon polaritons. The third and the fourth sections are devoted to technologies to incorporate the respective metallic nanostructures sustaining either localized or propagating surface plasmon polaritons into a solar cell. We finalize the chapter with conclusions on the broader impact of surface plasmon polaritons for photovoltaics. The chapter is written from the point of view of scientists working theoretically and numerically in the field of photon management; hence, emphasis is mainly put on those aspects, respectively.

5.2 Plasmon Polaritons and Their Use in Solar Cells

A plasmon is a harmonic oscillation of a free electron gas at a certain carrier density-dependent frequency in a metal. If such an oscillation resonantly couples to an electromagnetic field, a hybrid excitation arises, consisting of an electronic and a photonic part. This excitation is termed plasmon polariton. Since such coupling occurs at the interface between a dielectric and a metal, the term surface plasmon polariton (SPP) was coined [4]. The major characteristic of an SPP is its confined field to the interface from which it spatially exponentially decays into both materials. Depending on the dimensionality of the spatial confinement, one distinguishes two types of SPPs. Roughly spoken, if the electromagnetic field is confined in one or two dimensions, one speaks of a propagating surface plasmon polariton (PSPP). For a confinement in all three spatial dimensions by means of a metallic nanoparticle, one speaks of a localized surface plasmon polariton (LSPP). It has to be stressed that for the excitation of SPPs stringent constraints have to be met by the system. Since PSPPs are propagating surface waves, a dispersion relation is imposed relating unambiguously the wave vector component along the propagation direction to the frequency [5]. By contrast, the excitation of an LSPP requires a particular frequency.

SPPs are strongly dispersive and exhibit a line width. Light interacting with a medium that supports LSPPs experiences strongly enhanced scattering as well as absorption effects within a narrow spectral domain close to SPP resonance clearly to be observed in the far-field. By contrast, an excitation of a PSPP results in steep changes of the far-field reflection/transmission characteristic of the exciting light. However, in both cases the near-field close to the respective surface is appreciably enhanced. As a first direct use of plasmonic structures in solar cells, this enhanced field might facilitate the emission of photoelectrons which can contribute to an external current [6, 7].

Predominantly, these properties can be used for the purpose of photon management in solar cells. With regard to the spectral redistribution of the incident light, LSPPs can be exploited, e.g., to enhance the efficiency of up-converters (the energy of several photons is collected to generate one photon with higher energy for which the solar cell material is absorptive) [8, 9], down-converters (one high energy photon is converted into several low energy photons to reduce thermalization losses) [10, 11], or down-shifters (the energy of photons is slightly decreased to avoid the poor spectral response of solar cells in the short wavelength range) [12]. This can be achieved by placing the converting material close to the metallic nanoparticles where the excited LSPP may enhance, e.g., the excitation or the emission efficiency [13–16]. The respective effects can be intuitively understood by thinking of the metallic nanoparticle as a resonator to which the external illumination couples. The enhanced local field amplitude is then just a direct consequence of the increased photon lifetime at resonance by virtue of coupling to a plasmonic excitation of the nanoparticle. This will increase the probability of their interaction with the adjacent converting centers; hence affecting the aforementioned quantities [17]. Employing such LSPPs is appealing for a large variety of different solar cell concepts (e.g., plain silicon solar cells [18] but also more advanced concepts such as intermediate band solar cells [19] or quantum-well solar cells [20]) but it is beyond the scope of this chapter to discuss these concepts in detail.

Therefore, we focus in this chapter on the second aspect mentioned earlier, that is the immediate use of the excitation of an SPP to control the mould of light inside the solar cell with the aim to enhance its probability of absorption. First results were reported for organic solar cells, which comprise either thin metallic films, sustaining the excitation of PSPPs [21, 22], or small metallic clusters, sustaining the excitation of LSPPs [23–25]. The use of organic solar cells is not surprising for five reasons. At first, the resonance width of SPPs matches well with the absorption characteristics of the organic materials. At second, the thickness of an organic solar cell compares to the spatial extension of an SPP (exponential decay length of the field). At third, excited carriers in organic solar cells have a low mobility and a very low diffusion length. Such thin cells require light trapping compulsory. At fourth, the low refractive index contrast in such cells basically prohibits the use of more traditional approaches for the photon management. And at fifth, and probably trivial, improving a low efficiency solar cell is simpler than improving a solar cell working already at its limit. In the context of SPPs, their beneficial effects are always accompanied by an elevated absorption in the metal, whereby this dissipation has to be overcompensated by any effect used. This is the more difficult the higher the efficiency of the considered solar cell is. For all such reasons, this field of research remains vivid [26–29].

However, the underlying physical principles and associated optical actions of SPPs are universal and not restricted to a particular solar cell. This led to increasing research activities with regard to other types of solar cells whose efficiency is yet well below their theoretical limit. Among others, thin-film solar cells made of amorphous (a-Si:H)[30] and/ormicrocrystalline silicon (μc-Si) attracted much attention in the past [31]. The use of these materials is appealing, since their fabrication requires low costs and low energy consumption and can be easily up-scaled. However, particularly for the a-Si:H solar cell, the efficicency is currently limited by the finite (and small) diffusion length of the holes. This renders it pointless to increase the thickness of the solar cell material such that the incoming light would be fully absorbed. For state-of-the-art a-Si:H solar cells, the thickness of the active layer amounts to 250–350 nm, which renders an efficiency of 9.5% feasible [32]. Increasing the absorption without changing the thickness can be reached by various means. Reducing reflection losses at the entrance facet [33], using grating couplers to excite guided modes in the absorber layer [34, 35], or randomly textured surfaces to scatter light into the absorbing layer [36–38] are only a few approaches to be named.

By concentrating on the example of thin-film solar cells, we are going to subsequently outline how to incorporate metallic nanoparticles or metallic surfaces into the solar cell, to detail geometries that can potentially be used in any solar cell and to provide quantitative estimates, at least for the selected solar cells, to which extent their incorporation is beneficial. To facilitate the application of these approaches also to other solar cells, we concisely provide in the succeeding section the analytical background that permits to grasp an intuitive impression on how to exploit SPPs in any solar cell and how to layout an initial guess for the solar cell design. The section is accompanied by an outline of numerical methods and strategies that can be applied for designing and optimizing a real solar cell.

5.3 Fundamentals

5.3.1 Localized Surface Plasmon Polaritons

5.3.1.1 Existence Conditions

Localized surface plasmon polaritons are excited in metallic nanoparticles [39]. Although the conditions for their excitation strongly depend on the particle shape [40–42], we assume in the following that the nanoparticle is spherical;

5 Surface Plasmon Polaritons in Metallic Nanostructures 135

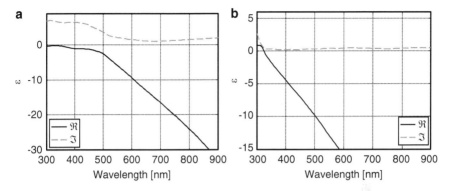

Fig. 5.1 Permittivity of Au (**a**) and Ag (**b**) as taken from [44] and used for all further considerations (Reprinted with permission from [44]. Copyrigth 1970. American Physical Society)

an assumption which permits for an analytical treatment. The sphere shall be characterized by a radius R. If the radius is much smaller than the wavelength of interest λ, the problem of plane wave scattering at the sphere can be solved in the quasi-static approximation [43]. In this approximation, phase variations of the incident electromagnetic field across the spatial domain of interest can be neglected. Thus, the problem reduces to an electrostatic one, where the field may be described by the scalar potential Φ being a solution to Laplace's equation ($\Delta \Phi = 0$). The sphere is furthermore assumed to be made of a metal, characterized by a dispersive permittivity $\epsilon_1(\omega)$, and embedded in a medium with permittivity $\epsilon_2(\omega)$. To excite an LSPP, the sphere's material has to have a negative real part of the permittivity like a metal, which can be described by the Drude model. The imaginary part of the permittivity should be small to ensure small damping of excited LSPPs. The material dispersion for the two best suited materials, gold (Au) and silver (Ag), is shown in Fig. 5.1 and was taken from literature [44]. It can be seen that the required metallic properties occur for Ag at wavelengths exceeding ≈ 340 nm and for Au beyond ≈ 540 nm, respectively. These wavelengths represent a lower bound for LSPPs excitation.

To predict the resonance wavelength of an excited LSPP, the Laplace equation is solved in spherical coordinates [43]

$$\frac{1}{r^2}\frac{\partial}{\partial r}\left(r^2\frac{\partial \Phi}{\partial r}\right) + \frac{1}{r^2 \sin\theta}\frac{\partial}{\partial \theta}\left(\sin\theta \frac{\partial \Phi}{\partial \theta}\right) + \frac{1}{r^2 \sin^2\theta}\frac{\partial^2 \Phi}{\partial \varphi^2} = 0. \quad (5.1)$$

The general solution to this equation reads as

$$\Phi(r, \theta, \varphi) = \sum_{n=0}^{\infty}\sum_{m=0}^{n}\left(a_{nm}r^n + \frac{b_{nm}}{r^{n+1}}\right)P_n^m(\cos\theta)\,e^{\imath m\varphi} \quad (5.2)$$

with $P_n^m(x)$ being the associated Legendre polynomials forming an orthonormal basis. Without loss of generality, we may assume a z-polarized electric field for illumination, its potential reads as $\Phi_0 = -E_0 z = -E_0 r P_1^0(\cos\theta)$. By writing the potential outside the sphere Φ_2 as a superposition of the incident Φ_0 and the scattered potential Φ_S, and the potential inside the sphere as Φ_1, the unknown coefficients a_{nm} and b_{nm} in the expansions of Φ_1 and Φ_S can be found by taking physical considerations into account and enforcing the usual boundary conditions. Boundary conditions are the continuity of the tangential component of the electric field and of the normal component of the electric displacement.

The solutions for the potentials read as

$$\Phi_1 = -\frac{3\epsilon_2}{\epsilon_1(\omega) + 2\epsilon_2} E_0 r \cos\theta \tag{5.3}$$

and

$$\Phi_2 = -E_0 r \cos\theta + \frac{\epsilon_1(\omega) - \epsilon_2}{\epsilon_1(\omega) + 2\epsilon_2} R^3 E_0 \frac{\cos\theta}{r^2}. \tag{5.4}$$

The associated electric field can be calculated from $\mathbf{E} = -\nabla\Phi$ for the outer region to be

$$\mathbf{E}_2 = E_0 \mathbf{e}_z + \frac{\epsilon_1(\omega) - \epsilon_2}{\epsilon_1(\omega) + 2\epsilon_2} \frac{R^3}{r^3} E_0 (2\cos\theta \mathbf{e}_r + \sin\theta \mathbf{e}_\theta). \tag{5.5}$$

By comparison, it can be seen that the scattered field corresponds to the field of a dipole having a moment

$$\mathbf{p} = \epsilon_2 \alpha E_0 \mathbf{e}_z. \tag{5.6}$$

Here, α is the polarizability, which is given by

$$\alpha = 4\pi R^3 \frac{\epsilon_1(\omega) - \epsilon_2}{\epsilon_1(\omega) + 2\epsilon_2}. \tag{5.7}$$

This equation for the polarizability is the major result of this section. Its value will dominate both the near- and the far-field response. It allows furthermore for an analytical discussion of the response of the system, which greatly simplifies the rational design of plasmonic systems used for solar cell applications.

Foremost it can be seen that in this quasi-static analysis the LSPP resonance wavelength depends only on the properties of both involved materials; namely, the denominator from (5.7) $\epsilon_1(\omega) + 2\epsilon_2$ shall attain values close to zero. The complete nullifying is prevented by the finite imaginary part of the metal that hinders the polarizability from taking unphysical singular values. When assuming a dielectric material with $\epsilon_2 = 2.25$ as the surrounding material, the resonance wavelength at which LSPP are excited corresponds to ≈ 400 nm for silver and ≈ 530 nm for gold. At this wavelength, the charge density in the metal oscillates resonantly and is driven by the external illumination. Both scattering strength and field enhancement being exploited in a solar cell for photon management are immediately affected by this polarizability.

5 Surface Plasmon Polaritons in Metallic Nanostructures

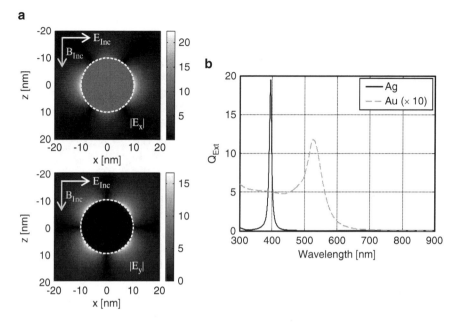

Fig. 5.2 (a) Amplitude around a silver nanosphere ($R = 10$ nm) upon illumination with a plane wave at the resonance wavelength. Fields are shown in a selected cross-section. (b) Spectrally resolved extinction cross section Q_{Ext} for a silver nanosphere ($R = 10$ nm) embedded in a dielectric host material

5.3.1.2 Quantities to Be Employed in a Solar Cell

The quantity that is primarily affected is the largely enhanced electromagnetic near field. It is seen from (5.5) in conjunction with (5.7) that the larger the polarizability the larger the near-field amplitude. This enhanced field directly causes the local absorption to be elevated. For illustrative purposes, Fig. 5.2 shows the near-field amplitude around a silver nanosphere with a radius of 10 nm surrounded by vacuum at the resonance wavelength.

The observable quantity in the far-field, which is beneficial for a solar cell is the scattering cross-section Q_{Sca}. It is defined as the ratio of the scattered intensity and the intensity of the incident field on the sphere averaged over time and reads as

$$Q_{Sca} = \frac{k^4}{6\pi}|\alpha|^2 \tag{5.8}$$

with $k = 2\pi/\lambda$ being the wave number. Correspondingly, the absorption cross section is given by

$$Q_{Abs} = k\Im(\alpha), \tag{5.9}$$

with $\Im(\alpha)$ being the imaginary part of the polarizability α. The extinction cross-section Q_{Ext} as the last quantity of interest is defined as $Q_{Ext} = Q_{Sca} + Q_{Abs}$.

An enhanced scattering of light into the solar cell will increase the optical path of photons and may excite guided modes in the solar cell for which the probability of absorption is unity. An example for the extinction cross section for the same silver sphere as considered above is shown in Fig. 5.2b. It can be seen that an LSPP is only excited in a narrow spectral domain. Moreover, from such simulation the resonance wavelength of LSPP excitation can be extracted, whereas the associated line width is a measure for the losses.

The above formulas may be used for a simple guess regarding the preferential nanoparticle size. Since for small particles absorption (linearly proportional to the polarizability) dominates over scattering (proportional to the square of the polarizability), metallic nanoparticles should be sufficiently large such that scattering prevails against absorption. For practical reasons, the diameter at which scattering starts to dominate over absorption can be roughly estimated to be 50 nm. Moreover, both quantities are resonantly enhanced whenever the polarizability diverges, hence at wavelengths around the LSPP resonance.

Finally, it has to be stressed that a rigorous analysis beyond the quasi-static approximation by relying on a direct solution of Maxwell's equations will reveal the following aspects:

- Increasing the particle size will red-shift the resonance wavelength and beyond a certain threshold it will also lower and broaden the resonance.
- Even in the quasi-static limit the resonance wavelength can be adjusted by changing the nanoparticle shape. As a rule of thumb, the longer the axis along the polarization direction of the incident beam the more the resonance is red-shifted at a constant volume of the nanoparticle [45].
- There are many options to control both the resonance wavelength and width. Some examples are the use of core-shell systems [46], structures made of coupled particles or particles with extremely involved geometries, which sustain the excitation of higher order plasmonic modes [47].
- The resonance wavelength can always be made sufficiently large, whereas a lower bound is given by the required negative real part of the permittivity.

In the discussions below, we will rely on the arguments obtained in this section in order to motivate the use of nanoparticles of a certain size.

5.3.2 Propagating Surface Plasmon Polaritons

5.3.2.1 Existence Conditions

The second type of plasmonic resonances that may be exploited in solar cells is the propagating surface plasmon polariton (PSPP) [48]. This hybrid state is bound to the interface between a dielectric and a metal or a sequence thereof, which propagates without changing its field profile. In a strict sense, this would require that all involved materials are lossless. The eigenmode itself is a guided mode and attains a sharp maximum at the interface and decays evanescently in both media, called for convenience the substrate and the cladding. The corresponding field profile is a rigorous solution to Maxwell's equations. For the following treatment, we assume that the normal of the interface points into z-direction, whereas the structure shall be invariant in y-direction. For a one-dimensional confinement, the PSPP is then characterized by a propagation constant $k_x(\omega)$ in the direction parallel to the interface.

The dispersion relation is generally given by providing the dependency of k_x on the frequency. The permittivity $\epsilon(\omega)$ of all media is expected to be frequency dependent. Although in most cases, the dielectric material might be assumed to have a constant permittivity free of dispersion, at least the metal is assumed to be dispersive. This permits to finally obtain the dispersion relation $k_x(\omega)$. For its derivation various methods exist. We rely here on a rather general method that derives the propagation constant of an arbitrary guided mode from the poles of the reflection coefficient R of the entire stack. The reflection coefficient of an arbitrarily stratified medium is obtained by using a transfer matrix technique as [49]:

$$R = \frac{(\alpha_s k_{sz} M_{22} - \alpha_c k_{cz} M_{11}) - \imath(M_{21} + \alpha_s k_{sz}\alpha_c k_{cz} M_{12})}{(\alpha_s k_{sz} M_{22} + \alpha_c k_{cz} M_{11}) + \imath(M_{21} - \alpha_s k_{sz}\alpha_c k_{cz} M_{12})}. \quad (5.10)$$

Here, k_{cz} and k_{sz} are the longitudinal wave vector components in the cladding (c) and the substrate (s). They are computed from the dispersion relation of the homogenous medium as

$$k_{c/s\,z} = \sqrt{\frac{\omega^2}{c^2}\epsilon_{c/s} - k_x^2}. \quad (5.11)$$

The stratified medium is characterized by the transfer matrix $\hat{\mathbf{M}}$ that is a polarization dependent 2×2 matrix for a given sequence of layers. $\alpha_{c/s}$ is a polarization-dependent parameter and reads as $\alpha_{c/s} = 1$ for TE polarized light (electric field component only tangential to the interface) and $\alpha_{c/s} = \frac{1}{\epsilon_{c/s}}$ for TM polarized light (magnetic field component only tangential to the interface).

For a guided or bound mode, the field in the substrate and the cladding has to be evanescent. It requires an imaginary longitudinal wave vector component, hence $k_x^2 > \frac{\omega^2}{c^2}\epsilon_{c/s}$ is required. By writing the imaginary longitudinal wave vector component as $k_{c/s\,z} = \imath\kappa_{c/s}$, the dispersion relation for a guided mode can be cast into the form

$$M_{11} + \alpha_s\kappa_s M_{12} + \frac{1}{\alpha_c\kappa_c}M_{21} + \frac{\alpha_s\kappa_s}{\alpha_c\kappa_c}M_{22} = 0. \quad (5.12)$$

For the sake of simplicity, we restrict our further considerations to the case of a single interface and assume the substrate to be metallic. It renders the transfer matrix $\hat{\mathbf{M}}$ to be the unit matrix. Nevertheless, this is not a general restriction but shall permit here only to derive all quantities of interest. With these assumptions, the dispersion relation (5.12) simplifies to

$$1 + \frac{\alpha_s \kappa_s}{\alpha_c \kappa_c} = 0. \tag{5.13}$$

Obviously, this equation has no solution for TE-polarized light because the evanescent propagation constants will always take positive values.

The solution for TM polarized light reads as

$$\frac{\kappa_c}{\epsilon_c} + \frac{\kappa_s}{\epsilon_s} = 0. \tag{5.14}$$

In general, solutions may only exist in a spectral domain where the permittivities of substrate and cladding have opposite signs, i.e., their frequency must be less than the plasma frequency of the metal. Plugging the expressions for the longitudinal wave vector component (5.11) into the solution for the resonance condition (5.14) provides the dispersion relation

$$k_x(\omega) = \frac{\omega}{c} \sqrt{\frac{\epsilon_s(\omega)\epsilon_c}{\epsilon_s(\omega) + \epsilon_c}}. \tag{5.15}$$

This equation relates the longitudinal wave vector component (propagation constant) of the PSPP to the frequency. Now the final constraints for the existence of PSPP can be read off. It does not suffice that the metal (the surface-active material) attains negative values but $\epsilon_s(\omega) + \epsilon_c < 0$ is required. By assuming metal–air interface and a damping free Drude model for the metal it holds for frequencies less than $\omega_p/\sqrt{2}$, where ω_p is the plasma frequency. This also gives the lower wavelength bound for the possible excitation of a PSPP. The dispersion relation is shown in Fig. 5.3a for the interface between air and a Drude metal.

It is evident that the wave vector of the PSPP is always larger than the wave vector in the involved dielectric. Consequently, special emphasis has to be put on the compensation of the transverse wave vector mismatch by exciting a PSPP. Different experimental schemes have been proposed where the so-called Otto and the Kretschmann configurations [50, 51] are the most popular ones. More advanced approaches rely on the use of a diffraction grating [52] but also isolated scatterers allow to excite a PSPP since they offer the entire wave vector spectrum.

5.3.2.2 Quantities to Be Exploited in a Solar Cell

The quantity, usually exploited in a solar cell, is the largely enhanced near-field amplitude. A snap shot of the magnetic field of an excited PSPP is shown in Fig. 5.3b. The field profile takes the form

5 Surface Plasmon Polaritons in Metallic Nanostructures

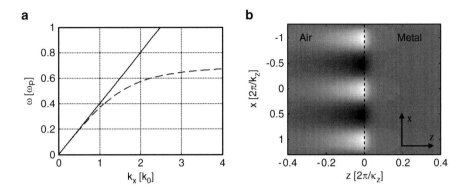

Fig. 5.3 (**a**) Dispersion relation for a plane wave propagating in free space (*black solid line*) and for a surface plasmon polariton propagating at the interface between a Drude metal and air (*gray dashed line*). (**b**) Snapshot of the magnetic field of a surface plasmon polariton propagating along this interface

$$H_y(x,z) = H_0 e^{ik_x x} e^{-\kappa_{c/s}|z|}. \tag{5.16}$$

The corresponding x- and z- component of the electric field are readily obtained employing the appropriate Maxwell's equation.

The observable quantities that provide indication of an excited PSPP are usually a strongly reduced reflection over a narrow spectral and angular domain upon using the indicated excitation techniques. Since the line width of a PSPP is usually much narrower than for the LSPP, the impact of photon management occurs likewise in a narrow spectral domain only. This is detrimental for a silicon-based solar cell because the entire spectrum of sun light has to be considered although main emphasis is put on wavelengths around the absorption edge. However, this spectral domain is suitably adapted to the absorption profile of other solar cell materials, such as, e.g., organic solar cell materials, where it may be beneficially exploited.

It remains to be mentioned that the above considerations apply only for the simplest scenario where PSPPs can be excited (planar interface between a metal and a dielectric). Arbitrary, more complex wave guides can be imagined in one- or two-dimensions, which can have an ever-increasing complexity in their geometrical cross section. As a simple extension, additional buffer layers between the absorber and the metal shall be mentioned. Nevertheless, the techniques to characterize the PSPPs remain the same. With an appropriate propagating eigenmode ansatz Maxwell's equations are analytically (or numerically) solved and the functional dependency of $k_x(\omega)$ is usually computed. In addition to the dispersion relation, the solution of the eigenvalue problem provides the field profiles of the eigenmodes, which allows to estimate the field overlap of the guided mode with the solar cell material. However, for current solar cells the shape of the eigenmodes is not of primary importance since only an external illumination allows for their excitation. The entire analysis of this coupling problem is usually fairly involved such that no analytical solutions are at hand and one has to resort to a numerical simulation of the problem. An overview of numerical techniques suitable for this problem is provided in the subsection below.

5.3.3 Simulation Techniques

Simulation techniques are indispensable tools for designing metallic nanostructures, applicable for photon management in solar cells and deviating from standard geometries outlined above. Moreover, quantitative predictions regarding the enhancement of the absorption or the solar cell efficiency are only possible with numerical techniques that solve Maxwell's equations rigorously because of the involved geometries of real cells. The usual computational procedure to estimate the enhancement of the absorption efficiency is as follows:

1. Compute with a numerical technique of choice the electromagnetic field at a given wavelength (most notably the electric field $\mathbf{E}(\mathbf{r},\omega)$) everywhere inside the solar cell.
2. With the electric field at hand, the locally resolved absorption is given by the divergence of the time averaged Poynting vector $\mathbf{S}(\mathbf{r},\omega)$ as $\mathbf{div}\langle\mathbf{S}(\mathbf{r},\omega)\rangle = -\frac{1}{2}\Im\{\omega\mathbf{P}(\mathbf{r},\omega)\mathbf{E}^*(\mathbf{r},\omega)\}$ with $\mathbf{P}(\mathbf{r},\omega)$ and ω being the dielectric polarization and the angular frequency of the light, respectively.
3. Integrate this local absorption over the entire spatial domain of interest, i.e. the light absorbing solar cell material, for obtaining the global absorption.
4. Weight the global absorption with the AM1.5G spectrum to obtain a measure how many photons are absorbed.
5. Iterate these steps over the entire relevant spectral domain to compute the total number of absorbed photons for a given angle of incidence.
6. If necessary, repeat the procedure above for the relevant angles of incidence for determining the characteristics of a nontracked solar cell.

Eventually, the number of absorbed photons has to be normalized with the number of absorbed photons in the same solar cell without photon management to evaluate the absorption enhancement. This quantity is sufficient to estimate the performance of a certain approach for the photon management. Moreover, the outlined procedure only allows to access the optical properties of the solar cell. An entire opto-electronical simulation in 3D would involve too many unknown electronical parameters to allow for an extraction of design guidelines. However, under the simplifying assumption that each absorbed photon generates one electron-hole pair and by assuming an external quantum efficiency of unity, the short-circuit current can be derived from the number of absorbed photons.

The key step in this procedure is the rigorous computation of the electromagnetic field, which is a very challenging task. For this purpose, various methods were developed in the past. They can be roughly divided into three categories.

The first category allows predominantly to analyze the scattering of light at single objects embedded in a dielectric environment. Techniques like the Discrete Dipole Approximation [53], the Multiple Multipole Method [54] and the Boundary Element Method [55] fall in this category. Although such methods are not widely used to quantify the absorption in a solar cell since only a single or a few elements are analyzed, the methods can be used to quantify the optical response from objects with complicated geometries and to optimize their plasmonic response to the absorption spectrum of the solar cell [56].

The second category allows to describe light propagation in and the diffraction of light at periodically arranged structures. Devoted techniques, like the Rigorous Coupled Wave Analysis [57], the Fourier Modal Method [58], or the Chandezon method [59], make intrinsically use of the periodicity by formulating Maxwell's equations in the spatial Fourier domain and solving them with a plane wave ansatz. The required perfect periodicity is met if structures are fabricated with a deterministic technique such as electron beam lithography. Moreover, consideration of such deterministic structures allows for a straightforward optimization of the structures' geometry. Therefore, they are perfectly suitable to estimate an upper limit for the absorption enhancement. In the following, we rely mostly on the assumption of such periodically arranged particles for this reason. Besides, the computational techniques are so sophisticated that they allow for an analysis of the considered system in a reasonable amount of time using state-of-the-art computational facilities.

The third category comprises general purpose tools, which can be applied to periodic structures as well as to single elements. Such methods solve Maxwell's equations either in the frequency domain, e.g. the Finite Element Method [60] or in the time domain, e.g. the Finite-Difference Time-Domain (FDTD) method [61–63]. They are always valid and a good choice in the optical simulation of a solar cell. Nonetheless, frequency domain methods seem to be preferred since the proper incorporation of the material dispersion is much simpler there when compared to time-domain methods. In the time domain, the material response function needs to be implemented by a superposition of several oscillators leading to a Drude–Lorentz type material dispersion. When tabulated material properties shall be considered, the number of oscillators that need to be taken into account is large, rendering the simulation computationally demanding.

5.4 Use of Localized Surface Plasmon Polaritons in Solar Cells

The direct use of LSPP excited in metallic nanoparticles is probably most appealing since the concept is rather simple. The metallic nanoparticles in the solar cell can be either fabricated on top [64] or on the bottom [65] of the cell by using either deterministic technologies [66] or various forms of self-organization [67, 68]. A sketch of a documented photovoltaic device in which metallic nanoparticles are

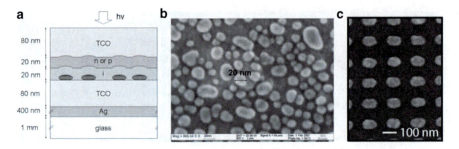

Fig. 5.4 (a) Sketch of the layout of an a-Si:H solar cell, which contains metallic nanoparticles and (b) a micrograph of the actual nanoparticles, which were fabricated by a self-organized process (Both figures were reprinted with permission from [65]. Copyright 2009. American Institute of Physics). (c) Alternatively, nanoparticles can be also fabricated by deterministic techniques to provide well-defined shapes. (Reprinted with permission from [66]. Copyright 2008. American Institute of Physics)

incorporated along with an image of fabricated nanoparticles using either a self-organized or a deterministic technique are shown in Fig. 5.4. The size of the metallic nanoparticles is usually adjusted such that the resonance matches the absorption edge of the solar cell material where its impact is most tremendous. Then the local near-field as well as the scattering cross section is significantly enhanced, which will lead to a much larger amount of light absorbed in the solar cell. Additionally, for a periodic arrangement of the metallic structures guided modes in the solar cell are potentially excited that will also cause a significantly elevated absorption. We wish to mention once again that also other effects, such as the enhanced light absorption at the plasmon resonance wavelength that leads to an enhanced photoemission of electrons from the metals, will contribute to an enhancement of the external current [65].

For nondeterministic approaches, the procedure to integrate metallic nanostructures consists in adjusting a fabrication parameter in the self-organization process and observing its effect either on the morphology of the generated metallic nanostructures [69] or directly on the performance of the solar cell [70]. This very parameter is then continuously tuned to optimize a quantity of interest, such as e.g. the light absorption or the efficiency. Whereas such approaches are low cost and suitable for up-scaling, they usually do not fully exploit the ability of LSPPs since a large portion of the fabricated metallic nanostructures act off-resonant and do not contribute to light trapping. This limits its use. Moreover, with the advance of large-scale nanofabrication technologies that allow to fabricate periodic patterns of nanoscopic particles at a macroscopic extent, more sophisticated design strategies were put forward to fully exploit their capabilities [71, 72]. Such an approach [73, 74] is detailed in this section. The analysis is performed to reveal fundamental guidelines universal to all solar cells that shall be optimized in their design.

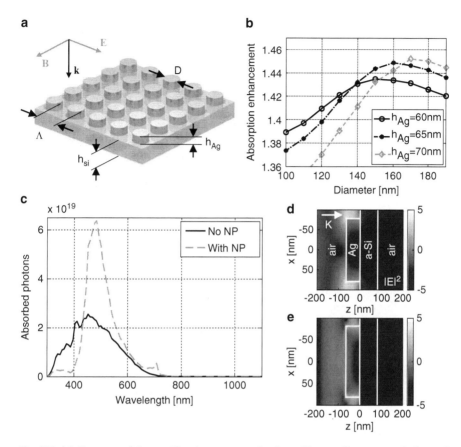

Fig. 5.5 (a) Geometry of the considered structure made of metallic nanodiscs on top of a layer of amorphous silicon. The parameters, which can be arbitrarily chosen in principle, are indicated. (b) Absorption enhancement in the silicon layer as a function of the size of the nanodiscs. The period was chosen to be 1.5× the nanodisc diameter D and the thickness of the silicon layer was chosen to be 80 nm. Surrounding material is assumed to be air. (c) Absorption spectrum of the structure weighted by the illuminating photon density, showing the largest absorption enhancement in (b). (d) Amplitude of the electric field on a logarithmic scale in the x–z plane. The incident electric field is x-polarized and the wavelength was chosen to be $\lambda = 484$ nm. Geometrical parameters of the system correspond to those shown in (c). (e) The same amplitude is shown but at a wavelength of $\lambda = 350$ nm, where the nanodiscs are non-resonant [74]

5.4.1 Thin-Film Solar Cells

The approach we consider at first consists of placing metallic nanoparticles with a predefined shape on top of the solar cell. The geometrical sketch of the structure is shown in Fig. 5.5. The choice of nanodiscs is motivated by the simple shape that reduces the number of free parameters, the polarization independent response of the structure and the ease in fabricating such nanodiscs. The solar cell is assumed to be made of a-Si:H, as motivated in the introduction. The nanodiscs are made of silver. It was chosen since, as we will see later, their LSPP resonance wavelengths are favorable when compared to gold. In all simulations, we rely on material properties as documented in literature [44, 75]. The structure is illuminated by a linearly polarized plane wave. The chosen C_4 symmetry excludes pronounced polarization-dependent resonances, which could affect the conclusions.

Depending on some selected geometrical parameters, the achievable absorption enhancement is shown in Fig. 5.5b. In this particular case, it is shown as a function of the size of the nanodiscs. Remaining parameters were motivated by supplementary simulations [74]. The period of 1.5× the nanodisc diameter D represents a compromise. On the one hand, the particles are sufficiently separated for a reduced coupling between neighboring nanodiscs, which would otherwise lead to excessive nonradiative losses, i.e., absorption. On the other hand, the nanodisc arrangement is yet sufficiently dense to show an appreciable absorption enhancement. Naturally, the larger the period the less the absorption enhancement. The thickness of the solar cell is also chosen by purpose since, at this thickness, the first Fabry–Perot resonance of the absorption enhancement occurs. For larger thicknesses similar maxima occur, though they are smeared out since an integration was performed over the entire relevant spectrum [15]. If the free spectral range of Fabry–Perot response becomes smaller, the resonant features disappear. Nevertheless, the conclusions do not depend on a specific thickness.

The spectrally resolved absorption weighted by the AM1.5g spectrum is shown in Fig. 5.5c. It can be seen that when compared to the LSPP resonance of the nanodiscs (which was estimated to be at 410 nm), the absorption is reduced at lower but enhanced at larger wavelengths. This is a universal feature. Since, as for all resonance phenomena, the response is in-phase with the driving field at frequencies smaller than the resonance, the field scattered by the nanodisc interferes constructively with the illumination. This increases the local field inside the solar cell and so the absorption. By contrast, at higher frequencies the scattered field is π out-of-phase with the illumination and interferes destructively. Evidently, this leads to the reduced absorption at smaller wavelengths [76,77]. For exactly this reason the use of gold nanoparticles on top of the solar cell is not beneficial, since the lowest feasible resonance wavelengths occur already above the absorption edge. This will lead to a reduction of the light absorption in a spectral domain, where the solar cell usually provides its largest contribution to the photo current. Nevertheless, a simple exit strategy would be the incorporation of nanoparticles on the back side, since short wavelength light is already sufficiently absorbed in the solar cell upon a single passage [72].

The resonant character can be recognized in looking at the amplitude distribution of the electric field in the solar cell. In Fig. 5.5d, e, the local amplitude of the electric field is displayed for a selected cross section. Only at resonance the field is significantly enhanced. Moreover, from the Fabry–Perot resonances of the field that occur inside the solar cell, it can be concluded that not only the enhanced near-field amplitude contributes to the absorption enhancement but also the strong scattering. Only for a much smaller solar cell thickness the enhanced near-field contributes predominantly, the absorption even tends to diverge for a vanishing thickness. However, this is trivial, since the absolute absorption also approaches zero and the absorption enhancement is not a useful quantity [74].

As the main result of this section, we have shown that, although with a slightly artificial solar cell geometry (no substrate and no superstrate), the absorption can be enhanced by a factor of ≈ 1.5 when compared to the same solar cell without the light trapping structure. This is observed by only placing the nanoparticles on top of the solar cell. This achieved absorption enhancement seems to be likewise feasible by considering other geometries [78]. However, it remains to be noted that in the design of a real solar cell, the entire geometry needs to be considered to provide reliable predictions. The enhancement as documented here should therefore merely be regarded as an optimal value to which the performance of a real system should be compared. Such a more feasible system is considered in the section below.

5.4.2 Tandem Solar Cells

The efficiency of a single junction a-Si:H cell can be enhanced by passing on to a tandem cell concept, where the a-Si:H top cell is series connected to a μc-Si bottom cell [79]. This bottom cell can usually be made such thick that the amount of absorbed photons in μc-Si equals that one in a-Si:H [80]. This renders the solar cell to be current matched. Since the voltages provided by each cell will add and the formerly unused long wavelength range can be exploited as well, the overall efficiency is enhanced. Due to the aforementioned thickness limitation of the a-Si:H cell, maximizing its current contribution is the primary goal in the cell design. Besides the methods mentioned already in Sect. 5.2, an additional intermediate reflecting layer (IRL) directly underneath the top cell helps to localize more light within the a-Si:H cell. The simplest approach consists of a layer made of a transparent conducting oxide [81] (TCO) whose refractive index is less than the indices of the surrounding silicon. Due to increased Fresnel reflection at both TCO interfaces, this TCO layer forms a Fabry–Perot cavity which features high back reflection for its resonance wavelengths. By optimizing the TCO thickness, the spectral domain of high reflection can be adjusted to match the absorption edge of the a-Si:H cell. However, off-resonance high back reflection can still be observed, especially in the long wavelength range, where reflected light cannot be absorbed anymore by the top cell. This is detrimental since it will then also not contribute to the current of the bottom cell. Therefore, more advanced schemes employ spectrally selective photonic crystals [82] or asymmetric (additional) structuring of the TCO layer [83, 84].

The incorporation of metallic nanoparticles as a spectrally selective IRL into a silicon-based tandem solar cell [85] shall be investigated in detail as a second example in this section. Such a cell is depicted in Fig. 5.6a, where silver nanodiscs are embedded in a dielectric ZnO IRL, whose permittivity was set to $\varepsilon = 4.0$. By

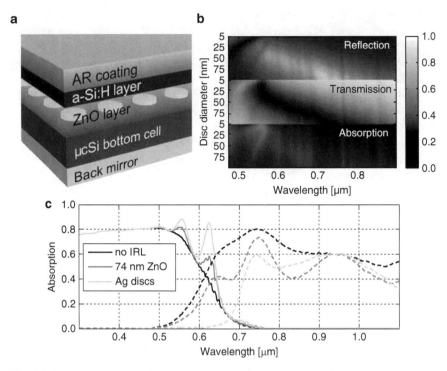

Fig. 5.6 Incorporation of metallic nanoparticles into a tandem cell (**a**). Spectral properties of the optimized nanodiscs when embedded in infinite ZnO (**b**). Absorption spectra of the top (*solid lines*) and the bottom cell (*dashed lines*) for different IRL concepts: no IRL, optimized ZnO layer, and optimized silver nanodiscs (**c**)

optimizing the shape and distribution of the nanodiscs, the LSPP resonance will be adjusted to the absorption edge of the a-Si:H top cell.

The spectral properties of 29-nm-thick Ag nanodiscs for varying diameter are shown in Fig. 5.6b assuming normal incidence of unpolarized light. The nanodiscs are arranged on a square lattice and embedded in infinite ZnO. To reveal the resonance wavelength, there is no need to take the silicon layers into account since the nanoparticles are buried in ZnO in the final cell such that the deviating dielectric surrounding of the silicon affects the resonance wavelength only negligibly. A constant metal fraction is realized by setting the period to 1.78× the diameter. As was detailed already in Sect. 5.3, the resonance wavelength can be tuned by the surrounding medium and the particle shape. The latter can also be observed in Fig. 5.6b, where the resonance is accompanied by enhanced absorption and reflection. Furthermore, the resonance wavelength cannot be shifted below 500 nm even for a vanishing particle size. If shorter resonance wavelengths are required, the surrounding medium has to have a smaller permittivity.

Thinner nanodiscs feature similar spectra (not shown), except for less pronounced resonances which also have a smaller line width. For particle sizes in excess, higher order LSPP resonances can be excited in thicker nanodiscs, which disturb the reflection by adding bands of high transmission (not shown). This imposes an upper limit on the particle size. Decreasing the period leads to stronger interparticle coupling and the non-radiative losses are enhanced. More light can be absorbed in the bottom cell by increasing the period at the cost of less back reflection into the top cell. Therefore, by properly choosing the metal fraction, the aspect ratio of the nanodics, and the permittivity of the IRL, the spectral position, as well as the strength of the back reflected light can be adjusted to match the spectral demands of the considered tandem cell. These parameters can be subject to optimization.

Figure 5.6c shows the absorption spectra of a-Si:H (solid lines) and μc-Si (dashed lines) for various IRL concepts. The black lines correspond to a cell, where no IRL is present. The top cell thickness was set to 350 nm, thus setting the spectral range, where the top cell absorption can be significantly enhanced to 500–750 nm. The bottom cell was 4.0 μm thick and followed by a perfectly reflecting substrate. Due to the coherent calculations and the finite thickness of the bottom cell, strong Fabry–Perot resonances would be observable in the μc-Si spectra. In order to simplify the assessment of the results, these oscillations were smeared out. The dark gray lines represent a cell, where the top cell current was maximized by adjusting the IRL thickness. Already an enhancement of 1.06 could be achieved. The light gray lines correspond to an IRL, which contains Ag nanodiscs arranged in a square lattice. For this purpose, the height and diameter of the nanodiscs, their lattice constant, as well as their distances to the top and the bottom cell were optimized. Such a tandem cell features a matched current, which is by a factor of 1.14 higher than that of a cell without any IRL at all. These optimized discs correspond to the spectral properties shown already in Fig. 5.6b.

5.5 Use of Propagating Surface Plasmon Polaritons in Solar Cells

As outlined already in Sect. 5.3, propagating surface plasmon polaritons can be excited at the interface of a metal and a dielectric. Since most solar cells do have a metallic backside reflector, it is an obvious choice to rely on this interface for the excitation since no additional actions need to be undertaken for its incorporation. However, it is by far not obvious that the excitation of PSPPs is beneficial, since the energy dissipated in the metal obviously will not contribute to an external current and is a source of losses. Since the primary contribution of this metallic backside is the strong reflection of light, the excitation of such PSPPs at the back contact is often suppressed either by a dielectric intermediate passivation layer or by an supportive structure, which shifts the PSPP resonance wavelength outside the spectral domain of interest [86].

To control the excitation of PSPPs various schemes are feasible and some of them are shown in Fig. 5.7. In general, a supportive structure is required, which

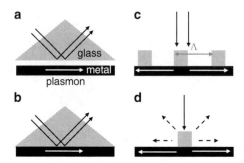

Fig. 5.7 Various principal configurations for the excitation of a propagating surface plasmon polariton. (**a**) Otto configuration, (**b**) Kretschmann configuration, (**c**) grating coupler, (**d**) isolated scatterer. The *black arrows* indicate the external illumination, whereas the white ones indicate schematically an excited PSPP

compensates the wave vector mismatch between the light impinging from free space and the PSPP; as outlined in Sect. 5.3. Moreover, the surface of the metallic structure itself might be corrugated, either periodically or randomly, which allows for the excitation of PSPPs as well [87].

In any case, the possible excitation of a PSPP constitutes an additional degree of freedom, which can be exploited in the solar cell. A careful analysis is required where pros and cons have to be checked against each other. Their balancing may result in pros an absorption enhancement for various types of solar cells; such as GaAs solar cells [88], organic solar cells [89], or also for amorphous silicon thin film solar cells [90]. All of them rely on the option to change the propagation direction of the incident light by 90° by redirecting it to propagate along the surface of the metal and hence along the direction, where the solar cell material is infinitely extended. This of course will significantly enhance the absorption of light in the solar cell material. Such an application is concisely detailed in this subsection.

5.5.1 Thin-Film Solar Cells

The experimental results we present in this section were taken from literature [22]. It is a simple, yet illustrative example how to take advantage of PSPPs in a solar cell. The solar cell consists of an organic material, namely copper phthalocyanine (CuPc) (layer thickness 33 nm). The material is absorptive in a narrow spectral domain and shows a double hump spectrum that peaks at 620 and 720 nm [91]. By using vacuum evaporation techniques, it is deposited on a thin aluminum layer (13 nm) that was deposited on a glass substrate. To excite a PSPP, a thin silver layer (21 nm) was finally deposited on the solar cell. The CuPc acts as a p-type semiconductor and a Schottky barrier is formed at the interface to the aluminium. The silver layer provides in addition to its plasmonic properties also the necessary ohmic contact

5 Surface Plasmon Polaritons in Metallic Nanostructures

to complete the solar cell [92]. By eventually attaching copper wires to the solar cell, the current–voltage characteristics could be measured [22]. The entire structure was mounted on a glass prism and illuminated through it under an angle such that light suffers from total internal reflection at the prism facet. The excitation scheme corresponds exactly to that as shown in Fig. 5.7b.

By using a white light Xe-lamp with a flat and broad spectrum at first, the normalized short circuit current was measured as a function of the angle of incidence and the polarization (see Fig. 5.8a) where normalization was done against the short circuit current at normal incidence. It can be seen that no enhancement was observed for s-polarized light (TE polarization) because PSPPs cannot be excited in this case (see Sect. 5.3). For p-polarization, the wave vector mismatch is provided at each wavelength for a slightly different angle. In the spectral domain of interest where CuPc is absorptive, this occurs at angles around 45° for the given configuration. Once a PSPP is excited, the short circuit current is significantly enhanced (by a factor of 1.7) since the optical path of the photons is prolonged when compared to their passage through the film at normal incidence. Moreover, as can be deduced from Fig. 5.8b, which shows the entire current-voltage characteristic of the cell for normal incidence and for an incidence angle of 45°, the power conversion efficiency is enhanced by a factor of 2.3.

However, it remains to be mentioned that such large enhancement was only possible because of the rather narrow absorption line width of CuPc. This permits

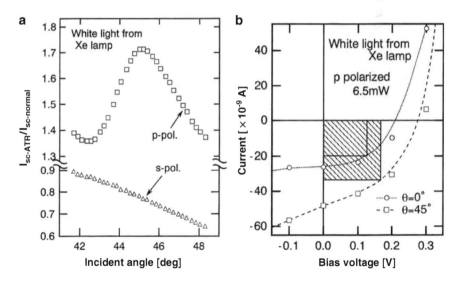

Fig. 5.8 Experimental characteristics of an organic solar cell that exploits PSPP excitation under white light illumination. (**a**) Normalized short circuit current as a function of the angle of incidence and the polarization. (**b**) Current–voltage characteristics of the solar cell for two different angles of incidence (Reproduced from [22] with permission. Copyright: Institute of Pure and Applied Physics, 1995)

for a proper match of the line width of the excited PSPP and hence a global increase of the solar cell efficiency. For other solar cell materials with broader absorption bands, the PSPP will only contribute to a narrow spectral domain and one has to take care that the dissipation in the extra metal film does not annihilate the positive contribution from the PSPP in a narrow spectral domain. Nevertheless, design strategies are documented in the literature where this is possible, though they do wait yet for their experimental implementation [90].

5.6 Conclusions

In summary, in this chapter we have shown that metallic nanostructures sustaining surface plasmon polaritons, being either localized or propagating, are an excellent tool for photon management. They facilitate the trapping of light inside the solar cell and provide an efficiency enhancement. The most appealing advantage of metallic nanostructures is their characteristics to act as an add-on to any existing solar cell concept. Physically, by exciting a surface plasmon polariton, the light is coupled to a charge density oscillation, where it remains bound to the surface, hence resonantly enhancing the interaction with any matter surrounding the metallic nanostructures. The only detrimental aspect is the unavoidable absorption of light inside the metal, which leads at first glance to a degradation of the solar cell performance. However, once such negative aspects are compensated by a careful design of the plasmonic structure, a benefit remains. Incorporation of metallic nanostructures is advantageous for any solar cell which has, by physical constraints, a finite thickness, which cannot be increased to ensure complete absorption of light. In this chapter, we have detailed applications of metallic nanostructures to thin-film solar cells made of amorphous and/or microcrystalline silicon as well as to organic solar cells. Nevertheless, before being ready for use in a solar cell various questions have to be answered. Such questions concern foremost the entire opto-electronical properties of solar cells containing metallic nanostructures, since, unfortunately, most cells are currently analyzed from an optical perspective only. Such a comprehensive theoretical/numerical analysis is currently beyond the available capabilities. Further experimental data is required underlining the potential of metallic nanoparticles for an efficiency enhancement. This relates to finding answers on such questions as what are possible surface effects occurring at the interface between a metal and a semiconducting material in a solar cell or if there are any long-term degradation effects of fabricated devices. It will be furthermore vital to verify the potential of such metallic nanostructures to act as an add-on to any solar cell, to exploit their possible use also in the context of more sophisticated solar cell concepts, such as organic bilayer [93], organic bulk heterojunction [94] or quantum dot solar cells [95], where they can likewise contribute to an efficiency enhancement within the framework of effects likewise described in this chapter. Only if such questions are answered in a satisfactory manner the exploitation of plasmonic effects will be also fostered in an industrial environment. If such level is reached, also end-consumer oriented issues have to be solved, e.g, if the integration of metallic structures into solar cells will pay off economically. If the benefit is worth the effort, then there should be no obstacle to see the implementation of metallic nanostructures emerging in the next generation of solar cells.

Acknowledgements We are indebted to numerous colleagues, coworkers, and project partners working in the field of photon management for numerous discussions and their advices on many practical aspects in this field of research. We thank Prof. S. Hayashi, Dr. C. Hägglund and Dr. E. Moulin for kindly providing us with some of the figures used in this chapter and most notably T. Kirchartz for his patience to discuss with us all photovoltaic aspects. We acknowledge the partial financial support of this work through the Deutsche Forschungsgemeinschaft (Nanosun and Nanosun 2), the Federal Ministry of Education and Research (Nanovolt) and the Thuringian State Government (SolLux). Some computations utilized the IBM p690 cluster JUMP of the Forschungszentrum in Jülich, Germany.

References

1. P. Würfel, *Physics of Solar Cells: From Basic Principles to Advanced Concepts*, 2nd edn. (Wiley, Berlin, 2009)
2. A. Martí, A. Luque, *Next generation photovoltaics*, 1st edn. (IOP Publishing Ltd, Bristol, 2004)
3. J. Nelson, *The Physics of Solar Cells*, 1st edn. (Imperial College Press, London, 2003)
4. S.A. Maier, *Plasmonics: Fundamentals and Applications*, 1st edn. (Springer, Berlin, 2007)
5. W.L. Barnes, A. Dereux, T.W. Ebbesen, Nature **424**, 824 (2003)
6. C. Daboo, M.J. Baird, H.P. Hughes, N. Apsley, M.T. Emeny, Thin Solid Films **20**, 9 (1991)
7. M. Westphalen, U. Kreibig, J. Rostalski, H. Lth, D. Meissner, Thin Solid Films **61**, 97 (2000)
8. A. Shalav, B.S. Richards, T. Trupke, K.W. Krämer, H.U. Güdel, Appl. Phys. Lett. **86**, 013505 (2005)
9. T. Trupke, M. A. Green, P. Würfel, J. Appl. Phys. **92**, 4117 (2002)
10. T. Trupke, M. A. Green, P. Würfel, J. Appl. Phys. **92**, 1668 (2002)
11. C. Strümpel, M. Mccann, G. Beaucarne, V. Arkhipov, A. Slaoui, V. Svrcek, C.D. Canizo, I. Tobias, Sol. Energ. Mater. Sol. Cell. **91**, 238 (2007)
12. E. Klampaftis, D. Ross, K. R. McIntosh, B. S. Richards, Sol. Energ. Mater. Sol. Cell. **93**, 1182 (2009)
13. J.N. Farahani, D.W. Pohl, H.-J. Eisler, B. Hecht, Phys. Rev. Lett. **95**, 017402 (2005)
14. S. Kühn, U. Håkanson, L. Rogobete, V. Sandoghdar, Phys. Rev. Lett. **97**, 017402 (2006)
15. F. Hallermann, C. Rockstuhl, S. Fahr, G. Seifert, S. Wackerow, H. Graener, G.v. Plessen, F. Lederer, Phys. Stat. Sol. (a) **205**, 2844 (2008)
16. R. Esteban, M. Laroche, J.-J. Greffet, J. Appl. Phys. **105**, 033107 (2009)
17. A. Luque, A. Martí, M.J. Mendes, I. Tobías, J. Appl. Phys. **104**, 113118 (2008)
18. B.S. Richards, Sol. Energ. Mater. Sol. Cell. **90**, 1189 (2006)
19. M.J. Mendes, A. Luque, I. Tobías, A. Martí, Appl. Phys. Lett. **95**, 071105 (2009)
20. D. Derkacs, W.V. Chen, P.M. Matheu, S.H. Lim, P.K.L. Yu, E.T. Yu, Appl. Phys. Lett. **93**, 091107 (2008)
21. S. Hayashi, K. Kozaru, K. Yamamoto, Solid State Commun. **79**, 763 (1991)
22. T. Kume, S. Hayashi, H. Ohkuma, K. Yamamoto, Jpn. J. Appl. Phys. **34**, 6448 (1995)
23. O. Stenzel, A. Stendal, K. Voigtsberger, C. von Borczyskowski, Sol. Energ. Mater. Sol. Cells **37**, 337 (1995)
24. O. Stenzel, S. Wilbrandt, A. Stendal, U. Beckers, K. Voigtsberger, C. von Borczyskowski, J. Phys. D Appl. Phys. **28**, 2154 (1995)

25. M. Westphalen, U. Kreibig, J. Rostalski, H. Lüth, D. Meissner, Sol. Energ. Mater. Sol. Cell. **61**, 97 (2000)
26. K. Tvingstedt, N.-K. Persson, O. Inganäs, A. Rahachou, I.V. Zozoulenko, Appl. Phys. Lett. **91**, 113514 (2007)
27. C. Hägglund, M. Zäch, B. Kasemo, Appl. Phys. Lett. **92**, 013113 (2008)
28. F.-C. Chen, J.-L. Wu, C.-L. Lee, Y. Hong, C.-H. Kuo, M.H. Huang, Appl. Phys. Lett. **95**, 013305 (2009)
29. D. Duché, P. Torchio, L. Escoubas, F. Monestier, J.-J. Simon, F. Flory, G. Mathian, Sol. Energ. Mater. Sol. Cell. **93**, 1377 (2009)
30. D.E. Carlson, C.R. Wronski, Appl. Phys. Lett. **28**, 671 (1976)
31. J. Meier, H. Keppner, S. Dubail, U. Kroll, P. Torres, P. Ziegler, J.A.A. Selvan, J. Cuperus, D. Fischer, A. Shah, Mater. Res. Soc. Symp. Proc. **507**, 139 (1999)
32. M.A. Green, K. Emery, Y. Hishikawa, W. Warta, Prog. Photovolt. Res. Appl. **17**, 85 (2009)
33. L. Schirone, G. Sotgiu, F.P. Califano, Thin Solid Films **297**, 296 (1997)
34. C. Heine, R.H. Morf, Appl. Opt. **34**, 2476 (1995)
35. L. Zeng, P. Bermel, Y. Yi, B.A. Alamariu, K.A. Broderick, J. Liu, C. Hong, X. Duan, J. Joannopoulos, L.C. Kimerling, Appl. Phys. Lett. **93**, 221105 (2008)
36. C. Rockstuhl, F. Lederer, K. Bittkau, R. Carius, Appl. Phys. Lett. **91**, 171104 (2007)
37. S. Fahr, C. Rockstuhl, F. Lederer, Appl. Phys. Lett. **92**, 171114 (2008)
38. C. Rockstuhl, S. Fahr, F. Lederer, T. Beckers, K. Bittkau, R. Carius, Appl. Phys. Lett. **93**, 061105 (2008)
39. G. Mie, Ann. Phys. **330**, 77 (1908)
40. J.P. Kottmann, O.J.F. Martin, D.R. Smith, S. Schultz, Phys. Rev. B **64**, 235402 (2001)
41. U. Hohenester, J.R. Krenn, Phys. Rev. B **72**, 195429 (2005)
42. C. Rockstuhl, M.G. Salt, H.P. Herzig, J. Am. Soc. Am. A **20**, 1969 (2003)
43. J.D. Jackson, *Classical Electrodynamics* (Wiley, NY, 1999)
44. P.B. Johnson, R.W. Christy, Phys. Rev. B **6**, 4370 (1970)
45. C.F. Bohren, D.R. Huffman, *Absorption and Scattering of Light by Small Particles* (Wiley, New York, 1983)
46. C. Hägglund, B. Kasemo, Opt. Express **17**, 11944 (2009)
47. Y.A. Akimov, W.S. Koh, K. Ostrikov, Opt. Express **17**, 10195 (2009)
48. H. Raether, *Surface Plasmons* (Springer, Berlin, 1988)
49. B.E.A. Saleh, M.C. Teich, *Grundlagen der Photonik*, 2nd edn. (Wiley, Berlin, 2008)
50. A. Otto, Zeitschrift für Physik **216**, 398 (1968)
51. E. Kretschmann, H. Raether, Z. Naturforsch. **23A**, 2135 (1968)
52. R.H. Ritchie, E.T. Arakawa, J.J. Cowan, R.N. Hamm, Phys. Rev. Lett. **21**, 1530 (1968)
53. B.T. Draine, P. J. Flatau, J. Opt. Soc. Am. A **11**, 1491 (1994)
54. C. Hafner, *The Generalized Multipole Technique for Computational Electromagnetics* (Artech House Inc., Boston, 1990)
55. J.M. Bendickson, E. Glytsis, T.K. Gaylord, J. Opt. Soc. Am. A **18**, 1487 (2001)
56. J.R. Cole, N.J. Halas, Appl. Phys. Lett. **89**, 153120 (2006)
57. M.G. Moharam, E.B. Grann, D.A. Pommet, T.K. Gaylord, J. Opt. Soc. Am. A **12**, 1068 (1995)
58. L. Li, J. Opt. Soc. Am. A **14**, 2758 (1997)
59. J. Chandezon, D. Maystre, G. Raoult, J. Opt. **11**, 235 (1980)
60. P. Monk, *Finite Element Methods for Maxwell's Equations* (Oxford University Press, NY, 2003)
61. A. Taflove, S.C. Hagness, *Computational Electrodynamics: The Finite-Difference Time-Domain Method* (Artech House Inc., Boston, 2005)
62. J.P. Berenger, J. Comp. Phys. **114**, 185 (1994)
63. A. Farjadpour, D. Roundy, A. Rodriguez, M. Ibanescu, P. Bermel, J.D. Joannopoulos, S.G. Johnson, G. Burr, Opt. Lett. **31**, 2972 (2006)
64. H.R. Stuart, D.G. Hall, Appl. Phys. Lett. **73**, 3815 (1998)
65. E. Moulin, P. Luo, B. Pieters, J. Sukmanowski, J. Kirchhoff, W. Reetz, T. Müller, R. Carius, F.-X. Royer, H. Stiebig, Appl. Phys. Lett. **95**, 033505 (2009)

66. C. Hägglund, M. Zäch, G. Petersson, B. Kasemo, Appl. Phys. Lett. **92**, 053110 (2008)
67. H.R. Stuart, D.G. Hall, Appl. Phys. Lett. **69**, 093103 (1996)
68. D. Derkacs, S.H. Lim, P. Matheu, W. Mar, E.T. Yu, Appl. Phys. Lett. **89**, 2327 (2006)
69. J. Springer, A. Poruba, L. Müllerova, M. Vanecek, O. Kluth, B. Rech, J. Appl. Phys. **92**, 1427 (2002)
70. K. Nakayama, K. Tanabe, H.A. Atwater, Appl. Phys. Lett. **93**, 121904 (2008)
71. K.R. Catchpole, A. Polman, Appl. Phys. Lett. **93**, 191113 (2008)
72. S. Mokkapati, F.J. Beck, A. Polman, K.R. Catchpole, Appl. Phys. Lett. **95**, 053115 (2009)
73. C. Rockstuhl, S. Fahr, F. Lederer, J. Appl. Phys. **104**, 123102 (2008)
74. C. Rockstuhl, F. Lederer, Appl. Phys. Lett. **94**, 213102 (2009)
75. O. Vetterl, F. Finger, R. Carius, P. Hapke, L. Houben, O. Kluth, A. Lambertz, A. Mück, B. Rech, H. Wagner, Sol. Energ. Mater. Sol. Cell. **62,** 97 (2000)
76. S.H. Lim, W. Mar, P. Matheu, D. Derkacs, E.T. Yu, J. Appl. Phys. **101**, 104309 (2007)
77. F.J. Beck, A. Polman, K.R. Catchpole, J. Appl. Phys. **105**, 114310 (2009)
78. R.A. Pala, J. White, E. Barnard, J. Liu, M.L. Brongersma, Adv. Mater. **21**, 3504 (2009)
79. A. Lambertz, A. Dasgupta, W. Reetz, A. Gordijn, R. Carius, F. Finger, in *Proceedings of the 22. European Photovoltaic Solar Energy Conference (EU PVSEC)*, Milan, Italy, 3–7 Sep 2007
80. A. Bielawny, C. Rockstuhl, F. Lederer, R.B. Wehrspohn, Opt. Express **17**, 8439 (2009)
81. P. Buehlmann, J. Bailat, D. Dominé, A. Billet, F. Meillaud, A. Feltrin, C. Ballif, Appl. Phys. Lett. **91**, 143505 (2007)
82. A. Bielawny, J. Üpping, P.T. Miclea, R.B. Wehrspohn, C. Rockstuhl, F. Lederer, M. Peters, L. Steidl, R. Zentel, S.-M. Lee, M. Knez, A. Lambertz, R. Carius, Phys. Stat. Sol. (a) **205**, 2796 (2008)
83. P. Obermeyer, C. Haase, H. Stiebig, Appl. Phys. Lett. **92**, 181102 (2008)
84. T. Söderström, F.-J. Haug, X. Niquille, V. Terrazzoni, C. Ballif, Appl. Phys. Lett. **94**, 063501 (2009)
85. S. Fahr, C. Rockstuhl, F. Lederer, Appl. Phys. Lett. **95**, 121105 (2009)
86. F.-J. Haug, T. Söderström, O. Cubero, V. Terrazzoni-Daudrix, C. Ballif, J. Appl. Phys. **104**, 064509 (2008)
87. R.H. Franken, R.L. Stolk, H. Li, C.H.M. van der Werf, J.K. Rath, R.E.I. Schropp, J. Appl. Phys. **102**, 014503 (2007)
88. V.E. Ferry, L.A. Sweatlock, D. Pacifici, H.A. Atwater, Nano Lett. **8**, 4391 (2008)
89. T. Wakamatsu, K. Saito, Y. Sakakibara, H. Yokoyama, Jpn. J. Appl. Phys. **34**, L1467 (1995)
90. N.C. Panoiu, R.M. Osgood Jr., Opt. Lett. **32**, 2825 (2007)
91. T. Kume, S. Hayashi, K. Yamamoto, Jpn. J. Appl. Phys. **32**, 1993 (1993)
92. G.A. Chamberlain, Sol. Cell. **8**, 47 (1983)
93. C.W. Tang, Appl. Phys. Lett. **48**, 183 (1986)
94. H. Hoppe, N.S. Sariciftci, J. Mater. Chem. **19**, 1924 (2004)
95. A.J. Nozik, Physica E **14**, 115 (2002)

Chapter 6
Non-Coherent Up-Conversion in Multi-Component Organic Systems

Stanislav Baluschev and Tzenka Miteva

Abstract The requirements for observing efficient energetically conjoined triplet–triplet annihilation (TTA) up-conversion (UC) in multi-component organic system are stated. The fundamental advantages of the TTA–UC regarding the other up-conversion techniques in the context of solar photonic applications are established. The device-architecture, optical and electrical characteristics of a photonic device comprised of a combination of upconvertor-device (UCd) and dye sensitized solar cell (DSSC) excited with sunlight are demonstrated.

6.1 Comparison Between Up-Conversion Processes

For photonic applications in the field of organic optoelectronics, of a great importance is the ability to blue shift the emission photons regarding the excitation photons by energy shift of 0.25 up to 2.5 eV. The main goal of the organic optoelectronics devices is to increase the efficiency of the organic photovoltaic cells by such means of UC-convertors or UC-sunlight concentrators. However, these specific applications modify radically the requirements of the UC-process used. Four principle requirements are discussed below.

First, the excitation intensity necessary for effective UC needs to be small – as low as some $W \cdot cm^{-2}$; such light intensities could be obtained by moderate concentration of the sunlight. However, if higher sunlight concentration is required, the next technical problem arising is the necessity of tracking of the Sun. Most tracking devices are complicated systems with complex management. A general rule of thumb is that the costs for sunlight concentration and tracking are comparable,

---

S. Baluschev (✉)
Max-Planck-Institute for Polymer Research, Ackermannweg 10, 55128 Mainz, Germany

Optics and Spectroscopy Department, Faculty of Physics, Sofia University
"St. Kliment Ochridski", 5 James Bourchier Blvd., 1164 Sofia, Bulgaria
e-mail: balouche@mpip-mainz.mpg.de

or may even exceed the cost of the photovoltaic device. Therefore, it is desirable to avoid or greatly simplify the necessity of both sunlight concentration and tracking. Once the excitation intensity problem is successfully solved by the organic UC-devices (UCd), a variety of futuristic solar devices will become possible.

Second, the spectral power density required for effective UC needs to be comparable with that of the terrestrial sunlight. For instance, at best the solar spectral irradiance (at global tilt of 37°) does not exceed $200\,\mu W\cdot cm^{-2}\,nm^{-1}$. In comparison, the spectral power density achievable by most of the laser systems is of the order of $W\cdot nm^{-1}$. Nevertheless, for the organic UCd efficient excitation by light with a low spectral power density is an ultimate requirement.

Third, the Sun is a non-coherent excitation source, therefore a priori the desired UC-process must efficiently utilize non-coherent photons.

Fourth, the UCd should not restrict the technological advantages of the organic solar cells already demonstrated, such as flexibility (robustness against mechanical distortions of the substrate material), small specific weight and easy up-scaling. Implicitly, the demonstrated possibilities for cost-effective fabrication of organic solar cells via ink-jet printing or roll-to-roll processing techniques have to be maintained.

Until now the examples of UC described in the literature such as simultaneous or sequential absorption of two or more photons with lower energy, second and higher harmonic generation of the fundamental wavelength and parametric processes have been commonly associated with the use of very high excitation intensities [1]. The required intensities are of the order of $MW\cdot cm^{-2}$ up to $GW\cdot cm^{-2}$. Additionally, except for the process of sequential absorption of two or more photons, all of them ultimately require excitation by *coherent* light sources (lasers).

In the context of photonic applications, special attention is deservedly placed upon the UC-systems based on rare-earth (RE) doped phosphors. The process of UC in ion-doped systems has been observed in various types of solids such as crystals and glasses – both in bulk and in waveguide forms [2], as well as in solutions containing nanocrystals or nanoparticles [3]. Furthermore, rare-earth ions have discrete energy levels with the potential for the UC process to occur from the individual ions alone. However, all RE-UC material systems have a common drawback, they put very high requirements on the spectral power density of the excitation source. As a consequence, efficient UC emission is experimentally shown only with laser excitation, never with sunlight. Additionally, these UC-systems can only produce UC emissions under moderate ($kW\cdot cm^{-2}$) to strong ($MW\cdot cm^{-2}$) excitation densities [2, 4, 5]. As the background to our hypothesis as to a possible solution to some of the aforementioned problems we propose a multicomponent system comprised of an emitter part (such as conjugated semiconductor polymers or aromatic hydrocarbon derivatives) and a sensitizer (such as metallated macrocycles). Conjugated polymers, such as the polyfluorenes (PF), have established themselves as functional materials in light-emitting diodes [6] and solar cells [7]. These polymers have good thermal and oxidative stability. Two-photon absorption (TPA) and its associated fluorescence have been also observed in PF [8]. Consequently, the

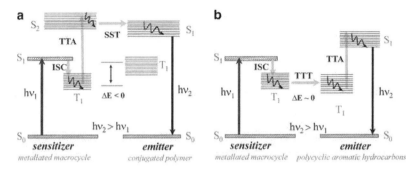

Fig. 6.1 Energetic schema of the process of TTA–UC in multi-component systems. (**a**) When the energy of the emitter triplet state is much higher than sensitizer triplet state (UC–*Type I*) and (**b**) by overlapping sensitizer and emitter triplet states (UC–*Type II*)

very appealing idea to enhance this UC emission led us to search for more efficient mechanisms and processes of low-energy photon excitation of PF.

Recently has been reported TTA–UC in thin films of metallated porphyrin macrocycles (MOEP), which act as sensitizers, blended in a matrix of blue emitting polymers with high fluorescence quantum yields, such as polyfluorenes [9–11] and polypentaphenylenes [12]. A schematic representation of the proposed process of TTA–UC is shown in Fig. 6.1a After the absorption of a photon in the singlet Q-band of the MOEP molecule, the long-lived triplet state of MOEP (Fig. 6.2b) is populated due to the efficient inter-system crossing (ISC) enhanced by spin-orbit coupling of the heavy metal centre [13]. This triplet state can be considered as a reservoir of excited states for subsequent energy transfer [14, 15]. Beyond the trivial depopulation channels for this triplet state, which include radiative decay (sensitizer phosphorescence) and non-radiative decay, there is an additional decay channel: the TTA-pathway. This up-conversion (UC) relaxation pathway, represented by the gray lines in Fig. 6.1a, is a consequence of a triplet–triplet annihilation process between two excited MOEP-molecules. One of the molecules returns to the ground state whilst the other is excited to a higher singlet state. This excitation is followed by an effective transfer of the MOEP higher singlet excitation to the excited singlet state of the emitter. Consequently, efficient matrix fluorescence is observed.

Solutions of PF2/6 (Fig. 6.2a) were prepared in dry toluene with a mass concentration of $10\,\text{mg}\,\text{ml}^{-1}$. For all sensitizers, the level of doping was kept constant –3 wt.%. Films of the MOEP/PF2/6 blends were prepared via spin-coating of the solutions onto quartz substrates in a nitrogen filled glove-box. In order to remove the residual amount of organic solvent, the samples were dried under low dynamic pressure ($\sim 10^{-1}$ mbar) and at a slightly elevated temperature $\sim 40\,°\text{C}$. The film thickness for all samples was adjusted to be $ca.$ 80 nm. The UC-efficiency measurements were conducted in a vacuum chamber with a dynamic vacuum higher than 10^{-4} mbar. An electronically controlled multi-stage Peltier-element was used to adjust the temperature of the sample. This active temperature control is obligatory, because

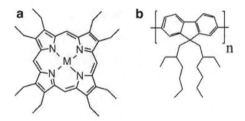

Fig. 6.2 (a) Structure of the sensitizer – (2,7,8,12,13,17,18-octaethyl-porphyrinato)M(II), M = Pt; Pd; Zn and Cu. (b) Structure of the emitter – poly(9,9-bis(2-ethylhexyl-fluorene-2,7-diyl) (PF2/6))

Fig. 6.3 The intensity dependence of the UC in PF 2/6 films, blended with 3% wt. MOEP on the excitation intensity, as follows: PtOEP (*squares*), PdOEP (*circles*), CuOEP (*hollow triangles*) and ZnOEP (*triangles*)

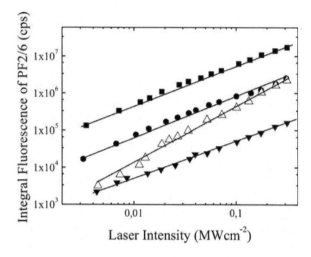

the TTA-rate between the sensitizers excited triplet states depends drastically on the matrix temperature. Obviously, controlling the substrate temperature only reduces the experimental variance, without completely resolving the temperature uncertainty – for instance, the local (across the excitation spot) temperature gradient, for which was not effectively compensated. As a pumping source an intracavity frequency doubled (532 nm) diode pumped *cw* – Nd:YAG laser was used.

The dependence of the integral UC-fluorescence on the excitation intensity is shown in Fig. 6.3. The solid lines are power law fits: $P_{UpConv} = a \cdot P_{exc}^b$ with $b = 1.06 \pm 0.02$ for the system PtOEP/PF2/6; $b = 1.09 \pm 0.02$ for the system PdOEP/PF2/6; $b = 1.42 \pm 0.05$ for the system CuOEP/PF2/6 and $b = 0.93 \pm 0.02$ for the system ZnOEP/PF2/6. Of particular note is the very broad region of excitation intensities, where the blends of MOEP/PF2/6 exhibit well approximated linear intensity dependence, more than 2 orders of magnitude wide! Furthermore, only the CuOEP–blend shows a steeply inclined intensity dependence. The intensity of the UC fluorescence of the most effective blend – PtOEP/PF2/6 – is more than 100 times greater than the intensity of the least effective one – the ZnOEP/PF2/6. This huge difference cannot be explained only by the difference between the optical densities of the samples at the wavelength of excitation ($OD_{PtOEP} \sim 10 \cdot OD_{ZnOEP}$). Instead, more importantly it is the difference between the ISC-coefficients: ISC_{ZnOEP} ~0.065 is almost 10 times less than the ISC_{PtOEP} ~0.5. This means, that even if both blends are absorbing the same number of photons, the PtOEP blend will store approximately 10 times more optical energy for the subsequent energy transfer. The second possible UC–pathway (Fig. 6.1b) is characterized by a triplet-triplet transfer (TTT) of the MOEP – triplet excitation directly to the emitter triplet state, followed again by effective TTA, but now between the triplet states of the emitter molecules. As a consequence of this process, one of the emitter molecules returns to the ground state, but the other emitter molecule is excited to the higher singlet state and strongly blue shifted emission is observed [16]. In order to evaluate the application potential of both possible UC-pathways, a direct comparison between the UC-efficiencies of model UC–compounds can be done. In the example model systems that follow the matrix emitter molecules were varied against a fixed sensitizer molecule. There are two reasons to use model compound systems. The first is that the energy position of the oligomer triplet level is higher (for OF7 ~ 2.16 eV [17]) than the energy of the triplet level of the corresponding polymer (for PF ~ 2.11 eV [17]). The triplet level of PtOEP is 1.91 eV. Therefore, the probability for triplet–triplet transfer between PtOEP and OF7 in a blended PtOEP/OF7 system will be more strongly reduced in comparison with a system containing PF (PtOEP/PF2/6 – blend) because the energy gap between the two corresponding triplet levels is broader. The second reason is that the fluorescence lifetime increases with decreasing chain length [17]. This behaviour can be explained by the decreased probabilities of intersystem crossing as the chain length decreases, hence more efficient UC is to be expected. To continue, the DPA molecule has a suitable triplet level (1.78 eV, [10]) with regards to the triplet level of the PtOEP. Therefore, TTT between these levels is highly probable. Finally, the DPA has a relatively long-lived triplet states at room temperature (~5 ms, [10]).

Solutions of the blue emissive matrix (OF7 or DPA, Fig. 6.4) were prepared in dry toluene with mass concentration of 10 mg/ml. In order to compare the absolute

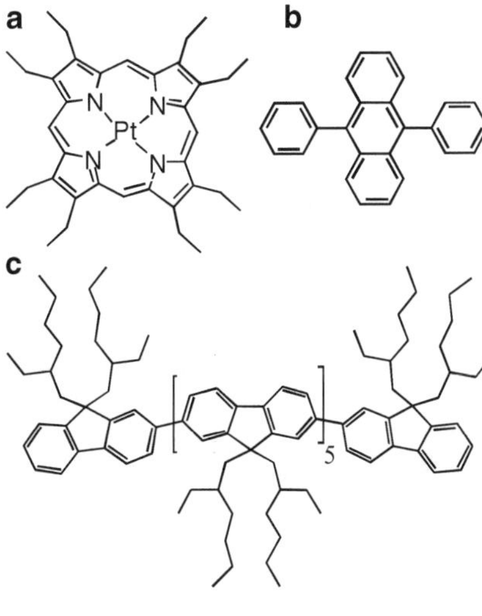

Fig. 6.4 (a) Structure of the sensitizer – PtOEP.(b) Structure of the emitter –9,10-diphenylanthracene (DPA) and (c) oligo((9,9-bis(2-ethylhexyl))fluorene heptamer (OF7)

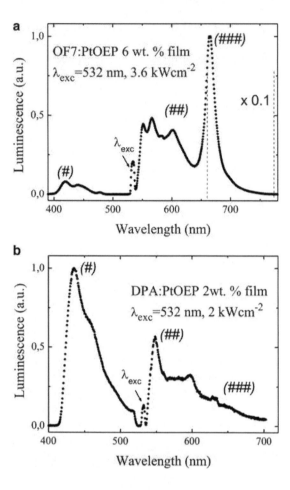

Fig. 6.5 (a) Luminescence spectra at room temperature of films of PtOEP/OF7 and (b) – PtOEP/DPA. For (a) and (b) the excitation wavelength was suppressed more than 10^{-6} times by using a notch filter ($\lambda = 532$ nm). Additional suppression only for (a), whereby the porphyrin phosphorescence was suppressed 10 times ($\lambda = 646$ nm ÷ $\lambda = 800$ nm)

number of emitted UC photons, optimal blending conditions for the both blends were used; these are as follows: 2 wt.% for the blend PtOEP/OF7 and 6 wt.% for the blend PtOEP/DPA. Thin films were prepared via spin coating of the solutions onto quartz substrates in a nitrogen filled glove box.

The *cw*-luminescence spectra for both model systems clearly shows the fluorescence of PtOEP (region ##, with maximum at $\lambda = 545$ nm, partially suppressed due to the notch filter used) and the phosphorescence of PtOEP (region ###, with maximum at $\lambda = 662$ nm), as well as the fluorescence of OF7 (Fig. 6.5a, region #, with maximum at $\lambda = 418$ nm) and the fluorescence of the DPA (Fig. 6.5b, region #, with maximum at $\lambda = 435$ nm). The strong dependence of the UC-efficiency on the studied UC-pathways is clearly manifested through comparing Fig. 6.5a, b. Recalling that the UC-pathway (Fig. 6.1b, UC-*Type II*) is characterized by a transfer of the triplet excitation of the PtOEP directly to the triplet state of the emitter molecules, followed by TTA, has much higher efficiency. This efficiency improvement is evident from a comparison of the total amount of UC-photons from the UC-*Type II* system, which is an order of magnitude higher regarding the number of UC-photons from the UC-*Type I* system. Additional evidence is that the phosphorescence emission of the PtOEP molecule in the case of UC-*Type II* is more than 2 orders of magnitude weaker than the phosphorescence observed from the UC-*Type I* (Fig. 6.1a). Moreover, the existence of strong sensitizer phosphorescence is further direct proof of the less effective use of the stored optical energy (Fig. 6.5a) in an UC-*Type I* system. Conversely, for the UC-*Type II* the sensitizer phosphorescence is almost quenched, which indicates very efficient use of the stored optical energy.

From the *cw*-luminescence spectra, also from the decay characteristics of the fluorescent species, one can understand the nature of the excitation pathways. The decay time of the delayed UC fluorescence of the studied model systems is drastically different. In Fig. 6.6, the decay time of the phosphorescence together with the delayed fluorescence for both UC channels are shown. Whereas the decay time of the PtOEP phosphorescence is comparable for both films, the decay time of the delayed fluorescence markedly differs by more than 2 orders of magnitude. For UC-*Type I*, the decay time of the UC fluorescence is defined by the decay time of the PtOEP-triplets. However, in the other UC-pathway, the decay time of the UC fluorescence (UC-*Type II*) is not dependent on the decay time of the PtOEP triplets, but is instead dependent on the decay time of the emitter triplets (at room temperature, the triplet life time for DPA is on the order of 5 ms [10]).

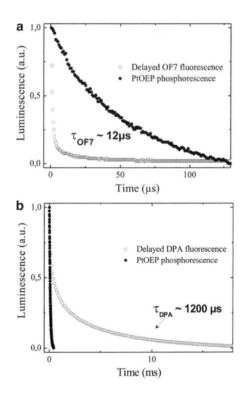

Fig. 6.6 Decay time of the delayed UC-fluorescence of OF7 ((**a**), *open circles*) and DPA ((**b**), *open circles*) and phosphorescence of PtOEP (*filled circles*, (**a**) and (**b**)). Excitation pulse duration 200 μs; intensity ∼3.6 kW·cm^{-2}

Fig. 6.7 (a)Dependence of the UC-efficiency of blend PtOEP/OF7 and (b) – PtOEP/DPA on the sensitizer concentration

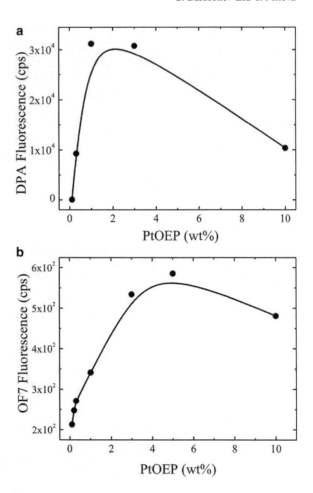

It is evident that in both UC-pathways an optimal doping concentration exists by which the built up triplet population of the sensitizer is used most efficiently by the emitter (Fig. 6.7). It is important to notice, that whereas for systems of UC-*TypeI* the optimum doping concentration requires relatively high doping levels, e.g. 6 wt.% (Fig. 6.7a), yet for UC-*TypeII* systems the optimal doping level is rather lower, and corresponds to doping concentration of 2 wt.% (Fig. 6.7b). From a practical point of view, the external efficiency of the UC process is of critical importance. The definition of the UC-efficiency will be done later. For now we need only mention that the Q.Y of the demonstrated systems is in order of $5 \cdot 10^{-3}$ for the solid-state film of PtOEP/OF7 (UC-*TypeI*), but for the solid-state film of PtOEP/DPA (UC-*TypeII*) is much higher $\sim 2 \cdot 10^{-2}$. A final remark is that the excitation intensity at which these example UC-systems are working is still relatively high – in order of $kW \cdot cm^{-2}$.

The results demonstrated (Figs. 6.3 and 6.5) can be summarized as follows: the process of TTA-supported UC in multi-component organic systems can work in solid state thin polymer (oligomer/single emitter) films with an overall efficiency comparable to the overall UC-efficiency of the earlier developed techniques such as UC in RE-ions doped glasses [2]. Unfortunately, the necessary intensity of the excitation light is also comparable [2] – as high as several kW·cm^{-2}. For instance, if we assume a central excitation wavelength of $\lambda = 700$ nm and spectral width of the UC-absorption in order of $\Delta\lambda = 10$ nm, then an intensity of 1 kW·cm^{-2} can only be obtained by focusing of the sunlight (AM 1.5) more than $8 \cdot 10^{+5}$ times! Technically such focusing of Sunlight is neither easy not inexpensive.

A possible solution of this problem would be if the necessary excitation intensity for observation of efficient UC could be drastically lowered. This is indeed the case when the studied TTA-supported UC is performed in liquid media. By increasing the local mobility of the molecules participating in the chain of multi-molecular interactions, the light intensity at which the TTA-UC process occurs is decreased drastically. A complete understanding as to exact mechanisms involved does not exist at the moment, but nevertheless the huge potential for applications can be already revealed. In order to distinguish the UC-process demonstrated in this work from the already known and described in [13] sensitized TTA-process, we will henceforth call the process energetically conjoined triplet–triplet annihilation-supported up-conversion process (TTA-UC). In the following sections, this process of TTA-UC will be elucidated. In contrast to the all earlier described UC-methods, the fundamental advantage of the TTA-UC is its inherent independence [18] on the coherence of the excitation light. Another principal advantage of this UC process is the very low intensity (as low as 20 mW·cm^{-2}) and extremely low spectral power density [19] (as low as 125 μW·nm^{-1}) of the excitation source required (indeed, we have demonstrated, that it can be the Sun [18, 19]). This therefore allows for the possibility of a step-wise UC of the green and red (IR-A) part of the terrestrial solar spectrum into blue wavelengths; this was first demonstrated in [20]. Subsequently, the excitation band for effective TTA-UC was extended to the NIR region [21] and then even to the IR-A region [22], whilst preserving all advantageous characteristics of the TTA-UC process – i.e. high quantum yield under non-coherent excitation and extremely low spectral power density of the excitation source (sunlight).

The characteristic parameters, contrasting the classical types of UC with the TTA-UC studied in this book, are summarized in the Table 6.1.

6.2 Requirements of an Organic System for Observation of Efficient TTA-UC

The process of TTA-UC (Fig. 6.1b) has to be considered as an inherently connected chain of three processes [23–26]: first intersystem crossing, then triplet–triplet transfer and finally triplet–triplet annihilation.

The first process in the chain is the intersystem crossing (ISC), which is strongly enhanced by the spin-orbit coupling to the metal centre of the metallated macrocycle

Table 6.1 Comparison between the optical parameters of the classical types of UC and the TTA-UC

"Classical"-type UC	TTA-UC (studied here)
− Coherent excitation required (except energy transfer UC (ETU) two-steps absorption)	+ Non-coherent excitation (inherently)
	Proof-of-principle shown experimentally in [18]
− High or very high excitation light intensity $\sim$kW·cm^{-2} (typically MW·cm^{-2})	+ Low excitation intensity $\sim$20 mW·cm^{-2}
	Proof-of-principle shown experimentally in [21]
− Extremely high spectral power density needed $\sim$1 W·nm^{-1}	+ Low spectral power density needed $\sim$100 μW·nm^{-1}
	Proof-of-principle shown experimentally in [19]
− Strongly restricted width of the absorption bands $\sim$GHz or up to $\sim$nm (only in the case of ETU)	+ Broad absorption band $\sim$20 or up to 200 nm (in case of senitizer ensembles)
	Proof-of-principle shown experimentally in [19, 20]

sensitizer molecules [23, 24]. The efficient ISC within the sensitizer molecules (for instance, the ISC-probability is almost 1 for the metallated porphyrins [13]) ensures a very efficient population of the sensitizer triplet level via single photon absorption.

On the other hand, the very weak ISC of the emitter molecules (its probability for various aromatic hydrocarbon derivatives has vanishing values [13]) substantially prohibits the depopulation of the excited emitter triplet states via phosphorescence and thus stores the created triplet population of the emitter for the process of TTA. Therefore, the first requirement of the molecular system is that a very large difference between the ISC-coefficients of the sensitizer and emitter molecules must exist:

$$c_{\text{sensitizer}}^{\text{ISC}} >> c_{\text{emitter}}^{\text{ISC}}. \tag{6.1}$$

The second process [25] in the chain is the transfer of the excitation of the sensitizer triplet to the emitter triplet (triplet–triplet transfer (TTT)). This process can be expressed through the equation:

$$T_{\text{sensitizer}}^{*} + S_{\text{emitter}}^{0} \longrightarrow S_{\text{sensitizer}}^{0} + T_{\text{emitter}}^{0}, \tag{6.2}$$

where the subscripts * and ^{0} identify the first excited triplet state, and the ground state, respectively. The efficiency of the TTT is determined by the extent of overlap of the sensitizer and emitter triplet manifolds, thus the second requirement of the molecular system is:

$$E_{\text{sensitizer}}^{\text{triplet}} \cong E_{\text{emitter}}^{\text{triplet}}. \tag{6.3}$$

The third process is the subsequent triplet-triplet annihilation (TTA), which in our system occurs mostly between the triplets of the emitter molecules [26].

$$2T_{\text{emitter}}^* \longrightarrow S_{\text{emitter}}^0 + S_{\text{emitter}}^* \longrightarrow 2S_{\text{emitter}}^0 + h\nu, \tag{6.4}$$

where the subscripts * and ^{0} identify the first excited triplet or singlet state, and the ground state, respectively. As a precondition for efficient energetically conjoined TTA-UC, there are two requirements related to the molecular structure of the emitter and the sensitizer molecules:

$$2E_{\text{emitter}}^{\text{triplet}} \geq E_{\text{emitter}}^{\text{singlet}}. \tag{6.5}$$

This requirement ensures that the sum-energy of two excited emitter triplet states is enough to populate the first excited singlet state of the emitter molecule, without using thermal energy.

The last requirement is related to the structure of the absorption spectrum of the sensitizer. In order to reduce the re-absorption of the UC emission generated via the sensitizer molecules in ground state, the so-called "transparency window" of the sensitizer molecule has to be sufficiently large. Serendipitously, metallated macrocycles such as porphyrins and phthalocyanines have a band-like absorption spectrum, with only more or less two strong bands – the Soret-band and the Q-band. Therefore, photons with energies lying far enough away from these two local absorption maxima will be hardly absorbed. For the energy positions of the absorption maxima of the sensitizer and the first excited singlet state of the emitter, it thus follows from the above the third requirement of the molecular system:

$$E_{\text{sensitizer}}^{\text{Soret-band}} > E_{\text{emitter}}^{\text{singlet*}} \gg E_{\text{sensitizer}}^{\text{Q-band}}. \tag{6.6}$$

In summary, the main outcome of these theoretically and experimentally determined requirements to the multi-component organic system is that: if the requirements specified in (6.1), (6.3), (6.5) and (6.6) are fulfilled, then it follows that it is likely that the requirements of (6.2) and 6.4 will be fulfilled also, and thus very efficient energetically conjoined TTA-UC will be observed.

Special attention must be drawn to the fact that all molecular energy levels, involved in the process of TTA-UC, are real molecular levels – no virtual energetic levels are involved. Therefore, the processes of internal energy conversion (i.e. thermalisation of the electronic states of the molecules involved) cannot be neglected. During the process of TTA-UC, there are three routes of internal energy conversion. The first of these internal energy conversion processes happens at the Q-band of the sensitizer molecule – the photons absorbed either at the Soret-band or in Q-band

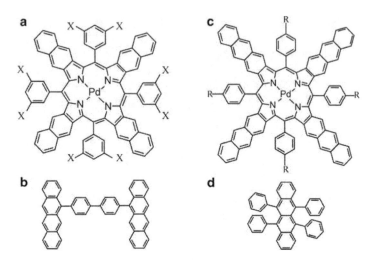

Fig. 6.8 (**a**): Structures of the sensitizer – *meso*-Tetraphenyl-octamethoxide-tetranaphtho[2,3] porphyrin Palladium (PdTNP, X = OMe) and (**b**): the emitter – 4,4'-bis(5-tetracenyl)-1,1'-biphenylene (BPBT); (**c**) Structures of the sensitizer – *tetrakis*-5,10,15,20-(*p*-methoxycarbonyl-phenyl)tetraanthra[2,3-*b*,*g*,*l*,*q*]porphyrin (PdTAP, R = COOH) and (**d**): the emitter rubrene

are equilibrated to the bottom of the Q-band. This energy conversion channel also includes the energy lost due to the process of ISC and the subsequent equilibration of the energy to the bottom of the sensitizer triplet state T_1. For instance, for the UC–couples PdTNP/BPBT (Fig. 6.8a, b) the energy lost caused from the first internal energy conversion channel can be estimated to be:

$$\Delta E = E_Q^{\text{sensitizer}} - E_{\text{Phophorescence}}^{\text{sensitizer}} \sim 0.39\,\text{eV}.$$

The second internal energy conversion channel represents the Stokes-shift at the excited triplet state of the emitter molecules. In fact, the emitter molecules must have vanishing ISC-coefficient (6.1). Therefore, characterisation of the energy position and spectrum of the emitter triplet state is not a trivial experimental or theoretical problem. Following the Mirror-Image Rule, one can estimate that the triplet state Stokes shift for the emitter molecule BPBT is approximately:

$$\Delta E_{\text{Stokes}}^{\text{emitterT}} = \sim 0.06\,\text{eV}.$$

The third internal energy conversion channel represents the Stokes-shift at the excited emitter singlet state. Therefore, emitter molecules with a small Stokes-shift are essential in order to minimize the energy-lost via this third channel. The Stokes-shift for the emitter BPBT is:

$$\Delta E_{\text{Stokes}}^{\text{emitterT}} = \sim 0.06\,\text{eV}.$$

As a consequence, these described processes of internal energy conversion lead to a noticeable loss of excitation energy; therefore, the up-converted emission a priori has a frequency lower than the doubled frequency of the excitation light.

6.3 Characterization of the TTA-UC

From a practical point of view, the external quantum efficiency of the up-conversion process is certainly of decisive importance. For organic photonic applications, such as UC-sunlight concentrators, a transparent, clear and rigorous definition of external quantum efficiency of the UC-process is needed. Therefore, the UC-efficiency definition must be free from possible uncertainties, and give lucid estimation of the expected UC-photon flux. On the other hand, the UC-efficiency definition must allow for a direct and meaningful comparison between the UC-efficiencies of different UC-methods at given experimental conditions, in order to choose the most appropriate method for a certain application. Of crucial importance is to be able to compare the UC-methods at realistic experimental conditions. For sunlight-UC, these conditions are: non-coherent excitation, excitation intensities of the order of mW·cm^{-2} and a spectral power density of the light source of the order of 200 µW·cm^{-2} nm^{-1}, that is, to use light parameters comparable to low concentrated sunlight. The classical definition for quantum yield (the JUPAC-definition) is stated, devised and derived for a single species. Therefore, it is not entirely directly applicable to more complicated, multi-species TTA-UC in organic systems. Nevertheless, that said, the main benefit of using such a rigorous quantum yield definition arises from the clear and transparent evaluation for the reader of the application potential of the TTA-UC.

For estimation of the quantum yield of the TTA-UC, we accepted the methodology commonly used for the determination of fluorescence quantum yields of single emitters. As such, we consider the TTA-UC as a one-step process with "absorption" – corresponding to the absorption of the sensitizer molecules – and "emission" – corresponding to the fluorescence of the emitter molecules (Fig. 6.9). The particulars of these processes, happening inside the "up-conversion black-box"

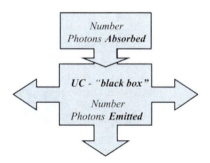

Fig. 6.9 Cartoon, depicting the UC-efficiency definition

are neglected. After this simplification of the TTA-UC, we compare the system under study to a reference standard of known quantum efficiency, together with other measurable parameters. The unknown quantum yield (Q_X) of the fluorescence, excited in UC regime is thus presented in (6.7):

$$Q_X = Q_R(E_X/E_R)(h\nu_X/h\nu_R)(n_X^2/n_R^2)(I_R/I_X)(A_R/A_X). \quad (6.7)$$

Here, E is the emission integral, I is the light intensity at the excitation wavelength, A is the absorbance of the solution at the excitation wavelength, n is the refractive index of the solvent at the emission wavelength, $h\nu$ is the energy of the excitation photons. The subscripts x and r, respectively, refer to the up-conversion system and to the optical standard. In order to minimize the influence of boundary effects, such as those caused by the diffusion of excited states onto the annihilation controlled process of up-conversion, we used for the excitation a collimated beam with diameter of $3.9 \cdot 10^{-3}$ m and an excitation intensity of the order of 10 mW·cm^{-2}. The quantum yield is an integral parameter describing the overall efficiency of a chain of events, starting with the absorption of a single photon in the Q-band of the sensitizer. The TTA-UC relies on the presence of the relatively long-lived triplet states of the sensitizer molecules. With this and the well-known fact that oxygen molecules are renowned to be very effective quenchers of the excited triplet states of the metallated macrocycles (such as the porphyrins – see for example [13]) then this predetermines the strong dependence of the up-conversion efficiency even upon a residual oxygen concentration in the order of ppm. So, correspondingly, all solutions investigated were prepared from degassed solvents and sealed in a nitrogen-filled glove-box. Therefore, the reported UC-efficiencies must be treated as only estimation values valid for the particular sensitize/emitter couple and for the given sample preparation techniques, but not as an absolute efficiency for this TTA-UC process.

6.3.1 Replacement of the Volatile Organic Solvent

In this chapter, a close connection to and consideration of the realistic experimental conditions for a practical working UC-device has been deserved. As such, it has to be mentioned that the UC-devices based on volatile organic solutions [20–22], although showing relatively high external quantum yields under realistic pumping conditions (low intensity non-coherent excitation), are not suitable for UC-applications. Sealing of these devices is neither efficient nor robust. As a consequence of the presence of organic solvent, fast aging of the sealing layer is observed, followed by ambient oxygen penetration. The result of molecular oxygen penetration is the strong reduction of the up-conversion efficiency. Additionally, photo-induced oxidation of the active molecules is observed.

Polymers with a high glass transition temperature have very good mechanical properties and high processability. Thus, the straightforward solution of the sealing problem, namely to use a polymer matrix embedding the UC-molecular couples has already been demonstrated in a number of UC-experiments [27–30]. Nevertheless, no reliable data about the efficiency of the demonstrated TTA-UC process were published, and more importantly, no UC-device excited with sunlight could be demonstrated. This failure supports the results (Figs. 6.6 and 6.9) and explanation of the TTA-UC mechanism demonstrated in this chapter. A possible explanation for the low UC-efficiency could be that the polymers with glass transition temperature significantly higher than room temperature have a very low degree of local viscosity, and as a consequence a low mobility of the UC-molecules. This aspect of the TTA-UC process certainly needs a deeper understanding. Importantly, this greatly limits the UC-applications, because in polymer films the efficiency of the TTA-UC is an order of magnitude lower than it is in organic solution – keeping all other experimental conditions identical [16,31]. Instead of using the aforementioned embedded encapsulation one can use a different solvent that does not destroy the sealing layers. Exchanging the volatile organic solvent (toluene) with a non-volatile (under normal conditions) trimer/tetramer mixture of styrene oligomers (Fig. 6.11d) satisfies all requirements for efficient and robust sealing: indeed, for such a system over a period of more than 100 days no observable change of the sealing material could be registered. More importantly, such UCd were able to work without any change in their efficiency over this period.

Unfortunately, the UC-dyes are not directly soluble in styrene matrix. Thus, a solvent-exchange procedure, sketched in Fig. 6.10d is used. The mother solution of the sensitizer and emitter is prepared in the conventional way, using as a solvent degassed HPLC-grade toluene. Moderate heating (up to T_1 ~40 °C) is necessary, when highly concentrated (on level of 10^{-3} M) solutions are prepared. The necessary amount of trimer/tetramer mixture of styrene oligomers is added

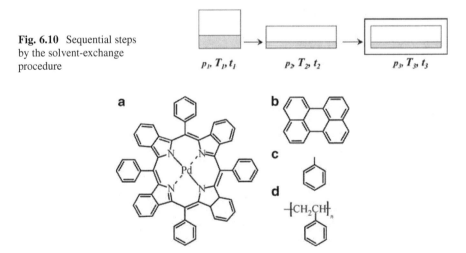

Fig. 6.10 Sequential steps by the solvent-exchange procedure

Fig. 6.11 Structures of (**a**) sensitizer – meso-Tetraphenyl-tetrabenzoporphyrin Palladium (PdTBP); (**b**) emitter Dibenz[de,kl]anthracene (perylene); (**c**) volatile solvent–toluene and (**d**) non-volatile solvent–styrene oligomers (PS400)

at the end of this first heating stage. At normal conditions, the styrene oligomer mixture (PS400) is a highly viscous liquid ($M_W \sim 400$ D). For the period of $t_2 \sim 20$–30 h, the sensitizer/emitter/toluene/styrene solution is dried in an Uhr-glass pot. The pressures (p_i, $i = 1 \div 3$) are the ambient pressure of the glove box ($p_1 = p_2$). The residual amount of toluene is removed by placing the Uhr-glass pot in a low dynamic pressure chamber. In this third stage of the solvent-exchange procedure, it is very important to apply in controlled manner a gradually increased temperature ramp, starting at room temperature ($\sim 20°C$) and slowly increasing up to ($T_3 \sim 50°C$) over a period of more than $t_3 \sim 5$–8 h. The pressure in the chamber must be kept constant, at a level of $p_3 \sim 1$-$2 \cdot 10^{-3}$ bar. Lower pressure (i.e. at a level of 10^{-4} bar) in the chamber, or stronger heating, will cause very fast evaporation of the volatile solvent and a situation close to "boiling" of the mixture, consequentially resulting in phase separation, or even the formation of emitter molecule microcrystals. Crucially, during the whole procedure all preparation stages must be done in an argon (or nitrogen) filled glove-box; the residual concentration of oxygen must be kept lower than 1–2 ppm throughout. The UC-device is prepared by the well-known "doctor blade" technique and sealed with glues, which do not need oxygen for their curing process. By following this procedure, a variety of non-volatile solvents or other than styrene oligomers could be used. With the following, a straightforward comparison of the UC-efficiency of particular UC-couple in volatile and non-volatile solvent will be done. The sensitizer, emitter and solvents used are shown in Fig. 6.11. The UC-efficiency will be calculated via the parameters and method defined in (6.7).

In Fig. 6.12a, comparison between the UC-efficiency of the UC-couple PdTBP/perylene dissolved either in a styrene oligomer mixture or in toluene is shown. It is evident, as expected, that the UC-efficiency in a volatile solvent is still higher than in a non-volatile one, but nevertheless already the efficiencies are

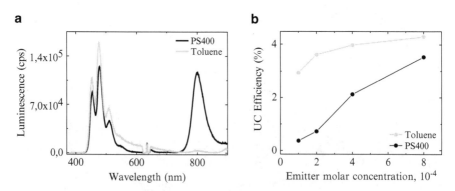

Fig. 6.12 (a) Luminescence spectra of the UC-couple PdTBP/perylene in toluene (*grey line*) and in PS400 (*black line*). Molar concentration of the active molecules $\sim 2 \cdot 10^{-5}$ M/$8 \cdot 10^{-4}$ M. Excitation intensity ~ 200 mW·cm^{-2}; spot diameter ~ 1090 μm; sample thickness ~ 400 μm. (**b**): Dependence of the UC-efficiency on the emitter molar concentration in toluene (the *grey points*) and in PS400 (the *black points*) obtained from both constant excitation intensity and constant sensitizer molar concentration ($2 \cdot 10^{-5}$ M)

comparable. In Fig. 6.12b, the dependence of the UC-efficiency on the molar ratio of sensitizer/emitter is shown. By increasing the molar ratio from 1:10 up to 1:40, the UC-efficiency in non-volatile solvent is increased proportionately much more than the efficiency in volatile solvent. Consequently, with a high molar ratio of sensitizer/emitter (on order of 1/40) the UC-efficiency in a non-volatile solvent approaches the efficiency in a volatile organic solvent. This result clearly shows that styrene oligomer mixture can replace the volatile organic solvent without significant loss in UC-efficiency. This fact opens the possibility to construct viable UCd working with low intensity coherent or non-coherent light (sunlight) and having good sealing characteristic and long lifetimes.

6.3.2 Materials and Methods

The specific behavior of the energetically conjoined TTA-UC will be demonstrated using examples of UC-couples shown in Fig. 6.13.

The absorption spectra of the sensitizer and emitters are shown in Fig. 6.14. The sensitizer has significant Q-band absorption (around $\lambda = 635$ nm). Of mention is that the absorption of all the emitter molecules is negligible at wavelengths close to the Q-band absorption of the sensitizer. Consequently, neither perylene and BPEA nor rubrene, singlet emission can be observed when these emitters are exposed to red excitation light in their pure solutions.

6.3.3 Transparency Window

Figure 6.15 is demonstrating the meaning of the transparency window – a requirement stated earlier in this chapter (6.6). The importance of a transparency window

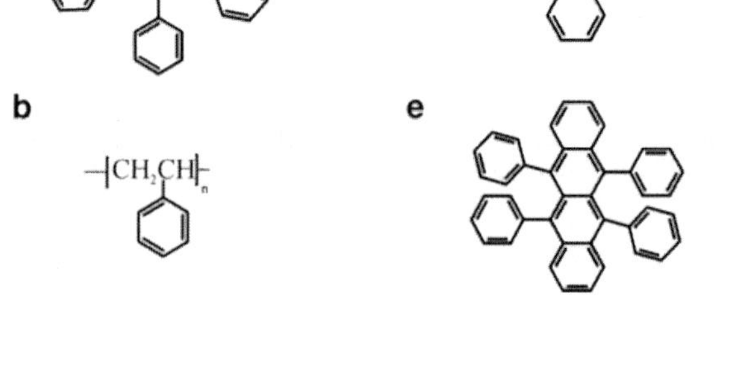

Fig. 6.13 Structures of the sensitizer – PdTBP (**a**); and the three emitters, as follows, (**c**) – perylene; (**d**) –9,10-Bis(phenylethynyl)anthracene (BPEA) and (**e**) –5,6,11,12-Tetraphenylnaphthacene (rubrene). The non-volatile solvent is PS400 (**b**)

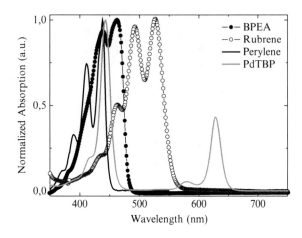

Fig. 6.14 Normalized absorption spectra of the sensitizer – PdTBP (*dark grey line*) and the emitters: perylene (*black line*), BPEA (*hollow circles*) and rubrene (*filled circles*) all in single toluene solutions

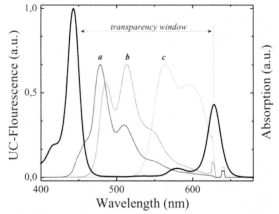

Fig. 6.15 UC-transparency window. Normalized absorption spectra of the sensitizer – PdTBP (*black line*) together with the normalized fluorescence spectra for sensitizer/emitter couples, as follows, PdTBP/perylene (line a), PdTBP/BPEA (line b) and PdTBP/rubrene (line c). All excited in UC-regime. Excitation wavelength $\lambda = 635$ nm. For convenience, the fluorescence spectra are normalized to a peak maximum of 0.6

is due to the inherent complexity of the UC-process in multicomponent organic systems. The complexity arises because simultaneously in the UC–medium there exist both excited triplet states of the sensitizer and emitter molecules as well as sensitizer molecules in ground state. All of these states can absorb the generated emitter fluorescence, either through the process of excited state absorption to higher lying triplet states, or by absorption from the ground state of the sensitizer. This last process is mainly responsible for the low efficiency of TTA–UC-based systems [32], where sensitizers with consolidated and continuous absorption spectrum, such as *Ir(ppy)₃* are used. In such systems, the singlet absorption band of the sensitizer completely overlaps the fluorescence spectrum of the emitter, causing a strong re-absorption of the UC light. Conversely, metallated macrocycles such as porphyrins and phthalocyanines have a band-like absorption spectrum, with more or less two strong bands – the Soret-band and the Q-band. For this reason in such systems it is possible to ensure that these absorption bands of the sensitizer do not overlap significantly the fluorescence spectrum of the emitter. This is evident from Fig. 6.15, where more than 90% of the emission bands of the chosen emitter molecules lay within the transparency window of the selected sensitizer molecule.

6.3.4 Triplet Harvesting

As a consequence of the well-pronounced heavy atom effect in metallated porphyrin macrocycles, the optical excitation is followed with a very effective population of the sensitizer triplet states [13]. On the other hand, there is large body of experimental evidence [13], that such deep lying triplet states decay mostly through non-radiative relaxation channels. On this point, we would now like to demonstrate the characteristic parameter of the energetically conjoined TTA: a phenomenon observed very rarely-namely, a system where non-radiative relaxation channels to the ground state can be effectively suppressed, thus allowing energy relaxation of the excited system via optical pathways. In other words, in multi-component systems, where the process of TTA-UC is effective and keeping all other experimental parameters the same, then the integral UC-fluorescence of the sensitizer/emitter couple will be larger than the integral sensitizer phosphorescence (in a single solution alone). We call this behaviour triplet harvesting.

The evidence for triplet harvesting behaviour is shown in Fig. 6.16. All four samples shown, described in Fig. 6.13, have equal optical density at the wavelength of excitation ($\lambda = 635$ nm). Therefore, for identical excitation intensities the amount of excited sensitizer triplet states in all samples is similar. The energy stored in the sensitizer triplet ensemble can be dissipated differently: in the case of pure sensitizer solution alone, only a small portion of the stored energy will be relaxed via phosphorescence; whereas in the case of UC-systems, this energy portion will be significantly larger. Figure 6.16b represents and compares the dependences of the integral sensitizer phosphorescence on the excitation intensity. It is evident that for the given relative concentrations of the sensitizer/emitter couples, the excitation intensity strength at which the total emitter fluorescence in the blended solutions exceeds the total phosphorescence of the pure solution of the sensitizer is different: for instance, in the case of PdTBP/perylene this intensity is 0.2 W·cm^{-2}, but in the case of PdTBP/rubrene the corresponding intensity is 5 W·cm^{-2}. For excitation intensities higher than 5 W·cm^{-2}, the total number of emitted fluorescent photons in blends exceeds the total number of phosphorescent photons of the pure solutions by more than 2–6 times (Fig. 6.16b). Taking into account the necessity that for the generation of one upconverted photon at least two excited triplet states of the

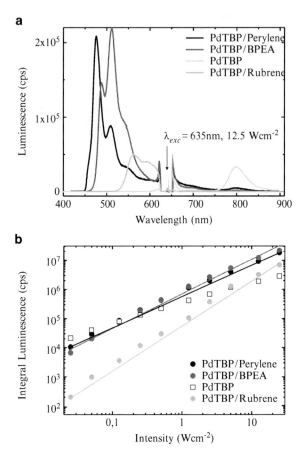

Fig. 6.16 (a) Comparison of the luminescence spectra of solutions containing only the sensitizer 10^{-4} M PdTBP (*grey line*)) and the three blended solutions, containing 10^{-4} M PdTBP/2×10^{-3} M emitter, as follows perylene (*black line*); BPEA (*dark grey line*) and rubrene (*light gray line*). The excitation intensity is 12.5 W·cm^{-2}. (**b**) Intensity dependence of the TTA-UC on the excitation light intensity. UC-couples as follows: PdTBP/perylene (*black circles*), PdTBP/BPEA (*dark grey circles*) and PdTBP/rubrene (*light grey circles*). *Solid lines* are power law fits. Excitation spot area 2 mm^2; $\lambda = 635$ nm

sensitizer are used, then the amount of non-radiatively decaying triplet states is reduced much stronger than could be estimated directly from Fig. 6.16b.

This is evidence for the efficacy of the "triplet harvesting" behaviour.

6.3.5 Intensity Depence

From the examples of UC shown in Fig. 6.16, the dependence of the integral UC-fluorescence upon the excitation intensity will now be discussed. The solid lines are power law fits: $P_{\text{UpConv}} = a \cdot P_{\text{exc}}^{b}$ with the following values for the UC-couples: for PdTBP/perylene $b = 0.97 \pm 0.05$, $b = 1.01 \pm 0.05$ for PdTBP/BPEA and $b = 1.37 \pm 0.05$ for PdTBP/rubrene. Please note, the region of excitation intensities where the couples PdTBP/perylene and PdTBP/BPEA show intensity dependence well approximated with linear function is more than 4 orders of magnitude wide! At a first glance, this behaviour of the studied TTA-UC is contradictory with the classical representation [13] of the TTA-process: indeed, the probability for TTA depends quadratically on the concentration of the excited emitter triplets. Actually, in the classical experiments for TTA, where emitter triplets were created through direct absorption of a UV-photon by the emitter singlet, followed from the strongly prohibited process of intersystem crossing in the emitter molecule, then there is a linear relationship between the created triplet states and the excitation intensity. As a consequence, classically a quadratic dependence between the excitation intensity and the intensity of the TTA-signal is observed. Furthermore, in the classical TTA-schema [13], delayed fluorescence is observed at excitation intensities, comparable with those for the non-linear optical processes (MW·cm^{-2} [1]) or UC in RE ion-doped systems (hundreds of kW·cm^{-2} [2]). In the presently reported schema of energetically-conjoined TTA-UC, realized in diffusion controlled oligomer matrices, the concentration of the excited emitter triplets is orders of magnitude higher than the corresponding concentration of excited emitter triplets states observed in classical experiments for TTA. Nevertheless, just as in the classical experiments, at high excitation intensities (and, hence high concentration of the excited triplet states) linear or sub-linear dependencies between the excitation intensity and the intensity of the generated UC-fluorescence are observed [33]. The crucial benefit of the energetically conjoined TTA-UC studied in this book is that a high concentration of excited triplet states is obtained at extremely low excitation intensities, actually more than 5 orders of magnitude lower than the intensities in the case of classical TTA. Finally, it must be noted that for the new TTA-UC schema the intensity regions within which the UC-signal of the specific UC-couple is linear versus the excitation intensity can vary. These dynamical properties are specific for each UC-couple, and furthermore may depend on the experimental parameters such as local temperature, total molar concentration of the optically active species (and their relative molar concentrations), viscosity of the oligomer/polymer matrix, and degree of re-absorption of the generated UC-emission. Therefore, as mentioned earlier, all the intensity dependencies of the energetically conjoined TTA-UC systems presented throughout this chapter should be treated as examples, certainly not as material parameters, independent from the actual experimental conditions.

6.3.6 Dependence on the Beam Diameter and Viscosity of the Optically Inactive Solvent

Earlier in that chapter, it was stated that the efficiency of the UC-process depends on the local mobility of the sensitizer and emitter molecules. With the following paragraphs, experimental results supporting this hypothesis will be demonstrated and discussed. This discussion will be open ended: it is certain that more questions will arise, than will have been answered. In addition, experimental data which lies beyond the realm of classical TTA-theory and -experiment will be demonstrated, further proving the uniqueness of the energetically conjoined TTA-UC.

Fig. 6.17 Dependence of the decay time of the UC-fluorescence on the molecular mass of the styrene-oligomer matrix mixture; UC-couple PdTBP/perylene; excitation wavelength $\lambda = 635$ nm; (**a**) -with large excitation beam (diameter $\sim 4{,}000\,\mu\text{m}$) and (**b**) -with small excitation beam (diameter $\sim 100\,\mu\text{m}$). Excitation intensity for both (**a**) and (**b**) is the same; room temperature; constant absolute molar concentration for the emitter and sensitizer molecules

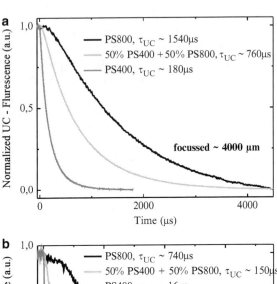
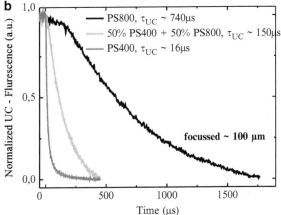

The decay time of the UC-emission mostly depends upon the strength of the TTA between the emitter triplets as shown in Fig. 6.17a. By gradually increasing the effective mass of the *oligo*-styrene mixture (and so gradually increasing the viscosity of the non-volatile solvent, working as optically inactive matrix) a strong increase of the decay time of the delayed UC-fluorescence is observed. This measured increase in the decay time is almost an order of magnitude greater as the effective molecular mass is changed from $M_w = 400$ D to $M_w = 800$ D.

The process of TTA-UC requires a high density of excited sensitizer and emitter triplet states. Therefore, all processes leading to the decrease of the local density of the excited triplet states are found to strongly influence not only the efficiency of the TTA-UC process, but also its characteristic dynamics. One such example of a non-emissive decay channel, leading to a strong decrease of the local density of the excited triplet states, is demonstrated in Fig. 6.17b.

The emitter triplet states are relatively long-lived. Consequently, the diffusion of triplet excitons through the borders of the optically addressed spot becomes significant. As a result, a large amount of the excited emitter triplet states are spread into the non-optically excited neighbouring area. These states are effectively lost to the process of TTA-UC. In Fig. 6.17b, the very strong dependence of the measured decay time of the UC-emission versus the diameter of the excitation beam is shown. Comparing the decay time in collimated beam ($d = 4{,}000\,\mu\mathrm{m}$) and focussed beam ($d = 100\,\mu\mathrm{m}$) a difference in the decay time of larger than order of magnitude is observed. It is important to notice that the initial densities of the sensitizer triplet states at the both measurements are equal. This example proves once more our statement that both the dynamical parameters and especially the efficiency of TTA-UC must be measured and compared at well-determined experimental conditions.

6.3.7 Molar Concentration Dependence

The dynamical characteristics of the energetically conjoined TTA-UC differs significantly from the already described sensitized TTA [13]. The classical theory predicts a simple relation between the decay time of the sensitizer phosphorescence and the decay time of the delayed emitter fluorescence, namely:

$$\tau_{\text{flourescence}}^{\text{delayed}} = \frac{1}{2\tau_{\text{phosphorescence}}}. \tag{6.8}$$

However, such a simple dependence is observed only for a defined sensitizer/emitter combination in specific experimental conditions namely: fixed molar ratio of sensitizer/emitter, sample temperature, total concentration of the active compounds and viscosity (or T_G) of the optically inactive matrix. In Fig. 6.18, the dynamical properties of the blend PdOEP/PF2/6 are shown.

The energetic schema for TTA-supported UC–*Type I* is presented in Fig. 6.1a. After optical excitation, the created sensitizer triplet ensemble is depopulated via radiative decay channel (sensitizer phosphorescence, region ### Fig. 6.5a) and non-radiative channels. Particularly, strong pronounced is the TTA between two excited MOEP-molecules. As a consequence, one of the PdOEP-molecules returns to the ground state, but the other PdOEP-molecule is excited to a higher excited singlet state (S_2). The energy of this S_2-state either could be effectively transferred to the excited singlet state (S_1) of the blue emitter (in this example PF2/6, thus UC-delayed fluorescence could be observed, region # Fig. 6.5a), or it can be relaxed inside the PdOEP-molecule, in the same manner, as blue photons, absorbed at the Soret-band of the neat porphyrin species: namely, it can create with some probability an excited triplet state or to participate the delayed fluorescence of the PdOEP molecule (region ## Fig. 6.5a). The dynamics of the observed delayed processes differs completely from the classical prediction: only for very low concentrations of the sensitizer molecule (0.1% wt.) is the duration of the delayed sensitizer fluorescence approximately 1/2 that of the decay time of the sensitizer phosphorescence (42 μs, and 90 μs, respectively). At this concentration, the decay

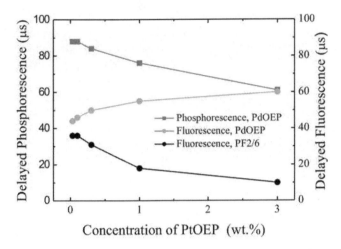

Fig. 6.18 Dependence of the decay time of the UC-fluorescence (*black filled circles*), the sensitizer fluorescence (*grey filled circles*) and sensitizer phosphorescence (*dark grey squares*) upon the concentration of the sensitizer. Conditions: constant sample temperature (room); solid-state film of the PdOEP/PF2/6; excitation wavelength $\lambda = 532$ nm. The structures of the active materials are shown in Fig. 6.2

time of the delayed emitter fluorescence (the UC-emission) is almost 3 times shorter (35 μs). However, with increasing sensitizer concentrations the decay time of the delayed emitter fluorescence decreases monotonically, approaching 10μs at 3% wt. doping level. On the contrary, in total disagreement with the classical behaviour, with increasing sensitizer concentrations the decay time of the delayed sensitizer fluorescence increases monotonically, consequently equalising with the sensitizer phosphorescence lifetime! In the example of PF2/6 doped with PdOEP, both of these decay times are around 62 μs at a 3% wt. doping level.

6.3.8 Molar Concentration Dependence

The UC-sunlight concentration devices are designed for outdoor applications. Therefore, for real-world applications of the overall UC-device, the temperature cannot be considered as constant; even more it is technically not undemanding to stabilize it at some temperature range. Taking into account the nature and large dynamic range of the sunlight intensity change during the day then technological solutions to the temperature problem becomes even more mandatory.

The examples given to show the dependence of the UC-efficiency upon the sample temperature shown in Fig. 6.19 verify the complex nature of the process of energetically conjoined TTA-UC. Indeed, upon the inclusion of the viscosity of the optically inactive non-volatile solvent as an experimental parameter, then the dependence of the UC-efficiency on the sample temperature can be completely

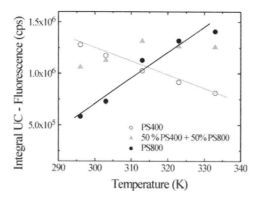

Fig. 6.19 The dependence of the integral UC-fluorescence in non-volatile solvents with different viscosities on the sample temperature. The UC-couple is PdTBP/Perylene (1/20·10⁻⁴ M, respectively) dissolved in styrene oligomer mixture of varying masses, as follows: $M_W = 400$ D (*black hollow circles*), $M_W = 600$ D (*grey triangles*) and $M_W = 800$ D (*black filled circles*). Conditions: excitation wavelength $\lambda = 635$ nm; excitation intensity ~ 80 mW·cm⁻², constant for all measurements; large excitation spot (3 × 4·10⁻³ m); sample thickness ~ 125 μm

changed: from that of decreasing efficiency with the increasing sample temperature at lower viscosities through to the absence of temperature dependence at a particular middling viscosity to the increasing efficiency with the increasing sample temperature at higher viscosities. This complex behaviour opens an additional "degree of freedom" by choosing the proper UC-system designed for certain applications.

In Fig. 6.20a, the common tendency of a decrease in the sensitizer phosphorescence with increasing sample temperature is shown. This dependence is common for all samples. Considering the ratio of the integral UC-fluorescence to the integral sensitizer phosphorescence, Fig. 6.20b, the physical picture becomes still clearer: as the TTA-probability increases with the sample temperature (TTA happens mostly between the emitter triplet states, UC-*Type II*) then the depopulation of the emitter triplet states becomes more efficient. Consequently, the channel for TTT (between the sensitizer triplet states to the emitter triplet states, Fig. 6.1b) becomes more efficient. As a result of combination of these two effects, the sensitizer triplet is almost completely quenched.

6.3.9 Temperature Dependence in Solid-State Films

In solid-state films, the UC-efficiency drops significantly with an increase in the sample temperature. This is mainly due to two parameters. The first parameter responsible for this efficiency decrease is the Q.Y. dependence of the polymer emitter itself on the sample temperature. As demonstrated in [34], the quantum efficiency of the fluorescence emission of a PF film is decreased only by 10% for

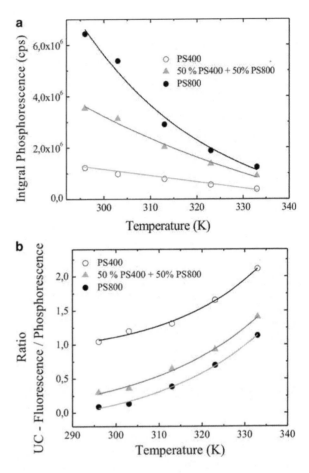

Fig. 6.20 Dependence of the integral sensitizer phosphorescence in non-volatile solvents with different viscosities on the sample temperature (**a**). Dependence of the ratio of the integral UC-fluorescence to the integral sensitizer phosphorescence (**b**). The same materials as in Fig. 6.19. The solid lines are as guidance to the eye

the temperature interval studied. Therefore, more than 2-fold decrease of the UC-efficiency (evident in Fig. 6.21) cannot be explained alone by the Q.Y. decrease of the emitter polymer. Instead additionally in solid-state films, with an increase of the sample temperature, the direct utilization of the sensitizer singlet states (S_2, Fig. 6.1a) via sensitizer singlet emission (from S_1, Fig. 6.1a) becomes more dominant. This singlet emission competes with the process of SST (Fig. 6.1a). As a consequence, the intensity of the UC-emission drops notably.

As an additional corroborating example to support the idea of how the studied energetically conjoined TTA-UC differs from the classical TTA [13] is shown in Fig. 6.22. Again, the decay time of the delayed sensitizer fluorescence equals the decay time of the sensitizer phosphorescence at elevated sample temperatures.

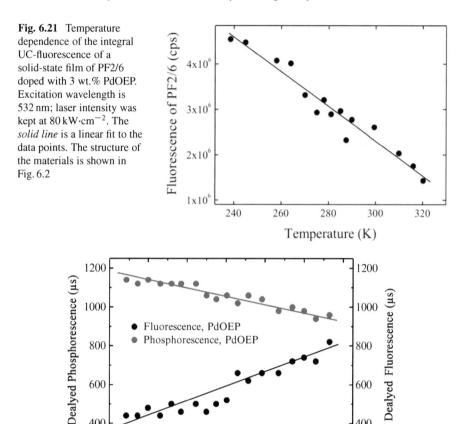

Fig. 6.21 Temperature dependence of the integral UC-fluorescence of a solid-state film of PF2/6 doped with 3 wt.% PdOEP. Excitation wavelength is 532 nm; laser intensity was kept at 80 kW·cm^{-2}. The *solid line* is a linear fit to the data points. The structure of the materials is shown in Fig. 6.2

Fig. 6.22 Dependence of the decay time of the delayed sensitizer-fluorescence (*black filled circles*) and sensitizer phosphorescence (*grey filled circles*) on sample temperature. Solid-state film of 1% wt. PdOEP in PS ($M_W \sim 6$ kD); excitation wavelength $\lambda = 405$ nm

6.4 Sunlight UC and Its Application for Organic Solar Cells

In the field of organic photovoltaic devices, major research efforts have been concentrated into the synthesis of new materials and the development new device concepts in order to increase the efficiency of photovoltaic operation [35]. An important and restricting feature of organic solar cells is the limited spectral range of optical absorption in organic dye molecules/polymers. The most efficient solar cells at present are harvesting photons mainly from the blue, green, and/or yellow parts of the solar spectrum. Therefore, it should be expected that they can benefit greatly from being combined with photon convertor devices and more specifically with photon up-convertor devices.

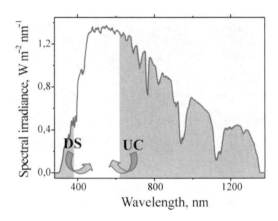

Fig. 6.23 Cartoon exploring the long sought for idea of DS (lower energy boundary 3.25 eV, corresponding to $\lambda = 380$ nm) and UC (higher energy boundary 2.0 eV, corresponding to $\lambda = 620$ nm)

Increasing the portion of the solar spectrum efficiently used by the organic photovoltaic devices has been a challenge for many research and development groups. Different concepts have been developed to overcome this challenge and one of them has been the use of photon conversion processes. Photon conversion processes such as down-shifting (DS), down-conversion (DC) and up-conversion (UC) aim at converting the solar spectrum, via luminescence re-emission, so that it matches better the absorption properties of the different photovoltaic devices (Fig. 6.23). The main goal of these conversion processes is the increase of the spectral power density of the Suns irradiation into the spectral region overlapping with the absorption region of the photovoltaic devices. As energy source for these conversion processes, the Suns irradiance, which is otherwise too far in the blue for the photovoltaic device to absorb (in the case of DS and DC) or too far in the red (in the case of UC), would be used. The utilization of photon conversion processes contrasts with the other concepts all of which focus on developing the photovoltaic devices themselves to better match the polychromatic solar spectrum, which is usually done either by widening the absorption spectrum of the devices or by building multi-junction devices [36, 37].

The conversion efficiency of a solar cell is an integral quantity, depending on:

$$\eta_{\text{IPCE}} = \int_{\lambda_2}^{\lambda_1} \zeta(\lambda) d\lambda. \tag{6.9}$$

Following the concept of photon conversion processes, it is possible to optimize both spectral parameters ($\omega(\lambda)$ and $\zeta(\lambda)$) independently of each other. As shown in Fig. 6.24, it is possible to have solar cells with comparable efficiencies, but with significant differences in the width of their IPCE-curves. In a hypothetic experiment, after efficient DS- and UC-processes, the energy of the solar spectrum would be concentrated to a region overlapping with spectral region of the solar cell with the large $\zeta(\lambda)$ (Fig. 6.22). The next advantage of the photon conversion techniques is that the conversion processes and devices could be considered and optimized

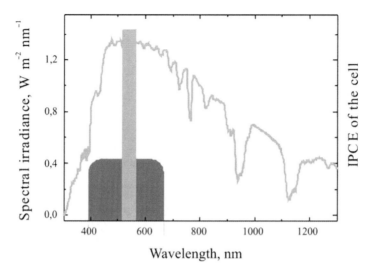

Fig. 6.24 Sketch exploring the idea of independent optimization of the solar cell spectral parameters: width of the IPCE curve – ($\omega(\lambda)$) – and spectral IPCE coefficient – $\zeta(\lambda)$ –. The *shaded zones* represent two solar cell devices with comparable efficiencies. The *light grey* zone – cell with narrow $\omega(\lambda)$ but very high $\zeta(\lambda)$–; the *dark grey* zone – a cell with broad $\omega(\lambda)$ and low $\zeta(\lambda)$–

independently, without affecting the particular physical properties of the operating photovoltaic material or device architectures. As an ensembled outcome, well-developed photon conversion devices would be combined with different existing photovoltaic devices.

6.4.1 Active UC-Media: The Molecular Couples

The extension of the π-conjugated system in the metallated porphyrins by annulated aromatic rings results in a noticeable red shift in the absorption and emission spectra. Thus, porphyrins, such as tetranaphthoporphyrins [38] and tetraanthraporphyrins [39] have expanded excitation spectra for UC down into the IR-A region of the visible spectrum.

Through attentive sample preparation (Fig. 6.10) also comprising of adjustment of the sensitizer/emitter couples the achievement of UC fluorescence excited by using ultra-low intensity (as low as 0.1 W·cm^{-2}) and extreme low spectral brightness (down to 100 μW·nm^{-1}) non-coherent sunlight was successfully demonstrated. A Dobsonian telescope (12″ Lightbridge, Meade Ins. Corp.) on the rooftop of the laboratory was used to collect the sunlight and couple it into an optical fiber (Multimode, 1,000 μm, $NA = 0.48$, Thorlabs Inc.). In order to reduce the thermal stress for the optical elements, the infrared tail of the solar spectrum (wavelengths longer than 850 nm) was rejected through the use of a large size (12″) interference filter (AHF Analysentechnik GmbH) before focusing onto the face of the optical

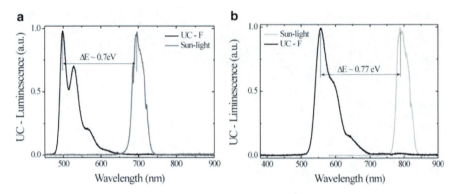

Fig. 6.25 (**a**): UC-fluorescence spectrum of PdTNP/BPBT (*black line*) excited by a portion of the solar spectrum (*grey line*) – near $\lambda = 700$ nm, $\Delta\lambda \sim 31$ nm FWHM; (**b**): UC-fluorescence spectrum of PdTAP/Rubrene (*black line*) excited by a portion of the Solar spectrum (*grey line*) – near $\lambda = 800$ nm, $\Delta\lambda \sim 42$ nm FWHM. For both samples: excitation intensity ~ 0.1 W·cm^{-2}

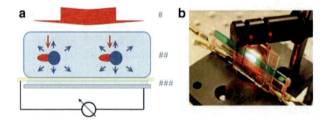

Fig. 6.26 (**a**) The schematic representation of the experimental combination of the UCd (##) and the DSSC (###). The red light (#) for exciting the UCd was either a laser diode with wavelength $\lambda = 635$ nm (intensity ~ 100 mW·cm^{-2}) or a portion of the solar spectrum ca. $\Delta\lambda \sim 30$ nm wide and with maximum at $\lambda = 624$ nm. (**b**) Photograph of the operating DSSC device (5 × 5 mm) with the UCd (the *green* – looking optical cell) and optical cell, for comparison containing only the transparent matrix material (PS400)

fiber. Finally, the necessary excitation sunlight band was selected by using a broadband interference filter at the output of the optical fiber (Fig. 6.25).

6.4.2 Active UC-Media: The Molecular Couples

In the experiments described below, the use of TTA-based UC-devices excited with sunlight in combination with dye sensitized solar cells (DSSC) will be demonstrated. The corresponding device-architecture (Fig. 6.26), its optical and electrical characteristics will be discussed. The structures of the sensitizer/emitter couple used are shown in Fig. 6.11.

Fig. 6.27 The absorption spectrum of the sensitizer (*black line*), luminescence of the system PdTBP/perylene excited in UC-regime (*light grey line*) and the IPCE-spectrum of the solar cell (*dark grey line*)

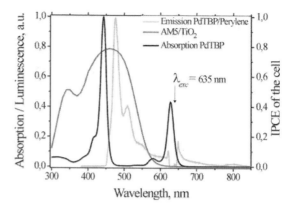

The IPCE spectrum of the DSSC used in the experiments is shown in Fig. 6.27. The optical properties of the UC-couple sensitizer/emitter were chosen to optimally match the excitation spectrum requirements of the organic dye used as sensitizer in the solar cell. As shown in Fig. 6.27, the UC-fluorescence emission of the emitter covers the second half of the IPCE-spectrum of the DSSC device. On the other hand, the Q-band absorption of the UC-sensitizer lies completely outside the IPCE-spectrum of the DSSC device; therefore, photons from the red light region will have negligible photon-to-electron conversion efficiency. The DSSC devices were constructed with a 10 μm thick porous TiO₂ layer, sensitized with dye absorbing down to 600 nm, a liquid electrolyte and had an illuminated active area of 0.25 cm². The same DSSC devices were used throughout all experiments, including excitation with simulated solar spectrum and, with or without UCd.

First, the solar cell was characterized with simulated sunlight (Sun-simulator, LOT Oriel Inc.) with an intensity of 100 mW·cm⁻². Under these conditions, the DSSC-device had a short circuit current density $J_{SC} = 14$ mA·cm⁻² and a power conversion efficiency of 6.4%. The IPCE curves of the DSSC devices show that the devices practically do not use light with wavelengths longer than 610 nm. The experimental set-up used for the measurements of the DSSC devices when working with upconverted red photons is shown schematically in Fig. 6.27. The combination of DSSC+UCd was characterised by using a laser diode as a light source; the illuminated area of the DSSC-device was ca. 0.125 cm² and the excitation intensity again 100 mW·cm⁻². For the next DSSC + UCd characterization, as light source portion of the solar spectrum collected with a telescope was used. This portion of the solar spectrum used as light source with an intensity of 100 mW·cm⁻² had width of ca. $\Delta\lambda = 30$ nm and peaked at $\lambda = 624$ nm. This intensity corresponds to a concentration of ca. 26 times of the Suns intensity at the location and time of the experiments (Mainz, Germany, August, 2 P.M., Blue sky). The current-voltage characteristics of the DSSC when working with UC photons are shown in Fig. 6.28. The measurements without the UCd were done by illuminating the DSSC through a device with the same geometry as the UCd but without any dyes, i.e. filled only with the transparent matrix material. In this case, without an

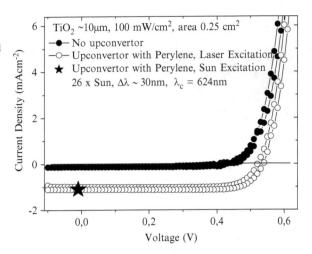

Fig. 6.28 The current voltage characteristics of the DSSC when working with red light, either without UCd (*filled circles*) or with UCd (*hollow circles*)

up-converter, the solar cell excited with $100\,mW\cdot cm^{-2}$ red light ($\lambda = 635\,nm$) has a very low ($<0.1\,mA\cdot cm^{-2}$) open-circuit photocurrent density (Fig. 6.28, filled circles current-voltage curve). When the DSSC was illuminated through the UCd device (Fig. 6.28), the measured short circuit current density was drastically increased up to $1.2\,mA\cdot cm^{-2}$ (at $100\,mW\cdot cm^{-2}$, Fig. 6.28, hollow circles current-voltage curve). Taking into account the relatively broad ($\Delta\lambda \sim 30\,nm$) UCd absorption band represented by the Q-band absorption of the sensitizer, the necessarily red-band excitation intensity of $100\,mW\cdot cm^{-2}$ is obtained as mentioned above with only 26-fold focusing (concentrating) of the Sun intensity at the location and time of the experiments. Such moderate sunlight focusing could be easily obtained in practical applications.

6.5 Conclusion

The UCd reported here have no limitation for size scaling; the contrary, namely very limited size scaling, is a common drawback for the UC-devices based on RE-doped crystalline glasses reported so far [40, 41]. The TTA-UCd device size would be limited mainly by the size of the optical substrates, although tiling would be fairly straightforward. Furthermore, for TTA-UCd across a broad range of pump intensities (practically, from $1\,mW\cdot cm^{-2}$ up to $10\,W\cdot cm^{-2}$) there is a nearly linear dependence of the UC-signal on the pump intensity, allowing to have effective and predictable UC-emission even at impermanent solar irradiation. The reported devices are also flexible; additionally, there is the possibility to use different non-volatile solvents in order to tune the total index of refraction of the device. The robustness of the materials allows efficient manufacturing techniques such as printing and roll-to-roll processing to be used. Further material advantage is that the emitters, used in the molecular systems showing efficient TTA-UC, are known as very stable and efficient optical down-shifters. This fact gives an additional advantage to the UCd based on TTA-UC: the incorporation of such devices within a luminescent solar concentrator (LSC) strategy [42] arises as a natural extension. The major benefit of the LSC is to concentrate the photon flux from a broad, strongly blue-shifted, solar band (commonly the UV-band) into the relatively narrow band of optimal photon-to-electron transfer efficiency, for any given solar cell dye. Due to the nature of the TTA-UC, a significant part of the red-shifted solar band will be concentrated into the same optimal wavelength region, as a consequence of the usage of the same emitter molecule for both of the DS- and UC-processes.

References

1. Y. Shen, *The Principles of Nonlinear Optics* (Wiley, New York, 2002)
2. F. Auzel, Chem. Rev. **104**(1), 139 (2004)
3. J.C. Boyer, L.A. Cuccia, J.A. Capobianco, Nano Lett. **7**(3), 847 (2007)
4. A. Shalav, B.S. Richards, M.A. Green, Sol. Energ. Mater. Sol. Cell. **91**(9), 829 (2007)
5. T. Trupke, A. Shalav, P. Wrfel, M.A. Green, Sol. Energ. Mater. Sol. Cell. **90**(18–19), 3327 (2006)
6. T. Miteva, A. Meisel, W. Knoll, H.G. Nothofer, U. Scherf, D.C. Mueller, K. Meerholz, A. Yasuda, D. Neher, Adv. Mater. **13**(8), 565 (2001)
7. R. Pacios, D.D.C. Bradley, J. Nelson, C. Brabec, Synthetic Metals **137**(1–3), 1469 (2003)
8. R. Schroeder, B. Ullrich, W. Graupner, U. Scherf, J. Phys. Condens. Matter **13**(16), L313 (2001)
9. P. Keivanidis, S. Baluschev, G.N.G.W. T. Miteva, A. Yasuda, Adv. Mater. **15**(24), 2095 (2003)
10. F. Laquai, G. Wegner, C. Im, A. Büsing, S. Heun, J. Chem. Phys. **123**(7), 074902 (2005)
11. S. Baluschev, P.E. Keivanidis, G. Wegner, J. Jacob, A.C. Grimsdale, K. Müllen, Appl. Phys. Lett. **86**(6), 061904 (2005)
12. S. Baluschev, J. Jacob, Y.S. Avlasevich, P.E. Keivanidis, T. Miteva, A. Yasuda, G. Nelles, A.C. Grimsdale, K. Müllen, G. Wegner, ChemPhysChem **6**(7), 1250 (2005)
13. M. Pope, C. Swenberg, *Electronic Processes in Organic Crystals* (Oxford University Press, New York, 1982)
14. J. Kalinowski, W. Stampor, M. J., M. Cocchi, D. Virgili, V. Fattori, P. Di Marco, Phys. Rev. B **66**(23), 235321 (2002)
15. M.A. Baldo, C. Adachi, S. Forrest, Phys. Rev. B **62**(16), 10967 (2000)
16. S. Baluschev, T. Miteva, B. Minch, V. Yakutkin, G. Nelles, A. Yasuda, G. Wegner, J. Appl. Phys. **101**(2), 023101 (2007)
17. D. Wasserberg, S.P. Dudek, S.C.J. Meskers, R.A.J. Janssen, Chem. Phys. Lett. **411**(1–3), 273 (2005)
18. S. Baluschev, T. Miteva, V. Yakutkin, G. Nelles, A. Yasuda, G. Wegner, Phys. Rev. Lett. **97**(14), 143903 (2006)
19. S. Baluschev, T. Miteva, V. Yakutkin, G. Nelles, S. Chernov, A. Aleshchenkov, A. Cheprakov, A. Yasuda, G. Wegner, Appl. Phys. Lett. **90**(18), 181103 (2007)
20. S. Baluschev, V. Yakutkin, T. Miteva, G. Wegner, T. Roberts, G. Nelles, A. Yasuda, S. Chernov, S. Aleshchenkov, A. Cheprakov, New J. Phys. **10**, 013007 (2008)
21. S. Baluschev, V. Yakutkin, T. Miteva, T. Avlasevich, S. Chernov, S. Aleshchenkov, G. Nelles, A. Cheprakov, A. Yasuda, K. Müllen, G. Wegner, Angewandte Chemie **46**, 7693 (2007)
22. V. Yakutkin, S. Aleshchenkov, S. Chernov, T. Miteva, G. Nelles, A. Cheprakov, S. Baluschev, Chem. Eur. J. **14**, 9846 (2008)
23. M. Kasha, Discussions Faraday Soc. **9**, 14 (1950)

24. R.L. Fulton, M. Gouterman, J. Chem. Phys. **35**(3), 1059 (1961)
25. A. Terenin, V. Ermolaev, Trans. Faraday Soc. **52**, 1042 (1956)
26. C.A. Parker, C.G. Hatchard, T.A. Joyce, Nature **205**(4978), 1282 (1965)
27. J. Mezyk, R. Tubino, A. Monguzzi, A. Mech, F. Meinardi, Phys. Rev. Lett. **102**(8), 087404 (2009)
28. A. Monguzzi, R. Tubino, F. Meinardi, J. Phys. Chem. A **113**(7), 1171 (2009)
29. J.A. OBrien, S. Rallabandi, U. Tripathy, M.F. Paige, R.P. Steer, Chem. Phys. Lett. **475**(4–6), 220 (2009)
30. R. Islangulov, J. Lott, C. Weder, F.N. Castellano, J. Am. Chem. Soc. **129**(42), 12652 (2007)
31. S. Baluschev, T. Miteva, B. Minch, V. Yakutkin, G. Nelles, A. Yasuda, G. Wegner, J. Appl. Phys. **102**(7), 076103 (2007)
32. D.V. Kozlov, F.N. Castellano, Anti-Stokes delayed fluorescence from metal-organic bichromophores. Chem. Commun. (24), 2860–2861 (2004). doi: 10.1039/b412681e
33. J.B. Birks, *Photophysics of Aromatic Molecules* (Wiley, New York, 1970)
34. M. Ariu, D.G. Lidzey, M. Sims, A.J. Cadby, P. Lane, D.D.C. Bradley, J. Phys. Condens. Matter **14**(42), 9975 (2002)
35. J.Y. Kim, K. Lee, N.E. Coates, D. Moses, T.Q. Nguyen, M. Dante, A.J. Heeger, Science **317**(5835), 222 (2007)
36. A. Luque, A. Martí, Phys. Rev. Lett. **78**(26), 5014 (1997)
37. M. Durr, A. Bamedi, A. Yasuda, G. Nelles, Appl. Phys. Lett. **84**(17), 3397 (2004)
38. O.S. Finikova, A.V. Cheprakov, P.J. Carroll, S.A. Vinogrado, J. Organ. Chem. **68**(19), 7517 (2003)
39. N. Kobayashi, W.A. Nevin, S. Mizunuma, H. Awaji, M. Yamaguchi, Chem. Phys. Lett. **205**(1), 51 (1993)
40. E. Downing, L. Hesselink, J. Ralston, R. Macfarlane, Science **273**(5279), 1185 (1996)
41. A. Rapaport, J. Miliez, M. Bass, A. Cassanho, H. Jensen, J. Display Technol. **2**(1), 68 (2006)
42. R. Koeppe, N.S. Sariciftci, A. Büchtemann, Appl. Phys. Lett. **90**(18), 181126 (2007)

Chapter 7
Next Generation Photovoltaics Based on Multiple Exciton Generation in Quantum Dot Solar Cells

Arthur J. Nozik

Abstract Next Generation solar cells based on Multiple Exciton Generation (MEG) in semiconductor quantum dots (QDs) are described. This application of QDs depends upon efficient MEG in QDs incorporated into PV cells, followed by efficient exciton splitting into free electrons and holes and their efficient separation and collection in the cell contacts to produce multiple free carriers per absorbed photon. Using time-resolved transient absorption, bleaching, photoluminescence and THz spectroscopy, MEG has been initially confirmed in several Group IV-VI, III-V, II-VI, and IV colloidal semiconductor QDs. Some controversy using these techniques have now been attributed to effects of the variable of the QD surface chemisty and under certain conditions to artifacts arising from long-lived trapping of photoinduced charge; in our opinion these controversies have been resolved and are discussed here. Furthermore, various photovoltaic cell architectures utilizing QDs have recently been constructed and the photocurrent and photovoltage characterisitics have been studied. These photocurrent measurements provide a more direct measurement of MEG since the photogenerated carriers are counted directly via the current, and they are very consistent with the QYs of MEG reported using the proper spectroscopic techniques; thus, these new photocurrent measurements confirm the existence of enhanced exciton and carrier multiplication in QDs. The past work and prognosis for QD-based Next Generation PV cells based on MEG are discussed.

A.J. Nozik (✉)
National Renewable Energy Laboratory Golden, 1617 Cole Blvd., Golden, CO 80401, USA

University of Colorado, Boulder, Department of Chemistry and Biochemistry, Boulder, CO 80309, USA
e-mail: Arthur.Nozik@nrel.gov

7.1 Introduction

Third Generation (also called Next Generation) Photovoltaic (PV) Solar Cells are defined as PV cells that have two characteristics: (1) a power conversion efficiency greater than the Shockley–Queisser limit of 32% [1], and (2) a very low cost per unit area. According to Green [2], in order to be classified as a Third Generation PV (TGPV) cell the value of the PV module cost/unit area divided by the watts delivered/unit area should yield a cost per peak watt (W_p) for the module of about \$0.20 US/W_p. Thus, for example, this goal would be achieved for a PV module having a cost of \$100 m^{-2} and an efficiency of 50% (500 watts m^{-2}); any combination of efficiency and areal cost yielding \$0.20/Wp would thus be classified as a TGPV cell. To obtain the yearly averaged energy cost for the PV system (\$/kWh), the balance of systems (BOS) cost (\$/m^2) needs to be added to the module cost, and then the annual-averaged capacity factor for the PV system together with other operating costs (e.g., interest rates, maintenance, taxes) need to be considered. A simple, but rough, conversion is to simply multiple the total \$ per peak watt cost(module + BOS) by 0.04–0.05 (depending upon geographical location) to obtain \$/kWh. If the module and BOS costs for the TGPV systems remain about equal, as they are today, then a \$0.20/W_p module cost would result in a total system cost of \$0.40/Wp and thus represents an energy cost of about \$0.02/kWh – a cost competitive with or even lower than that of energy from coal.

One route to achieve TGPV cells that is now being extensively investigated is to use semiconductor nanostructures in PV cells. Nanostructures of semiconductor materials exhibit quantization effects when the electronic particles of these materials are confined by potential barriers to very small regions of space. The confinement can be in 1 dimension (producing quantum films, also termed quantum wells in the early 1980s as the first examples of quantization in nanoscale materials), in two dimensions (producing quantum wires or rods), or in three dimensions (producing quantum dots (QDs)). Some authors refer to these three nanostructure regimes as 2D, 1D, or 0D, respectively, although these terms are not as precise. Nanostructures of other classes of materials, such as metals and organic materials, are also possible, but the present discussion will be limited to semiconductor nanostructures.

Nanostructures of crystalline materials are also referred to as nanocrystals (NCs); and this term includes a variety of nanoscale shapes with the three types of spatial confinement, including spheres, cubes, rods, wires, tubes, tetrapods, ribbons, discs and platelets [3]. The first six shapes are being studied for renewable energy applications, but the focus here will be on the use of spherical semiconductor NCs (QDs).

One ubiquitous feature of all present PV cells is that photons having energies greater than the semiconductor bandgap create free carriers or excitons that have energies in excess of the bandgap; these carriers or excitons are called "hot carriers" or "hot excitons". This excess electron energy is kinetic free energy and is lost quickly (ps to sub-ps time scales) through electron–phonon scattering, thus converting the excess kinetic energy into heat. The free carriers or excitons then occupy the lowest energy levels (the bottom and top of the conduction and valence bands, respectively) where they can be removed to do electrical work or lost through radiative or non-radiative recombination. In 1961, Shockley and Queisser [1] (S-Q) calculated the maximum possible thermodynamic efficiency of converting solar irradiance into electrical free energy in a PV cell assuming: (1) complete carrier cooling, and (2) that the only other free energy loss mechanism was radiative recombination. This detailed balance calculation in the radiative limit yields a maximum thermodynamic efficiency of 31–33%, depending upon the details of the AM1.5 solar spectrum utilized, with optimum bandgaps between about 1.1–1.4 eV.

One way used presently to reduce the energy loss due to carrier cooling is to stack a series of semiconductors with different bandgaps in tandem with the largest bandgap irradiated first followed by decreasing bandgaps. In the limit of a large number of different multiple bandgaps matched to the solar spectrum, the conversion efficiency can reach 67% at one-sun intensity. However, in practice only 2–3 bandgaps are used because for these multijunction PV cells most of the gain in efficiency is obtained with 3 bandgaps; after that there are diminishing returns. Detailed Balance calculations show that with 2 bandgaps the maximum efficiency is 43%, with 3 it is 48%, with 4 it is 52%, and with 5 it is 55%.

7.2 Solar Cells Utilizing Hot Carriers for Enhanced Conversion Efficiency

In 1982, thermodynamic calculations [4] showed for the first time that the same high conversion efficiency of solar irradiance into free energy in a tandem stack of different bandgaps can be also be obtained by utilizing the total excess kinetic energy of hot photogenerated carriers in a single bandgap semiconductor before they cooled to the lattice temperature through electron-phonon scattering; in the limit of a carrier temperature of 3,000 K the conversion efficiency also reaches 67%, the same value as for a tandem PV cell with a multiple stack of bandgaps matched to the solar spectrum. This can be achieved by transporting the hot carriers to carrier-collecting contacts with appropriate work functions (either into an electrolyte redox system in a photoelectrochemical fuel producing cell [5] or a solid state ohmic contact in a PV cell [6], before the carriers cool. These cells are called hot carrier solar cells [2,5–8].

Another approach to beneficially utilize hot electron–hole pairs is to use their excess kinetic energy to produce additional electron–hole pairs, and thus increase the possible photocurrent. However, this approach yields a lower maximum conversion efficiency of about 45% at one-sun intensity. This lower efficiency occurs because to satisfy energy conservation the photon energies between 1 and 2 E_g are lost through electron-phonon scattering and produce heat. In bulk semiconductors, this process is called impact ionization [9] and is an inverse Auger type of process. However, impact ionization (I.I.) cannot contribute to improved quantum yields (QYs) in present solar cells based on bulk semiconductors such as Si, CdTe, $CuIn_xGa_{1-x}Se_2$, or III–V semiconductors in a multi-junction, tandem structure because the maximum QY for I.I. does not produce extra carriers until photon energies reach the ultraviolet region of the spectrum ($h\upsilon > 3.5$ eV), where solar photons are absent. In bulk semiconductors, the threshold photon energy for I.I. exceeds that required for energy conservation alone because crystal momentum (k) must also be conserved [9]. Additionally, the rate of I.I. must compete with the rate of energy relaxation by phonon emission through electron–phonon scattering. It has been shown that in bulk semiconductors the rate of I.I. becomes competitive with phonon scattering rates only when the kinetic energy of the electron is many multiples of the bandgap energy (E_g) [10–12]. In bulk semiconductors, the observed transition between inefficient and efficient I.I. occurs slowly; for example, in Si the I.I. efficiency was found to be only 5% (i.e. total quantum yield = 105%) at $h\upsilon \approx 4$ eV ($3.6\,E_g$), and 25% at $h\upsilon \approx 4.8$ eV ($4.4\,E_g$) [9, 13].

7.3 Quantum Dots, Multiple Exciton Generation, and Third Generation Solar Cells

Because of spatial confinement of electrons and holes in quantum dots: (1) the e^-–h^+ pairs are correlated and thus exist as excitons rather than free carriers, (2) the rate of hot electron and hole (i.e., exciton) cooling can be slowed because of the formation of discrete electronic states [3], (3) momentum is not a good quantum number and thus the need to conserve crystal momentum is relaxed [3], and (4) Auger processes are greatly enhanced because of increased e^-–h^+ Coulomb interaction [14–17]. Because of these factors, it has been predicted [14–17] that the production of multiple e^-–h^+ pairs will be enhanced in QDs compared to bulk semiconductors; the threshold energy ($h\upsilon_{th}$) for electron hole pair multiplication (EHPM) should be reduced and the EHPM efficiency, η_{EHPM} (defined as the number of excitons produced per additional bandgap of photon energy above the EHPM threshold photon energy) is expected to be enhanced. In Qs we label the formation of multiple excitons Multiple Exciton Generation (MEG) [11]; free carriers can only be subsequently collected as separated electrons and holes upon the dissociation of the excitons in various PV device structures. Very recent reports [18–20] propose that the high energy photons required for MEG produce separated e^-–h^+ pairs (ie, free carriers) in either isolated QDs or in electronically-coupled QD arrays very rapidly (femtosecond time scale or essentially immediately); thus, it is presently unclear whether carrier multiplication and separation in QD arrays first require dissociation of thermalized excitons in individual QDs, followed by inter-QD transport, or whether these processes occur essentially instantaneously upon photon absorption in the QD array.

The possibility of enhanced MEG in QDs was first proposed in 2001 [14, 15] and the original concept is shown in Fig. 7.1. Figure 7.2 presents S-Q detailed balance calculations in the radiative limit for conventional solar cells compared to QD solar cells exhibiting various MEG characteristics regarding the threshold photon energy ($h\upsilon_{th}$) and η_{EHPM} [22]. The nature of the MEG characteristics corresponding to the various S-Q calculations presented in Fig. 7.2 are shown in Fig. 7.3. The M_{max}

7 Next Generation Photovoltaics Based on Multiple Exciton Generation

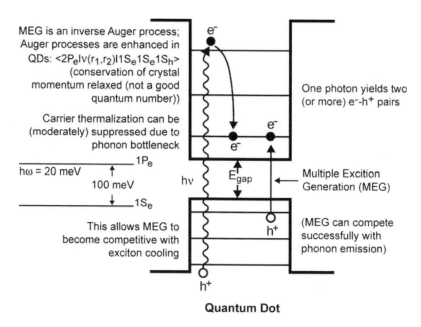

Fig. 7.1 Multiple exciton generation (MEG) in semiconductor quantum dots. Reproduced with permission from [21]

characteristic is a staircase function where the threshold for MEG is $2 E_g$ with N excitons being produced when the photon energy increases to N times E_g, and the QY remains constant between $N E_g$ and $(N + 1) E_g$. The characteristics labelled $L(N)$ have thresholds equal to $N E_g$ and the increase in QY after the threshold is linear with a slope related to the MEG efficiency. It has been shown [22] very recently that the threshold photon energy for MEG to occur and its efficiency, η_{EHPM}, are related by the expression:

$$h\nu_{th}/E_g = 1 + 1/\eta_{EHPM}. \tag{7.1}$$

This recent paper [22] also presents a rigorous derivation of why the appropriate parameter to use when comparing the efficiency of MEG in QDs vs. impact ionization in bulk materials is $h\nu/E_g$ and not just the absolute photon energy $h\nu$. When $h\nu/E_g$ is used, the linear slope of plots of MEG QY vs. $h\nu/E_g$ is the MEG efficiency, η_{EHPM} [22]. The use of just $h\nu$ in such plots was proposed in a prior publication [23] and led to the invalid conclusion that there is no difference between MEG in QDs and I.I. in bulk semiconductors. Our counter-view is clearly supported by Fig. 7.4, which is based on experimental data for the MEG QY obtained as a function of photon energy for bulk PbSe [23] and PbSe QDs [22]. From Fig. 7.4, the EHPM efficiency for PbSe QDs is calculated to be twice that of bulk PbSe, the threshold photon energy for EHPM is calculated to be much lower (about 3–4 E_g

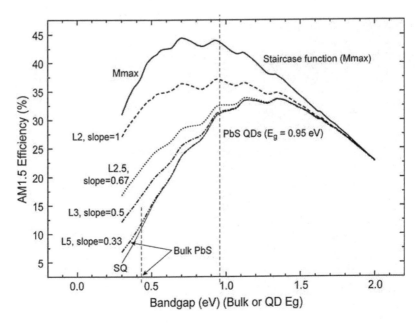

Fig. 7.2 Shockley–Queisser detailed balance thermodynamic calculations for PV power conversion efficiencies for bulk semiconductors and for QD PV cells with various MEG characteristics (M_{max}, L2, L2.5, L3, L5). The definition of the MEG characteristics is described in the text and is shown in Fig. 7.3. The slope for the designated $L(N)$ characteristic is the MEG efficiency (additional excitons per bandgap of photon energy above the MEG threshold energy); N is the threshold photon energy in bandgap units. Reproduced with permission from [22]

vs. about 6 E_g), and the rate of EHPM vs the rate of hot exciton cooling is calculated to be 3 times faster in PbSe QDs compared to bulk material [22].

Multiexcitons have been initially detected using several spectroscopic measurements [17, 21–26]. More recently [27, 28], MEG and subsequent carrier multiplication has been measured in the photocurrent of QD-based solar cells. The photocurrent measurements are more direct since the electron population is counted via the photocurrent; nevertheless, the photocurrent results confirm the validity and accuracy of careful spectroscopic measurements. The first spectroscopic method used was to monitor the signature of MEG using transient (pump-probe) absorption (TA) spectroscopy [24]. The multiple exciton analysis relies on time-resolved TA data taken as a function of the photon excitation (pump) energy. In one type of TA experiment, the probe pulse monitors the interband bleach dynamics with excitation across the QD bandgap [17, 24]; whereas in a second type of experiment the probe pulse is in the mid-IR and monitors the intraband transitions (e.g. $1S_e$–$1P_e$) of the newly created excitons [17, 21, 25, 26]. In the former case, the peak magnitude of the initial early time (3 ps) photoinduced absorption change created by the pump pulse together with the faster Auger decay dynamics of the generated multiexcitons and the resultant TA signal after the extra excitons have decayed (>300 ps), are related to the number of excitons created. In the latter case, the TA dynamics of the

7 Next Generation Photovoltaics Based on Multiple Exciton Generation

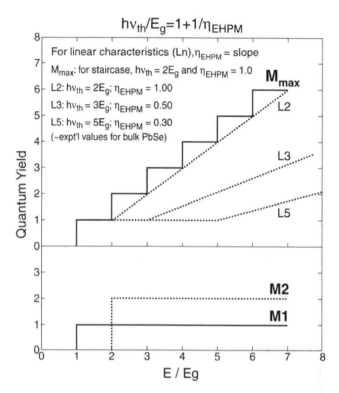

Fig. 7.3 Various MEG characteristics. The linear characteristic is labeled $L(N)$, where N is the MEG threshold energy in bandgap units (from [22]), and the threshold and the MEG efficiency are related through (7.1) in the text. The staircase characteristic (M_{max}) yields the highest conversion efficiency

photoinduced mid-IR intraband absorption after the pump pulse is monitored and analyzed. In [17, 21–26], the transients are detected by probing either with a probe pulse exciting across the QD bandgap, or with a mid-IR probe pulse that monitors the first $1S_e$–$1S_p$ intraband transition; both experiments yield the same MEG QYs.

The first experimental report of exciton multiplication was presented by Schaller and Klimov [24] for PbSe NCs and they reported an excitation energy threshold for the efficient formation of two excitons per photon at 3 E_g. Subsequent work reported that efficient MEG occurs also in PbS [17] and in PbTe QDs [29]. Additional experiments observing MEG have now been reported for PbSe [30, 31], CdSe [32, 33], PbTe [29], InAs [34], Si [35], InP [36], CdTe [37] and CdSe/CdTe core-shell QDs [38]. For InP QDs, the MEG threshold was 2.1 E_g and photocharging (see next paragraph for significance of QD charging for MEG) was claimed not to be present in the QD samples [36]. For the CdSe/CdTe QDs, time-resolved photoluminescence (TRPL) was used to monitor the effects of multiexcitons on the PL decay dynamics to determine the MEG QY. The timescale for MEG has been reported to be <100 fs [17]. This ultrafast MEG rate is much faster than the

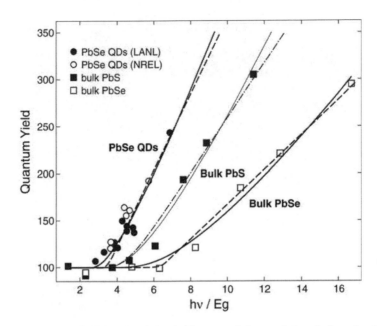

Fig. 7.4 Comparison of efficiency and threshold energy of electron–hole pair formation in QDs vs bulk IB-VI QDs. For QDs, data from NREL and the Los Alamos National Laboratory are equivalent. Effects due to photocharging have been eliminated in these experiments by flowing or stirring the colloidal QD samples. Reproduced with permission from [22]

hot exciton cooling rate produced by electron–phonon interactions, and MEG can therefore beat exciton cooling and become efficient.

However, a few published reports [39–41] could not reproduce some of the reported early positive MEG results (for example, InAs) [39–41], or if MEG was indeed observed the efficiency was claimed to be much lower [23, 42], and in one report MEG was claimed to be only equivalent to impact ionization in bulk materials [42]. Thus, some controversy has arisen about the efficiency of MEG in QDs.

The reason for this inconsistency has been attributed to the influence of QD surface treatments and surface chemistry on MEG dynamics compared to cooling dynamics [3, 21, 22, 25, 26, 43], and in some cases to the effects of surface charge produced during transient pump-probe spectroscopic experiments to determine MEG quantum yields [44, 45]. In the latter case, long-lived charge could produce trions (two electrons and a hole or two holes and an electron) in the QDs after the absorption of an additional photon in the QDs in a pump-probe experiment. Trions can also undergo fast Auger recombination compared to single excitons and thus could confound the fast early time decay of transient absorption or bleaching signals that are the signature of MEG, and lead to over-estimation of the MEG QY [44, 45]. However, recent work [30, 36, 46] shows that charging effects may not always be significant, since they depend upon the specific QD surface chemistry, photon fluence, photon energy, and QD size [46]. In any case, the possibility of photocharging effects can be eliminated in MEG experiments based on time-resolved TA or PL spectroscopy by flowing or stirring the colloidal QD suspension to refresh the sample volume of QDs being probed [44–46]. To be absolutely certain about MEG QY values determined by pump-probe spectroscopy, the possibility of charging needs to be examined for all experiments done on static solutions or solid-state films, and experiments need to be done to ensure trion or trapped charge-influenced decay is not affecting the determined MEG QYs.

It is noted that in addition to many reported MEG effects in semiconductor QDs, MEG has also been recently reported in single wall carbon nanotubes [47, 48] and in PbSe quantum rods [49]. Theoretical considerations [50] suggest that MEG should be enhanced in nanotubes and nanorods compared to QDs. This has been attributed in part to stronger Coulomb coupling and the absence or reduction of surface states in nanotubes and nanorods. Further research is underway on these new and interesting direction for exciton multiplication in various nanostructures other than spherical semiconductor QDS, and in general advances in theory and additional experiments to better understand MEG in various nanostructures are expected.

7.4 Configurations of QD Solar Cells

The two fundamental pathways for enhancing the conversion efficiency (increased photovoltage [4, 51] or increased photocurrent [14, 15, 52, 53]) can be accessed, in principle, in three different QD solar cell configurations [3]; these configurations are shown in Fig. 7.5 and they are described below. However, it is emphasized that these potential high efficiency configurations are theoretical and to date no actual enhanced *power conversion efficiencies* significantly greater than present high efficiency PV solar cells have yet been reported. Notwithstanding, hot electron and hot exciton effects to increase the photovoltage and photocurrent in various semiconductor structures have indeed been reported. In one recent publication [54] on hot electron effects in a nanoscopically thin Si solar cells, a very small increase in photovoltage ($\sim$17 mV) due to hot electron collection was reported in an amorphous nanoscale $p–i–n$ Si cell [54]. Furthermore, two other recent publications have reported hot electron effects in QD-sensitized single crystals of TiO_2. In one publication [55], PbSe QDs were deposited on single crystal rutile and the process of hot electron injection from the QDs into the TiO_2 conduction band and its timescale (<50 fs) were measured using ultrafast second harmonic generation. In a second paper [28], a monolayer of PbS QDs was deposited on single crystal anatase TiO_2 in a photoelectrochemical cell and the measured photocurrent QY at $h\upsilon = 3.2\ E_g$ was twice that measured below the MEG threshold ($2.5\ E_g$). In a third recent publication [27], both external quantum yields (photocurrent/incident photon flux) and internal quantum yields (photocurrent/absorbed photon flux) above 100% were reported for photocurrent in solar cells based on PbSe QDs. These are encouraging results, and the potential payoff of success in highly efficient MEG in QD-based or nanostructured solar cells justifies continued research in this area.

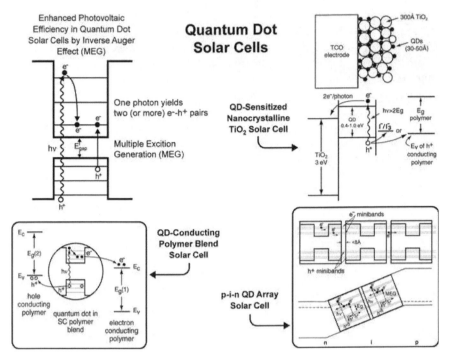

Fig. 7.5 Various architectures for QD solar cells. Reproduced with permission [14]

7.4.1 Photoelectrodes Composed of Quantum Dot Arrays

In this configuration (Fig. 7.3 bottom right), the QDs are formed into an ordered 3-D array with inter-QD spacing sufficiently small such that strong electronic coupling occurs to allow long-range electron transport. If the QDs have the same size and are aligned, then this system is a 3-D analog to a 1-D superlattice and the miniband structures formed therein [56]. The moderately delocalized but still quantized 3-D states could be expected to produce MEG. Also, the slower carrier cooling and delocalized electrons could permit the transport and collection of hot carriers to produce a higher photopotential in a PV or photoelectrochemical cell.

Significant progress has been made in forming 3-D arrays of both colloidal [57] and epitaxial [58] IV–VI, II–VI and III–V QDs. The former two systems have been formed via evaporation, crystallization, or self-assembly of colloidal QD solutions containing a reasonably uniform QD size distribution. Although the process can lead to close-packed QD films, they exhibit a significant degree of disorder. Concerning the III–V materials, arrays of epitaxial QDs have been formed by successive epitaxial deposition of epitaxial QD layers; after the first layer of epitaxial QDs is formed, successive layers tend to form with the QDs in each layer aligned on top of each other [58, 59]. Major issues are the nature of the electronic states as a function of inter-dot distance, array order vs disorder, QD orientation and shape, surface states, surface structure/passivation, and surface chemistry. Transport properties of QD arrays are also of critical importance.

7.4.2 Quantum Dot-Sensitized Nanocrystalline TiO$_2$ Solar Cells

This configuration (Fig. 7.3 top right) is a variation of the type of photovoltaic cell that is based on dye-sensitization of nanocrystalline TiO$_2$ layers [60–62]. In this PV cell, dye molecules are chemisorbed onto the surface of 10–30 nm-size TiO$_2$ particles that have been sintered into a highly porous nanocrystalline 10–20 μm TiO$_2$ film. Upon photoexcitation of the dye molecules, electrons are very efficiently injected from the excited state of the dye into the conduction band of the TiO2, affecting charge separation and producing a photovoltaic effect.

For the QD-sensitized cell, QDs are substituted for the dye molecules; they can be adsorbed from a colloidal QD solution [63] or produced in-situ [64–67]. Successful PV effects in such cells have been reported for several semiconductor QDs including InP, InAs, CdSe, CdS, and PbS [63–67]. Many reviews of this approach are published [68–72]. Possible advantages of QDs over dye molecules are the tunability of optical properties with size and better heterojunction formation with solid hole conductors. Also, as discussed here, a unique potential capability of the QD-sensitized solar cell is the production of quantum yields greater than one by MEG.

7.4.3 Quantum Dots Dispersed in Organic Semiconductor Polymer Matrices

Recently, photovoltaic effects have been reported in structures consisting of QDs forming intimate junctions with organic semiconductor polymers. In one configuration, a disordered array of CdSe QDs is formed in a hole-conducting polymer–MEH-PPV {poly[2-methoxy, 5-(2-ethyl)-hexyloxy-p- phenyleneviny-lene]} [73]. Upon photoexcitation of the QDs, the photogenerated holes are injected into the MEH-PPV polymer phase, and are collected via an electrical contact to the polymer phase. The electrons remain in the CdSe QDs and are collected through diffusion and percolation in the nanocrystalline phase to an electrical contact to the QD network. Initial results show relatively low conversion efficiencies [73, 74], but improvements have been reported with rod-like CdSe QD shapes [75] embedded in poly(3-hexylthiophene) (the rod-like shape enhances electron transport through the nanocrystalline QD phase) and recently [76] with newer polymers (PCDTBT, Konarka) that allow for better electrical properties (3.13%, NREL-certified). In another configuration [77], a polycrystalline TiO$_2$ layer is used as the electron conducting phase, and MEH-PPV is used to conduct the holes; the electron and holes are injected into their respective transport mediums upon photoexcitation of the QDs.

A variation of these configurations is to disperse the QDs into a blend of electron and hole-conducting phases, such as conducting polymers (Fig. 7.3 bottom left) and fullerene derivatives. This scheme is the inverse of light emitting diode structures based on QDs [78–82]. In the PV cell, each type of carrier-transporting phase would have a selective electrical contact to remove the respective charge carriers. A critical factor for success is to prevent electron–hole recombination at the interfaces of the two polymer blends; prevention of electron–hole recombination is also critical for the other QD configurations mentioned above.

All of these possible configurations for QD solar cells are being investigated in various laboratories and reviews are available [83]. Although reasonably high and reliable MEG efficiencies have been reported, including photocurrent QYs >100% from QDs bound to single crystal TiO_2 surfaces at photon energies of $3.2\,E_g$ [28] and QD arrays [27], no QD-based PV device has yet shown an enhanced power conversion efficiency due to MEG effects. Further research is necessary to establish these charge separation dynamics in the various QD solar cell configurations.

7.5 Schottky Junction and *p–n* Junction Solar Cells Based on Films of QD Arrays

To date, MEG has been studied in NCs of the lead salts [17, 24, 29, 84], InAs [41, 85, 86], CdSe [39, 87] and Si [35] using several time-resolved and quasi-CW spectroscopies; negative results were reported for InAs [41] and CdSe [39]. It is therefore important to establish whether significant MEG photocurrent can be collected from a NC solar cell [85, 88–92]. A simple, all-inorganic metal/NC/metal sandwich cell has been reported [93] that produces a large short-circuit photocurrent ($\sim$25 mA cm^2) by way of a Schottky junction at the negative electrode. The PbSe NC film, deposited via layer-by-layer (LbL) dip coating, yields an AM1.5G power conversion efficiency of 2.1%. This NC device produces one of the largest short-circuit currents of any nanostructured solar cell, without the need for sintering, superlattice order or separate phases for electron and hole transport. The report by Sambur, Novel, and Parkinson [28], described above, is the first result showing a QY > 1 for photocurrent in a photovoltaic device (a photoelectrochemical cell based on semiconductor – liquid junctions); the IQE value is 160%, but is uncorrected for reflection and absorption outside the QDs, and is twice the QY value below the MEG threshold photon energy.

The PbSe NC film device was fabricated by depositing a 60–300 nm-thick film of monodisperse, spheroidal PbSe NCs onto patterned indium tin oxide (ITO) coated glass using a layer-by-layer dip coating method, followed by evaporation of a top metal contact. In this LbL method [93], a layer of NCs is deposited onto the ITO surface by dip coating from a hexane solution and then washed in 0.01 M 1,2-ethanedithiol (EDT) in acetonitrile to remove the electrically insulating oleate ligands that originally solubilizes the NCs. Large-area, crack-free and mildly conductive ($\sigma = 5 \cdot 10^{-5}$ S·cm^{-1}) NC films result. The NCs pack randomly in the films are partially coated in adsorbed ethanedithiolate, and show p-type DC conductivity under illumination [93]. Working devices were also fabricated from PbS and CdSe NCs, which indicates that the approach is not restricted to EDT-treated PbSe NCs and that it should be possible to improve cell efficiency by engineering the surface of the NCs to attain longer carrier diffusion lengths.

Multiple lines of evidence suggested that the photogenerated electron–hole pairs in the device are separated by a Schottky barrier at the evaporated metal contact, and also recently observed in films of PbS NCs [94].

However, changing the contact metal from gold to calcium ($|\Delta\phi_m| = 2.3$ eV) resulted in only a 0.15 V increase in V_{OC}, which suggested that the surface Fermi level is pinned and the barrier height is relatively independent of the metal. Schottky barrier formation is often due to defects formed at an interface by deposition of a metal [95]. Direct evidence for the Schottky junction was obtained by capacitance-voltage (C-V) measurements on complete cells [93]. The location of the Schottky junction was determined by comparing the EQE spectra from cells of different thickness [93].

This film concept has been further extended into heterojunction nanocrystalline p-n PV devices with a p-type PbS or PbSe QD layer acting as the main absorber, while n-type ZnO was the window layer. This was the first demonstration that the polarity of charge can be inverted and the junction be relocated to the light-incident side of the device [96]. Other groups [97] have employed ZnO to increase the Voc of the device, yet only smaller PbSe NCs with a large bandgap can effectively inject charge across the PbSe/ZnO barrier. Recent results show a much enhanced air stability of devices when employing ZnO below a PbS layer deposited in air with Au as the back electrode. In this configuration, the maximum certified (by NREL) device efficiency was 4.4% which was stable in air for months [98]. The same p-n solar cell structure using n-type PbSe and n-ZnO very recently was shown [27] to yield internal quantum efficiencies above 100% starting at a photon energy of 2.7 eV and reached 130% at a photon energy of 3.4 eV; the bandgap of the PbSe QD for these efficiency results was 0.72 eV.

7.6 Conclusions

Many classes of semiconductor quantum dots (also called nanocrystals) have been synthesized, including Groups II–VI, III–V, IV–VI, IV and their alloys, various inter-Group and intra-Groups core-shell configurations, and various nanocrystal shapes.

The application of QDs photovoltaics solar cells to enhance their conversion efficiencies is a promising and increasingly active field of research. Such cells are termed "Next Generation" or "3rd Generation" solar cells. One approach to enhance efficiency in QD-based PV cells compared to conventional bulk semiconductor-based-PV is to create efficient MEG from a large fraction of the photons in the solar spectrum. Enhanced MEG quantum yields have now been confirmed by several groups in isolated colloidal QDs of PbSe, PbS, PbTe, Si, InP, CdTe, core-shell CdSe/CdTe, and in QD solar cells based on QD arrays of PbS, PbTe, and PbSe. It is noted that MEG has also been reported in single-wall carbon nanotubes and PbSe quantum rods.

The most common initial method for determining the MEG QY has been ps to ns time-resolved spectroscopy (transient absorption, bleaching, photoluminescence, and THz); steady-state photocurrent spectroscopy measurements at solar intensities have now also been used to show MEG and the results agree with the spectroscopic measurements. Discrepancies in the literature for reported MEG QY values for PbSe and CdSe are explained by variations in the surface chemistry of the QDs and in some cases the effects of charging of QDs when photogenerated electrons or holes are trapped at the surface producing a charged QD core. After accounting for these variations and effects, MEG has been confirmed in the many QD materials discussed here, and for these materials have threshold photon energies for MEG ranging from 2.1 E_g to 3 E_g, and total QYs at 3.0 E_g, for example, ranging from 120% to 200%. With these MEG characteristics the improvement in PV power conversion efficiency is relatively minor; to achieve significant increased power conversion efficiency the MEG threshold needs to be close to 2 E_g. The threshold photon energy for MEG has been shown to be related to the MEG efficiency (number of extra excitons produced per bandgap of photon energy excitation beyond the MEG threshold energy).

Three generic types of QD solar cells that could utilize MEG to enhance conversion efficiency can be defined: (1) photoelectrodes composed of QD arrays that form either Schottky junctions with a metal layer, a hetero p–n junction with a second NC semiconductor layer, or the i-region of a p–i–n device; (2) QD-sensitized nanocrystalline TiO2 films; and (3) QDs dispersed into a multiphase mixture of electron – and hole-conducting matrices, such as C60 and hole conducting polymers (like polythiophene or MEH-PPV), respectively. Additional research and understanding is required to realize the potential of MEG to significantly enhance solar cell performance.

Acknowledgements During preparation of this manuscript, the author was supported by the Center for Advanced Solar Photophysics, an Energy Frontier Research Center funded by the U.S. Department of Energy, Office of Science, Office of Basic Energy Sciences. Vital contributions to the research described here were made by my colleagues at NREL: Matt Beard, Joey Luther, Matt Law, Justin Johnson, Mark Hanna, Randy Ellingson, Aaron Midgett, Tavi Semonin, Jianbo Gao, Qing Song, Jim Murphy, and Sasha Efros at the Naval Research Laboratory.

References

1. W. Shockley, H.J. Queisser, J. Appl. Phys. **32**(3), 510 (1961)
2. M.A. Green (ed.), *Third Generation Photovoltaics* (Bridge Printery, Sydney, 2001)
3. A.J. Nozik, M.C. Beard, J.M. Luther, M. Law, R.J. Ellingson, J.C. Johnson, Chem. Rev. **110**(11), 6873 (2010)
4. R.T. Ross, A.J. Nozik, J. Appl. Phys. **53**(5), 3813 (1982)

5. A.J. Nozik, Ann. Rev. Phys. Chem. **29**, 189 (1978)
6. M.C. Hanna, Z.H. Lu, A.J. Nozik, Future Generat. Photovoltaic Technol. **404**, 309 (1997)
7. A. Martí, A. Luque (eds.), *Next Generation Photovoltaics: High efficiency through Full Spectrum Utilization* (Institute of Physics, Bristol, 2003)
8. http://gcep.stanford.edu/research/factsheets/hotcarriersolorcell_results.html
9. M. Wolf, R. Brendel, J.H. Werner, H.J. Queisser, J. Appl. Phys. **83**(8), 4213 (1998)
10. J. Bude, K. Hess, J. Appl. Phys. **72**(8), 3554 (1992)
11. H.K. Jung, K. Taniguchi, C. Hamaguchi, J. Appl. Phys. **79**(5), 2473 (1996)
12. D. Harrison, R.A. Abram, S. Brand, J. Appl. Phys. **85**(12), 8186 (1999)
13. O. Christensen, J. Appl. Phys. **47**(2), 689 (1976)
14. A.J. Nozik, Phys. E-Low-Dimensional Syst. Nanostructures **14**(1-2), 115 (2002)
15. A.J. Nozik, Ann. Rev. Phys. Chem. **52**, 193 (2001)
16. A. Shabaev, A.L. Efros, A.J. Nozik, Nano Lett. **6**(12), 2856 (2006)
17. R.J. Ellingson, M.C. Beard, J.C. Johnson, P.R. Yu, O.I. Micic, A.J. Nozik, A. Shabaev, A.L. Efros, Nano Lett. **5**(5), 865 (2005)
18. M. Aerts, C.S.S. Sandeep, Y. Gao, T.J. Savenijie, J.M. Schins, A.J. Houtepen, S. Kinge, D.A. Siebbeles, Nano Lett. **11**, 4485–4489 (2011)
19. W.D.A.M. de Boer, M.T. Trinh, D. Timmerman, J.M. Schins, L.D.A. Siebbeles, T. Gregorkiewicz, Appl. Phys. Lett. **99**, 53126–53128 (2011)
20. B. Cho, W.K. Peters, R.J. Hill, T.L. Courtney, D.L. Jonas, Nano Lett. **10**, 2498–2505 (2011)
21. A.J. Nozik, Chem. Phys. Lett. **457**(1–3), 3 (2008)
22. M.C. Beard, A.G. Midgett, M.C. Hanna, J.M. Luther, B.K. Hughes, A.J. Nozik, Nano Lett. **10**(8), 3019 (2010)
23. J.J.H. Pijpers, R. Ulbricht, K.J. Tielrooij, A. Osherov, Y. Golan, C. Delerue, G. Allan, M. Bonn, Nat. Phys. **5**(11), 811 (2009)
24. R.D. Schaller, V.I. Klimov, Phys. Rev. Lett. **92**(18), 186601 (2004)
25. M.C. Beard, R.J. Ellingson, Laser Photon. Rev. **2**(5), 377 (2008)
26. H.W. Hillhouse, M.C. Beard, Curr. Opin. Colloid Interface Sci. **14**(4), 245 (2009)
27. O. Semonin, J.M. Luther, S. Choi, Y.U. Chen, J. Gao, A.J. Nozik, M.C. Beard, Science **334**, 1530 (2011)
28. J.B. Sambur, T. Novet, B.A. Parkinson, Science **330**, 63 (2010)
29. J.E. Murphy, M.C. Beard, A.G. Norman, S.P. Ahrenkiel, J.C. Johnson, P.R. Yu, O.I. Micic, R.J. Ellingson, A.J. Nozik, J. Am. Chem. Soc. **128**(10), 3241 (2006)
30. M.T. Trinh, A.J. Houtepen, J.M. Schins, J. Piris, L.D.A. Siebbeles, Nano Lett. **8**(7), 2112 (2008)
31. M.B. Ji, S. Park, S.T. Connor, T. Mokari, Y. Cui, K.J. Gaffney, Nano Lett. **9**(3), 1217 (2009)
32. R.D. Schaller, M.A. Petruska, V.I. Klimov, Appl. Phys. Lett. **87**(25), 253102 (2005)
33. R.D. Schaller, M. Sykora, S. Jeong, V.I. Klimov, J. Phys. Chem. B **110**, 25332 (2006)
34. R.D. Schaller, J.M. Pietryga, V.I. Klimov, Nano Lett. **7**(11), 3469 (2007)
35. M.C. Beard, K.P. Knutsen, P.R. Yu, J.M. Luther, Q. Song, W.K. Metzger, R.J. Ellingson, A.J. Nozik, Nano Lett. **7**(8), 2506 (2007)
36. S.K. Stubbs, S.J.O. Hardman, D.M. Graham, B.F. Spencer, W.R. Flavell, P. Glarvey, O. Masala, N.L. Pickett, D.J. Binks, Phys. Rev. B **81**(8) (2010)
37. Y. Kobayashi, T. Udagawa, N. Tamai, Chem. Lett. **38**(8), 830 (2009)
38. D. Gachet, A. Avidan, I. Pinkas, D. Oron, Nano Lett. **10**(1), 164 (2010)
39. G. Nair, M.G. Bawendi, Phys. Rev. B **76**(8) (2007)
40. J.J.H. Pijpers, E. Hendry, M.T.W. Milder, R. Fanciulli, J. Savolainen, J.L. Herek, D. Vanmaekelbergh, S. Ruhman, D. Mocatta, D. Oron, A. Aharoni, U. Banin, M. Bonn, J. Phys. Chem. C **112**(12), 4783 (2008)
41. M. Ben-Lulu, D. Mocatta, M. Bonn, U. Banin, S. Ruhman, Nano Lett. **8**(4), 1207 (2008)
42. G. Nair, S.M. Geyer, L.Y. Chang, M.G. Bawendi, Phys. Rev. B **78**(12) (2008)
43. M.C. Beard, A.G. Midgett, M. Law, O.E. Semonin, R.J. Ellingson, A.J. Nozik, Nano Lett. **9**(2), 836 (2009)

44. J.A. McGuire, J. Joo, J.M. Pietryga, R.D. Schaller, V.I. Klimov, Accounts Chem. Res. **41**(12), 1810 (2008)
45. J.A. McGuire, M. Sykora, J. Joo, J.M. Pietryga, V.I. Klimov, Nano Lett. **10**(6), 2049 (2010)
46. A.G. Midgett, H.W. Hillhouse, B.K. Huges, A.J. Nozik, M.C. Beard, J. Phys. Chem. C **114**(41), 17486 (2010)
47. N.M. Gabor, Z.H. Zhong, K. Bosnick, J. Park, P.L. McEuen, Science **325**, 1367 (2009)
48. S.J. Wang, M. Khafizov, X.M. Tu, M. Zheng, T.D. Krauss, Nano Lett. **10**(7), 2381 (2010)
49. P.D. Cunningham, J.E. Boercker, E.E. Foos, E.E. Lumb, A.R. Smith, J.G. Tischler, J.S. Melinger, Nano Lett. **11**, 3476-3481 (2011)
50. A.C. Bartnik, Al.L. Efros, W.K. Koh, C.B. Murray, F.W. Wise, Phys. Rev B **82**, 195313 (2010)
51. D.S. Boudreaux, F. Williams, A.J. Nozik, J. Appl. Phys. **51**(4), 2158 (1980)
52. S. Kolodinski, J.H. Werner, T. Wittchen, H.J. Queisser, Appl. Phys. Lett. **63**(17), 2405 (1993)
53. P.T. Landsberg, H. Nussbaumer, G. Willeke, J. Appl. Phys. **74**(2), 1451 (1993)
54. K. Kempa, M.J. Naughton, Z.F. Ren, A. Herczynski, T. Kirkpatrick, J. Rybczynski, Y. Gao, Appl. Phys. Lett. **95**(23), 23121 (2009)
55. W.A. Tisdale, K.J. Williams, B.A. Timp, D.J. Norris, E.S. Aydil, X.Y. Zhu, Science **328**(5985), 1543 (2010)
56. R.D.J. Miller, G. McLendon, A.J. Nozik, W. Schmickler, F. Willig, *Surface Electron Transfer Processes* (VCH Publishers, New York, 1995)
57. C.B. Murray, C.R. Kagan, M.G. Bawendi, Ann. Rev. Mater. Sci. **30**, 545 (2000)
58. M. Sugawara (ed.), *Self-Assembled InGaAs/GaAs quantum dots* (Academic, San Diego, 1999)
59. Y. Nakata, Y. Sugiyama, M. Sugawara, in Semiconductors and Semimetals, vol. 60, ed. by M. Sugawara (Academic, San Diego, 1999)
60. A. Hagfeldt, M. Grätzel, Accounts Chem. Res. **33**(5), 269 (2000)
61. J. Moser, P. Bonnote, M. Grätzel, Coord. Chem. Rev. **98**, 3183 (1994)
62. M. Grätzel, Progr. Photovoltaics **8**(1), 171 (2000)
63. A. Zaban, O.I. Mićić, B.A. Gregg, A.J. Nozik, Langmuir **14**(12), 3153 (1998)
64. R. Vogel, P. Hoyer, H. Weller, J. Phys. Chem. **98**(12), 3183 (1994)
65. H. Weller, Berichte Der Bunsen-Gesellschaft-Phys. Chem. Chem. Phys. **95**(11), 1361 (1991)
66. D. Liu, P.V. Kamat, J. Phys. Chem. **97**(41), 10769 (1993)
67. P. Hoyer, R. Könenkamp, Appl. Phys. Lett. **66**(3), 349 (1995)
68. P.V. Kamat, K. Tvrdy, D.R. Baker, J.G. Radich, Chem. Rev. **110**(11), 6664 (2010)
69. P.V. Kamat, J. Phys. Chem. C **111**(7), 2834 (2007)
70. P.V. Kamat, J. Phys. Chem. C **112**(48), 18737 (2008)
71. M. Grätzel, Accounts Chem. Res. **42**(11), 1788 (2009)
72. S. Kumar, G.D. Scholes, Microchim. Acta **160**(3), 315 (2008)
73. N.C. Greenham, X.G. Peng, A.P. Alivisatos, Phys. Rev. B **54**(24), 17628 (1996)
74. N.C. Greenham, X.G. Peng, A.P. Alivisatos, Future Generat. Photovoltaic Technol. (404), 295 (1997)
75. W.U. Huynh, X.G. Peng, A.P. Alivisatos, Adv. Mater. **11**(11), 923 (1999)
76. S. Dayal, N. Kopidakis, D.C. Olson, D.S. Ginley, G. Rumbles, Nano Lett. **10**(1), 239 (2010)
77. A.C. Arango, S.A. Carter, P.J. Brock, Appl. Phys. Lett. **74**(12), 1698 (1999)
78. B.O. Dabbousi, M.G. Bawendi, O. Onitsuka, M.F. Rubner, Appl. Phys. Lett. **66**(11), 1316 (1995)
79. V. Colvin, M. Schlamp, A.P. Alivisatos, Nature **370**, 354 (1994)
80. M.C. Schlamp, X.G. Peng, A.P. Alivisatos, J. Appl. Phys. **82**(11), 5837 (1997)
81. H. Mattoussi, L.H. Radzilowski, B.O. Dabbousi, D.E. Fogg, R.R. Schrock, E.L. Thomas, M.F. Rubner, M.G. Bawendi, J. Appl. Phys. **86**(8), 4390 (1999)
82. H. Mattoussi, L.H. Radzilowski, B.O. Dabbousi, E.L. Thomas, M.G. Bawendi, M.F. Rubner, J. Appl. Phys. **83**(12), 7965 (1998)
83. A.J. Nozik, J. Miller (eds.) Chem. Rev. **110**, 6433–6936 (2010
84. J.M. Luther, M.C. Beard, Q. Song, M. Law, R.J. Ellingson, A.J. Nozik, Nano Lett. **7**(6), 1779 (2007)

85. X.M. Jiang, R.D. Schaller, S.B. Lee, J.M. Pietryga, V.I. Klimov, A.A. Zakhidov, J. Mater. Res. **22**(8), 2204 (2007)
86. J.J.H. Pijpers, E. Hendry, M.T.W. Milder, R. Fanciulli, J. Savolainen, J.L. Herek, D. Vanmaekelbergh, S. Ruhman, D. Mocatta, D. Oron, A. Aharoni, U. Banin, M. Bonn, J. Phys. Chem. C **111**(11), 4146 (2007)
87. R.D. Schaller, M. Sykora, J.M. Pietryga, V.I. Klimov, Nano Lett. **6**, 424 (2006)
88. D.H. Cui, J. Xu, T. Zhu, G. Paradee, S. Ashok, M. Gerhold, Appl. Phys. Lett. **88**(18), 183111/1 (2006)
89. K.P. Fritz, S. Guenes, J. Luther, S. Kumar, N.S. Saricifitci, G.D. Scholes, J. Photochem. Photobiol. a-Chem. **195**(1), 39 (2008)
90. S.A. McDonald, G. Konstantatos, S.G. Zhang, P.W. Cyr, E.J.D. Klem, L. Levina, E.H. Sargent, Nat. Mater. **4**(2), 138 (2005)
91. L.M. Qi, H. Colfen, M. Antonietti, Nano Lett. **1**(2), 61 (2001)
92. A.A.R. Watt, D. Blake, J.H. Warner, E.A. Thomsen, E.L. Tavenner, H. Rubinsztein-Dunlop, P. Meredith, J. Phys. D-Appl. Phys. **38**(12), 2006 (2005)
93. J.M. Luther, M. Law, M.C. Beard, Q. Song, M.O. Reese, R.J. Ellingson, A.J. Nozik, Nano Lett. **8**(10), 3488 (2008)
94. J.P. Clifford, K.W. Johnston, L. Levina, E.H. Sargent, Appl. Phys. Lett. **91**(25) (2007)
95. R.B. Schoolar, J.D. Jensen, G.M. Black, Appl. Phys. Lett. **91**(9), 620 (1977)
96. K.S. Leschkies, T.J. Beatty, M.S. Kang, D.J. Norris, E.S. Aydil, Acs Nano **3**(11), 3638 (2009)
97. J.J. Choi, Y.F. Lim, M.B. Santiago-Berrios, M. Oh, B.R. Hyun, L.F. Sung, A.C. Bartnik, A. Goedhart, G.G. Malliaras, H.D. Abruna, F.W. Wise, T. Hanrath, Nano Lett. **9**(11), 3749 (2009)
98. J.M. Luther, J.B. Gao, M.T. Lloyd, O.E. Semonin, M.C. Beard, A.J. Nozik, Adv. Mater. **22**(33), 3704 (2010)

Chapter 8
Fundamentals of Intermediate Band Solar Cells

Antonio Martí and Antonio Luque

Abstract Intermediate band solar cells aim to exploit the energy of below bandgap energy photons. They are based on materials that are characterised by the existence of an additional electronic band (intermediate band) located in between the conduction and valence band. An optimised IBSC has near the same limiting efficiency potential (63.2%) than a triple junction solar cells but without requiring tunnel junctions to connect the single gap solar cells. This chapter reviews its fundamental theory and introduces the different approaches that are being followed towards its implementation: quantum dots, the insertion of suitable impurities into a semiconductor host at sufficiently high densities (bulk approach) and the molecular approaches.

8.1 Introduction

Figure 8.1 sketches the basic structure of and intermediate band solar cell (IBSC). It consists of the so-called intermediate band material sandwiched between conventional p- and n-type semiconductors (emitters). Later on in this chapter we will describe how the IBSC is capable of preserving the *output voltage of the cell*. In this respect, we advance that the emitters, by preventing the IB region from contacting the metallic external contacts, allows for this preservation.

The intermediate band material is characterised by the existence of a set of energy levels located inside the semiconductor bandgap and separated from the conduction band (CB) and valence band (VB) by a null density of states. We designate this collection of energy levels as *band* to emphasise that we want them

A. Martí (✉)
Instituto de Energía Solar, Universidad Politécnica de Madrid, Avenida de la Complutense 30, 28040 Madrid, Spain
e-mail: antonio.marti@upm.es

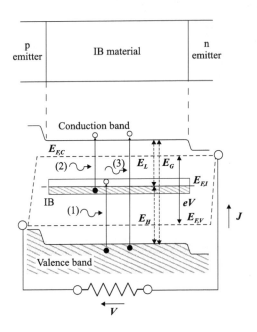

Fig. 8.1 General structure of an intermediate and solar cell (Reprinted with permission from [4]. Copyright 2008. Elsevier)

to behave differently from conventional *deep centres*. This differentiation mainly relies on the fact that, while deep centres introduce non-radiative recombination in the semiconductor, an ideal *band* only introduces radiative recombination (and, perhaps, Auger recombination). This immediately raises the question about how a collection of deep centres can be turned into a *band*. The answer to this question has been provided by Luque et al. from fundamental basis in [1]. Basically, a collection of deep centres will evolve into a band when its density is sufficiently high (typically beyond the Mott transition, $6 \times 10^{19}\,\mathrm{cm}^{-3}$, under some standard conditions) and the electron wavefunction becomes extended. Under this fundamental view, any impurity responsible of a deep centre in a conventional semiconductor can produce an intermediate band if it can be diluted into it at this high concentration without producing additional defects or clusters. Experimental results confirming these predictions have been recently obtained in silicon samples implanted with Ti [2]. Notice that, traditionally, it has been considered that the formation of an electronic band as a consequence of the occurrence of the Mott transition only had the implication on the fact of allowing carrier transport through the band; after the analysis in [1], we also accept that the formation of the band inhibits non-radiative recombination through the deep centre. In this way, recombination to and from the IB becomes no different than recombination between standard conduction and valence bands in semiconductors.

The intermediate band (IB) splits the semiconductor bandgap, E_G, into two subbandgaps: E_L and E_H. E_L denotes the lowest bandgap without implying necessarily it is located above or below the IB. E_H denotes the highest bandgap. For simplification purposes, the reference level for determining these gaps is usually taken at the middle of the IB. The IB allows two below bandgap energy photons creating a single electron-hole pair as follows: one below bandgap energy photon (photon 1) pumps an electron from the VB to the IB and a second one (photon 2) pumps an electron from the IB to the CB. For these two process being possible, it is necessary that the IB is partially filled with electrons so it can host both empty states to receive electrons from the VB as filled states to supply electrons to the CB. This semifilling can be achieved: (a) naturally, because the IB material intrinsically exhibits a metallic IB (several examples are given later in this chapter); (b) by compensating doping (that is, by adding acceptors to a material that naturally exhibits a completely filled band or donors to a material that naturally exhibits an empty band) and (c) by the operating conditions of the cell (for example, under strong illumination [3]). In addition to these below bandgap absorption, the conventional absorption of photons (photon 3) by means of transitions from the VB to the CB exists.

The absorption of below bandgap photons described above allows increasing the photogenerated current of the cell when compared to the photocurrent that would be generated by a single gap cell without IB. However, increasing the photocurrent of a solar cell is not sufficient to increase its efficiency if its output voltage is not preserved. In the case of the IBSC, this "voltage preservation" means that, although the IB introduces additional recombination (even if this is radiative), the trade-off between increased photo-current vs voltage loss becomes favourable. In fact, in the case of the IBSC, this trade-off can become so favourable that allows increasing the limiting efficiency from 40.7% (single gap solar cells) to 63.2% (both figures at maximum light concentration) [5, 6]. This last figure is similar to the limiting efficiency of a triple junction solar cell and is obtained for optimised bandgaps of $E_L = 0.71$ eV, $E_H = 1.24$ eV and $E_G = 1.95$ eV (Fig. 8.2). It is worthwhile mentioning, for example, that, operating at one sun, the increment in photocurrent produced by introducing an IB in a semiconductor host with $E_G < 1.14$ eV would not compensate the voltage loss even if the recombination introduced by the IB is only radiative [7]. The IBSC limiting efficiencies have been reviewed recently [8] and it has been found that, as the light concentration decreases, the insertion of selective reflectors increases the limiting efficiency from those previously calculated.

In the IBSC, due to the existence of a null density of states between the CB and the IB and between the IB and the VB, carrier relaxation within bands is assumed to be a much faster process than carrier relaxation between bands. Therefore, each band has associated its own quasi-Fermi level ($E_{F,C}$, $E_{F,I}$ and $E_{F,V}$ for the CB, IB and VB, respectively). The output voltage of the cell, V, is given by the split between electron and hole quasi-Fermi levels:

$$eV = E_{F,C} - E_{F,V} \qquad (8.1)$$

being e the electron charge. The highly doped p emitter determines the hole quasi-Fermi level and the highly doped n emitter, the electron quasi-Fermi level. From the plot in Fig. 8.1, it is easily understood that the output voltage of the IBSC is limited

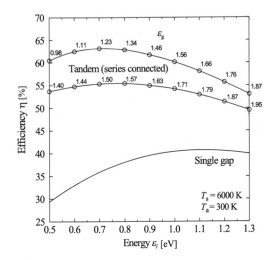

Fig. 8.2 Limiting efficiency of an IBSC as a function of its bandgaps. The value of the low bandgap, E_L (ϵ_1 in the plot), is given in the x-axis and the value of the total bandgap, E_G (ϵ_g in the plot), is marked on the plot. The efficiency limit is compared to that of a single gap solar cell and a tandem of two solar cells connected in series. In this case, the figure on the plot indicates the bandgap of the top cell and the figure in the x-axis, the value of bandgap of the bottom cell (Reprinted with permission from [5]. Copyright 1997. American Physical Society)

by the total bandgap of the cell, E_G, and not by the sub-bandgaps (E_L nor E_H). It is in this context that we mean that the IBSC has the potential to *preserve the output voltage* of the cell.

It is worthwhile mentioning that the emitters have not to be necessarily made of the same semiconductor than the material that hosts the intermediate band although, probably, should be lattice matched to it to avoid the introduction of non-radiative recombination. Actually, the emitters can have a larger bandgap (they can limit the output voltage of the cell if made with a semiconductor with lower bandgap). Effectively doping these large bandgap emitters, however, can become challenging [9].

The idea of using intermediate energy level to boost the efficiency of a solar cell was first considered by Wolf [10] in 1961. The approach was apparently disregarded until in 1997 Luque and Martí [5] calculated its limiting efficiency using detailed balance arguments and proposed [6] the use of single gap emitters to achieve the necessary electron and hole quasi-Fermi level split in order to preserve the output voltage. By now, the theory has been reviewed several times [11, 12], and it is not our intention to review it again exhaustively in this chapter but to provide a self-contained guide to the other chapters in this book that have been written in relation to this concept and provide some highlights of the international research about the concept.

8.2 Intermediate Band Solar Cell Model

The ideal performance of the IBSC is calculated according to detailed balance arguments. Taking again Fig. 8.1 as reference, this means, for example, that the total number of electrons, G_{CI} that per unit of area and time are pumped from the intermediate band to the conduction band is given by:

$$G_{CI} = H_S \int_{E_L}^{E_H} b(\epsilon, T_S, 0)\, d\epsilon + (H_C - H_S) \int_{E_L}^{E_H} b(\epsilon, T_C, 0)\, d\epsilon, \quad (8.2)$$

where H_S is the étendue [13] of the sun per unit of cell area, including the possible use of a concentrator ($H_S = \pi$ at maximum concentration), H_C is the étendue per unit of cell area with which photons can enter into the cell ($H_C > H_S$) and

$$b(\epsilon, T, \mu) = \frac{2}{h^3 c^2} \frac{\epsilon^2}{\exp\left(\frac{\epsilon-\mu}{kT}\right) - 1}. \quad (8.3)$$

Within this definition, $b(\epsilon, T_S, 0)$ corresponds to the spectral photon flux from the sun per solid angle when assumed as a black body at temperature T_S and $b(\epsilon, T_C, 0)$ are the thermal photons received from the ambient assumed at the cell temperature T_C. The use of (8.2) assumes that total photon absorption exists for photons with energy between E_L and E_H.

Detailed balance arguments [5] imply that, associated with the generation process described by (8.2), a radiative recombination process between the CB and the IB, at a rate of R_{CI} electrons per second and unit of area, also exists being given by:

$$R_{CI} = H_C \int_{E_L}^{E_H} b(\epsilon, T_C, \mu_{CI})\, dx \quad (8.4)$$

being $\mu_{CI} = E_{F,C} - E_{F,I}$ the split of quasi-Fermi levels between the conduction band and the intermediate band. The use of 8.2 and 8.4 also implies *photon selectivity*, which means that each photon can only be absorbed be one kind of transition (in the case being illustrated here, if a photon can be absorbed through a transition from the IB to the CB, it cannot be absorbed by means of a transition from the VB to the IB nor from the VB to the CB). Readers interested in a model for the IBSC that avoids photon selectivity can consult [14]. Similar equations, with similar notation, hold for transitions from the VB to the IB and from the VB to the CB so that the current voltage ($J(V)$) characteristic of the cell is given by:

$$J(V) = e\left(G_{CI} + G_{CV} - R_{CI} - R_{CV}\right), \quad (8.5)$$

where the following conditions have to be satisfied:

$$G_{CI} - R_{CI} = G_{IV} - R_{IV} \qquad (8.6)$$

$$eV = \mu_{CI} + \mu_{IV} = \mu_{CV}. \qquad (8.7)$$

The first equation means that no current is extracted from the IB while the second simply states the relationship between external voltage and electrochemical potentials.

A few comments are in order before finishing this section. First, it must be realised that the operation of the IBSC does not imply the simultaneous concurrence of two photons in order to pump an electron from the VB to the CB. In other words, the electron that is pumped from the CB to the IB has not to be the same one being pumped from the IB to the CB.

It is clear that there is a limitation for the maximum energy bandwidth that the IB can have (for example, it cannot extend itself to the conduction and valence band since then there would not be IB). The main fundamental limitation to this bandwidth comes from the appearance of stimulated emission. In this respect, it has to be fulfilled that the distance between band edges from the IB and the CB ($CB_L - IB_H$, see Fig. 8.3) and from the IB and the VB ($IB_L - VB_H$) have to be higher than the corresponding quasi-Fermi level splits ($CB_L - IB_L > E_{F,C} - E_{F,I}$ and $IB_L - BV_H > E_{F,I} - E_{F,V}$). The maximum IB energy bandwidth becomes then dependent on the injection level. In this respect, for example, it has been calculated [15] that the optimum IBSC can tolerate an IB bandwidth of ≈ 100 meV for operation at 1,000 suns and ≈ 700 meV for operation at 1 sun.

In the model described above, the quasi-Fermi levels $E_{F,C}$ and $E_{F,V}$ are allowed to shift but the IB quasi-Fermi level, $E_{F,I}$ is assumed to be fixed at its equilibrium position. The physical background for this assumption is to consider that the density of states at the IB is sufficiently high as to provide sufficient charge with an

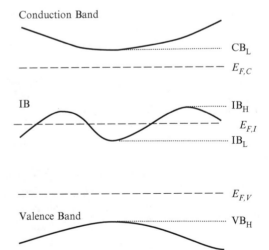

Fig. 8.3 Drawing showing the upper and lower limits of the bands together with the quasi-Fermi levels

infinitesimal displacement. In practical terms [16], this means that for operation at 1,000 suns the density of states at the IB can be one order of magnitude lower than the density of states at the CB and VB. A model has been developed [17] that allows shifting also the IB quasi-Fermi level.

As mentioned, there is no current extracted from the IB. This means that, in principle, carriers at the IB have not necessarily to have a high mobility what is compatible with an IB with a thin bandwidth (a thin IB has associated a high effective mass and, therefore, low mobility). Nevertheless, it has been advanced [18] that some mobility in the IB might be required in practical devices, in particular when carrier generation through the IB is not uniform. This would be, for example, the case in which the absorption associated with the VB → IB transition is stronger than the one associated to the IB → CB transition. In this case, the IB should transport carriers from the top part of the cell (where VB → IB would mainly take place) to the rear part of the device (where IB → CB) transitions take place.

The limiting efficiency of the IBSC has been calculated assuming ideal photon selectivity. It has been shown [14] that when no ideal photon selectivity exists, the limiting efficiency can be recovered as the absorption coefficient associated with the VB → CB transition is much stronger than the one associated with the VB → IB transition and this, indeed, is much stronger than the one associated to the IB → CB transition. If the IBSC is implemented with a material in which the absorption coefficients have such differences in strength, it is foreseen [19] that light management architectures will have to be used in order to confine the light that is absorbed the weakest. Among the concepts that are being studied with this application in mind they are the use of diffracting grids [20, 21] and plasmon resonance [22, 23] (Fig. 8.4). In this respect, Chap. 5 in this book expands the plasmon theory.

Along the following sections, several systems will be identified as intermediate band material candidates. The systems can be grouped in *quantum dot* systems and *bulk systems*. In all of them, the term *intermediate band* will be used indistinguishably. However, the nature of the IB is different in one system from the other. Hence, in the QD case, the IB arises from the energy levels associated with the quantum confinement of the electrons, typically in the CB. However, it is not necessary (in an ideal system) that this confined energy levels form an actual *band* in the sense of being connected from one dot to the next by tunnelling (although it might be a desirable property according to the discussion above related to the non-uniform generation of light). In this regard, these states in individual quantum dots might be already sufficiently extended as to inhibit non-radiative recombination without requiring connection with neighbouring quantum dots. In bulk systems, however, the formation of a *band* in the sense of allowing electrical conductivity through the band might be necessary. This is so because, conduction through the IB, when formed through the insertion of some kind impurities, is associated with the occurrence of a Mott transition and, according to our discussion in [1], the occurrence of a Mott transition is also related to the inhibition of non-radiative recombination, which is the actual property for the band seek under the IBSC approach.

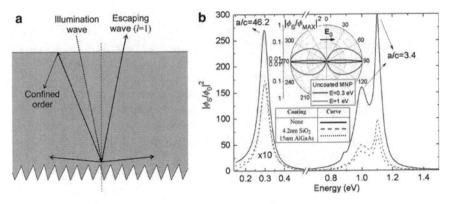

Fig. 8.4 (a) Illustration of the concept of using diffracting grids to increase the optical path of those photons that are absorbed weakly (Reprinted with permission from [20]. Copyright 2008. American Institute of Physics). (b) Illustration of the aspect ratio (a/c) of a silver nanoparticle that would be required to enhance the near field in the range of 0.3 or 1.1 eV. These are the optical ranges, for example, where the subbandgap absorption of an InAs/GaAs quantum dot IBSC are located (Reprinted with permission from [23]. Copyright 2009. American Institute of Physics)

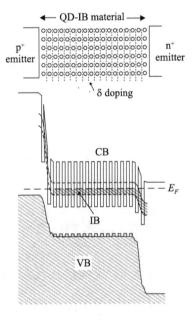

Fig. 8.5 Intermediate band solar cell implemented with quantum dots (Reprinted with permission from [27]. Copyright 2006. Elsevier).

8.3 Quantum Dot Intermediate Band Solar Cells

Figure 8.5 illustrates an IBSC implemented with quantum dots (QD-IBSC). In this cell, conventional p- and n-type emitters sandwich an array of QDs where the IB arises from the confined energy levels of the electrons in the conduction band [24]. The electron confinement is due to existence of a conduction band offset between the dot and barrier material. We have chosen the conduction band to illustrate the formation of the IB because, typically, electrons in the CB have a lower effective mass than holes what facilitates a higher separation between energy levels and, therefore, the formation of an IB separated from the conduction band by a null density of states.

Quantum dots and not wells are preferred to implement the IBSC because, on the one hand, quantum wells do not provide a true null density of states between the intermediate band and the conduction band but it is actually a continuum. On the other hand, what it would be the absorption of light from the IB to the CB is forbidden in a quantum well for front illumination [25]. The reader interested in a first approach to the design constrains of an IBSC implemented with quantum dots can consult [26].

InAs quantum dots in GaAs barriers were the first IBSC prototypes where the principles of operation of the IBSC, such as the existence of three separated quasi-Fermi levels [28] and the absorption of two below bandgap energy photons could be tested [29]. Chapter 10 in this book gives more details about the growth process of these structures.

Several groups in the world have manufactured by now QD-IBSCs [30–35]. Their work shows results in common that allow extracting the following lessons related to the practical implementation of the QD-IBSC [36].

- The QD-IBSC prototypes have not exceed the limiting efficiency of their GaAs reference counterparts (cells with nearly identical structure but without quantum dots). In this respect, Fig. 8.6 shows a typical current-voltage characteristic of a QD-IBSC compared with a GaAs reference. Results have been obtained by Hubbard et al. from the Rochester Institute of Technology and NASA Glenn Research Centre [31]. On the one hand, this result is expected since, even in the ideal case, in an IBSC implemented with GaAs as host material, when operated at one sun, the current gain does not significantly exceed the voltage loss [4]. In fact, the voltage loss observed is similar to what the ideal model predicts when only ideal radiative recombination with unity radiation emittance exists. However, in this case, it must be due to undesirable non-radiative recombination since, if it were radiative, it should be accompanied by a significant increase in photocurrent that the ideal model for unity absorbance (as required by the Kirchhoff law when the emittance is one) also predicts. Either way, the result is very good and demonstrates that strain balanced layers of QDs introduce very small non-radiative recombination.

  Considered globally, the main reason for the general lack of efficiency enhancement in the different QD-IBSCs manufactured, as will be emphasised next, is likely the poor performance of the IB → CB transition both in terms of absorbing photons as to issue a clear bandgap between the IB and the CB. As mentioned, QD-IBSCs can be accompanied by additional non-radiative recombination provided by the host structure [37], particularly stronger when strain uncompensated layers are grown [38].

- The quantum efficiency of the prototypes manufactured show production of photocurrent for below bandgap energy photons what, at first sight, could be

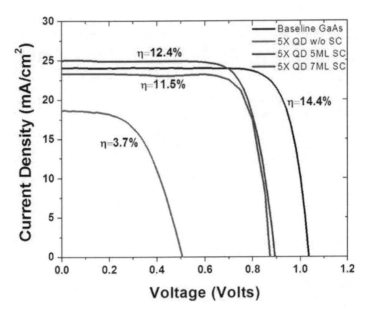

Fig. 8.6 Current-voltage characteristics of QD-IBSC (with different number of QD layers and with and without GaP strain compensating layers) compared with a GaAs reference (baseline, 14.4%) (Reprinted with permission from [39]. Copyright 2008. IEEE). The 3.7% efficiency curve corresponds to the cell without strain compensating layers and the 11.5% and 12.4% curves to the cells containing 7 and 5 monolayers (ML), respectively

regarded as an evident success of the theory (see Fig. 8.7). However, most of the quantum efficiency measurements have been carried out (we assume) illuminating the devices with monochromatic light (that is, using a single light source) and measuring the output the current. If this is the case, notice from the ideal theory that no sub-bandgap current should be produced (since the second photon is missed and also assuming that, due to the existence of ideal photon selectivity, a photon that pumps an electron from the VB to the IB is not able to pump an electron from the IB to the CB). A careful analysis reveals that the extraction of photocurrent for below bandgap energy photons without the concurrence of the second photon is still possible if, instead, the electron is thermally pumped from the IB to the CB [12]. This certainly produces cells with increased photocurrent (although small due to the poor QD light absorption) but cannot provide an increment in efficiency due to thermodynamic reasons [40]. Essentially, when thermally pumped, electrons take their energy from phonons at the lattice temperature but since these have no electrochemical energy, no work (and therefore, efficiency) can be gained from them. A different case would be that in which the electron in the IB would gain its energy from an electron recombining from the IB to the VB (impact ionisation). Under this circumstance, efficiency enhancement over single gap solar cells is possible [41]. In fact, it has been experimentally proven that, when thermal and tunnel scape are effectively blockade (by decreasing the temperature and increasing the barrier thickness)

8 Fundamentals of Intermediate Band Solar Cells

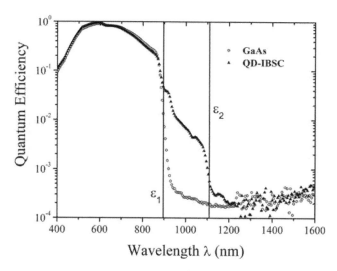

Fig. 8.7 Representative quantum efficiency of a QD-IBSC compared with a GaAs reference cell (Reprinted with permission from [17]. Copyright 2006. American Institute of Physics)

open-circuit voltage close to the total bandgap E_G are obtained [42] (see also Sect. 10.6 in Chap. 10).

- Absorption provided by the quantum dots, in particular to what it refers to the IB → CB transitions is weak. It is this weakness precisely what prevents, in another field, the achievement of efficient quantum dot infrared photodetectors and increases the need for achieving efficient light management techniques that increase the absorption of light in this energy range. It is foreseen that success in enhancement of the intraband IB → CB absorption in QDs will lead also to a new generation of QD infrared photodetectors. Actually, some recent results in this field have begun to be analysed from the perspective of IBSC operation [43].
- In spite of using QDs, the bandgap between the IB and the CB is not free of additional states corresponding to excited states [44]. Chapter 9 in this book shows detailed calculations of the states appearing in between the IB and the CB. The presence of additional levels introduce additional paths for recombination [37]. The use of concentrated light has been predicted as a means to minimise the impact on the performance of the cell of these additional levels [4] but no clear experimental confirmation in this respect has been obtained yet.

8.4 Thin-Film Intermediate Band Solar Cells

The insertion of transition elements in I–III–VI$_2$ compounds has been predicted to produce an intermediate band [45]. Figure 8.8 summarises the limiting efficiency of thin-film intermediate band solar cells [7] candidates where the position of the

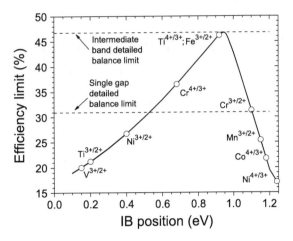

Fig. 8.8 Limiting efficiency of intermediate and solar cells implemented on CuGaS$_2$ on the basis of the insertion of several transition elements (Reprinted with permission from [7]. Copyright 2008. American Institute of Physics)

IB has been calculated on the basis of an elemental model that assumes that the position of the energy level introduced by the transition element is independent of the host when referred to the vacuum level [9] (the model has been proven to apply to III–V and II–VI compounds and was extrapolated to I–III–VI$_2$'s). Related to these predictions, researchers at the Helmholtz Centrum Berlin (HZB) [46] have inserted Ti in CuInS$_2$ as a first step towards its insertion in a more optimum system as it is CuGaS$_2$. Their results have shown an efficiency enhancement when compared with the host cell without Ti. However, this enhancement cannot be attributed to the existence of an intermediate band but perhaps to an improved (an unexpected) passivation of the grain boundaries. Thin-film intermediate band solar cells (TF-IBSC) will be discussed in detail in Chap. 11 of this book.

8.5 InGaN Intermediate Band Solar Cells

As mentioned, one hint for the implementation of bulk IBSCs goes through the identification of suitable impurities that, when inserted at low concentration, introduce deep centres at the appropriate position into the semiconductor. However, for the practical implementation of this approach, these impurities have to be inserted at sufficiently high concentration as to exceed Mott's transition [1] without forming clusters or introducing additional defects. Impurity concentrations beyond their solubility limit can be achieved by using non-equilibrium annealing techniques such as laser pulsed melting (PLM). In this respect, in Chap. 13, the experimental work related to the insertion of Ti in silicon by ionic implantation with the aim of creating an IB, and the subsequent annealing using PLM will be described.

However, it would be ideal if one could identify elements that introduce a deep centre at low concentrations but that also can be incorporated into the semiconductor host at sufficiently high concentrations without the need of using non-equilibrium

Fig. 8.9 Estimated location of Mn acceptor level as a function of the In content in $In_{1-x}Ga_xN$ (Reprinted with permission from [47]. Copyright 2009. Elsevier)

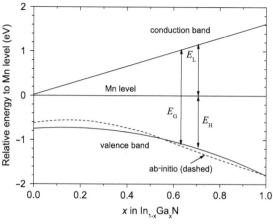

Fig. 8.10 Limiting efficiency of intermediate and solar cells implemented on $In_{1-x}Ga_xN$:Mn. Calculations are for maximum concentration for AM1.5 Direct spectrum and for the sun assumed as a blackbody at 6,000 K (Reprinted with permission from [47]. Copyright 2009. Elsevier)

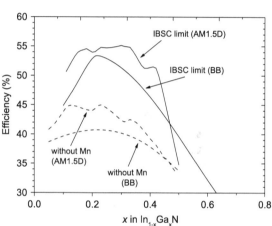

techniques. In this respect, and motivated by the high solubility of Mn in GaAs and the fact that this element introduces a deep centre in GaN (Fig. 8.9), it has been proposed [47] that InGaN:Mn can lead to a suitable IBSC (Fig. 8.10). Chapter 12 in this book discusses further this approach.

8.6 Highly Mismatched Alloys

Highly mismatched alloys are semiconductors in which the insertion of a small amount of an element of high electronegativity produces a strong modification of its fundamental properties. Among this properties we count a significant modification of their bandgap structure. The effect is due [48] to the interaction between localised states associated with the high electronegativity atoms and the extended states of the host semiconductor matrix and is described within the theory framework known as

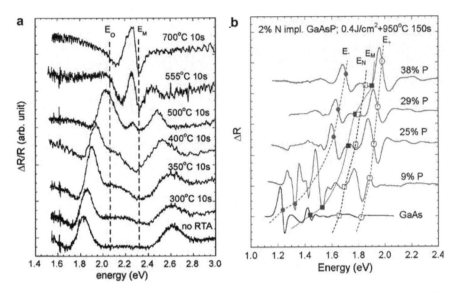

Fig. 8.11 Examples of formation of an intermediate band in (**a**) $Zn_{0.88}Mn_{0.12}Te$ implanted with O (Reprinted with permission from [48]. Copyright 2003. American Physical Society); and (**b**) $GaAs_{1-y}P_y$ for different y contents as measured using photoreflectance techniques (Reprinted with permission from [50]. Copyright 2006. American Institute of Physics)

band anticrossing [49]. As a result, the conduction band splits into two subbands of different parabolicity what can lead to the formation of a band separated from the conduction band and, therefore, to the formation of an IB. Examples reported are diluted II–VI semiconductors [48] and GaNAsP quaternary alloys [50]. In these systems, the IB formation was first detected using photoreflectance techniques (Fig. 8.11). Finally, researches at the Lawrence Berkeley National Laboratory, Rose Street Labs Energy and Sumika Electronic Materials have reported an operation IBSC based on GaNAs [51]. [52] has studied the potential of several II–VI systems with N incorporated as intermediate band solar cell candidates.

8.7 ZnTe:O Intermediate Band Solar Cells

Solar cells manufactured by the insertion of O in ZnTe, grown by molecular beam epitaxy (MBE) by researchers at the University of Michigan, [53, 54] constitute the first case of manufactured bulk intermediate band solar cell (although with an efficiency below 1%). The ZnTe:O cell shows, not only the production of subbandgap photocurrent but also this current is produced as a consequence of the incidence of two photons (Fig. 8.12) and is measured at room temperature. This system is also a particular case of highly mismatched alloy.

8 Fundamentals of Intermediate Band Solar Cells

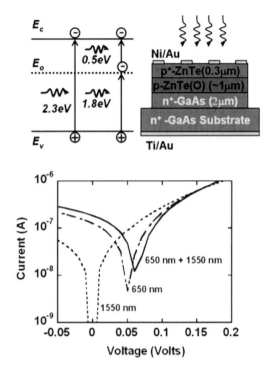

Fig. 8.12 (*Up*) structure and bandgap diagram of a ZnTe:O cell. (*Bottom*) current-voltage characteristics for illumination with single monochromatic beams and with two simultaneous monochromatic beams able to pump electrons from the VB to the IB and from the IB to the CB (Reprinted with permission from [53]. Copyright 2009. American Institute of Physics)

8.8 Molecular Approach

Daukes and Schmidt have proposed a molecular approach to the IBSC [55] (Fig. 8.13). Under this approach, two sensitisers would collect below bandgap energy photons through singlet ground → singlet excited transitions. The two singlet excited states would then transit to two triplets and then, through a triplet-triplet annihilation, the energy would be transferred to an emitter electrically connected to the external load. The absorption spectrum of the sensitisers can be tailored to satisfy the photon selectivity conditions required by the intermediate band solar cell. These authors have also suggested that this approach would be compatible with dye sensitised solar cells [56].

8.9 In$_2$S$_3$:V Intermediate Band Material

Vanadium inserted in In$_2$S$_3$ is the first example of intermediate band material, first predicted by ab-initio calculations [57] and later synthesised (in powder form). The experimental evidence of the existence of an intermediate band comes from light absorption measurements where transitions are observed at the energy thresholds predicted by the theory [58] (Fig. 8.14).

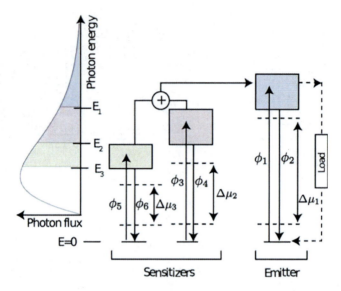

Fig. 8.13 Illustration of the molecular approach to the IBSC (Reprinted with permission from [55]. Copyright 2008. American Institute of Physics). The transitions marked with ϕ indicate the absorption (*up*) recombination processes (*down*)

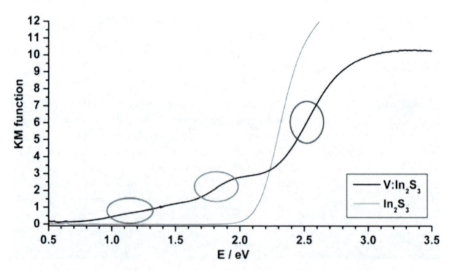

Fig. 8.14 Absorption related plot of V:In$_2$S$_3$ compared with a In$_2$S$_3$ reference without V inserted (Reprinted with permission from [58]. Copyright 2008. American Chemical Society). The *circles* represent, from *left* to *right*, absorption related to the IB → CB, VB → IB and VB → CB transitions, respectively

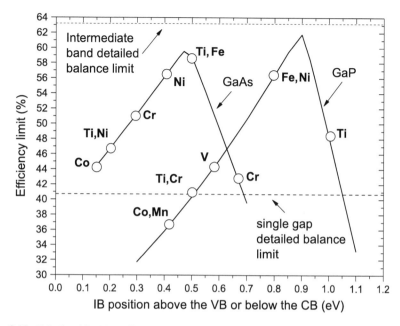

Fig. 8.15 Calculated limiting efficiency of an intermediate band solar cell made from the insertion of several transition elements in GaAs and GaP semiconductor hots (Reproduced with permission from [61]. Copyright 2009. IEEE)

8.10 Other Intermediate Band Solar Cell Systems

Several other intermediate band materials have been suggested as result of some kind of ab-initio calculations, as for example, those based on the insertion of Ti and Cr in GaAs or GaP [59,60]. In this respect, Fig. 8.15 shows the efficiency predictions [61] when the IB is assumed to arise from the deep centre associated with the corresponding transition metal assuming the impurity could be incorporated into the semiconductor at concentrations exceeding the Mott transition. Other systems have been predicted by the insertion of Ti and V in $MgIn_2S_4$ and In_2S_3 (Sect. 8.9) [57], Fe in AlP [62], Cr in ZnS and ZnTe [63].

8.11 Conclusions

Intermediate band solar cells are solar cells based on materials that have an intermediate electronic band located between the conduction and valence band. They have a limiting efficiency that exceeds that of a single gap solar cells due basically to the fact that they allow the absorption of below bandgap energy photons and the subsequent production of photocurrent with an output voltage still limited by the total bandgap. Several strategies have been proposed to take the concept to practice. Among them, InAs/GaAs quantum dots and ZnTe:O have been able to produce actual cells in which the absorption of two below bandgap energy photons has been demonstrated. In InAs/GaAs QDs, open-circuit voltages close to the semiconductor bandgap have been measured (at low temperature) showing that output voltage can be preserved in IBSCs. In QDs, the poor light absorption associated with the IB $\to$ CB transition, together with the presence of additional energy levels in between these two bands, is pointed out as the main difficulty to achieve actual high efficiency devices. Several other systems (diluted semiconductor oxides, GaNAsP quaternary alloys and In_2S_3:V) have synthesised intermediate band materials although, apparently, the lack of suitable emitters or their proper synthesis in crystalline form has prevented the manufacturing of actual solar cells. In addition, many other systems (for example, InGaN:Mn, molecular IBSC, AlP:Fe, GaAs or GaP with transition elements inserted, ZnS and ZnTe with Cr inserted) have been theoretically predicted.

Acknowledgements Authors acknowledge the financial support of the European Project IBPOWER (211640) to carry out their research on intermediate band solar cells.

References

1. A. Luque, A. Martí, E. Antolín, C. Tablero, Phy. B-Condens. Matter **382**(1–2), 320 (2006). doi: 10.1016/j.physb.2006.03.006
2. E. Antolín, A. Martí, J. Olea, D. Pastor, G. González-Díaz, I. Martil, A. Luque, Appl. Phys. Lett. **94**(4), 042115 (2009)
3. R. Strandberg, T.W. Reenaas, J. Appl. Phys. **105**(12), 124512 (2009). doi: 10.1063/1.3153141
4. A. Martí, E. Antolín, E. Cánovas, N. López, P. Linares, A. Luque, C. Stanley, C. Farmer, Thin Solid Films **516**(20), 6716 (2008); Proceedings on Adv. Mater. and Concepts for Photovoltaics EMRS 2007 Conference, Strasbourg, France
5. A. Luque, A. Martí, Phys. Rev. Lett. **78**(26), 5014 (1997)
6. A. Luque, A. Martí, Progr. Photovoltaics **9**(2), 73 (2001)
7. A. Martí, D.F. Marrón, A. Luque, J. Appl. Phys. **103**(7), 073706 (2008). doi: 10.1063/1.2901213
8. R. Strandberg, T.W. Reenaas, Appl. Phys. Lett. **97**(3), 031910 (2010). doi: 10.1063/1.3466269. http://link.aip.org/link/?APL/97/031910/1
9. M.J. Caldas, A. Fazzio, A. Zunger, Appl. Phys. Lett. **45**(6), 671 (1984)
10. M. Wolf, Sol. Energ. **5**(3), 83 (1961)
11. A. Martí, L. Cuadra, A. Luque, in *Intermediate Band Solar Cells*, Next Generation Photovoltaics: High Efficiency through Full Spectrum Utilization, ed. by A. Marti, A. Luque. (Institute of Physics Publishing, Bristol, 2003), pp. 140–162
12. A. Martí, C.R. Stanley, A. Luque, *Intermediate Band Solar Cells (IBSC) Using Nanotechnolgy* (Elsevier, Amsterdam, 2006), chap. 17
13. R. Winston, W.T. Welford, *Optics of Non Imaging Concentrators* (Academic, San Diego, 1979)
14. L. Cuadra, A. Martí, A. Luque, IEEE Trans. Electron Dev. **51**(6), 1002 (2004). doi: 10.1109/TED.2004.828161
15. L. Cuadra, A. Martí, A. Luque, Modelling of the adsorption coefficient of the intermediate band solar cell, in *Proceedings of the 16th European Photovoltaic specialist Conference held in Glasgow*. (Taylor & Francis Group, Oxford, 2000)

16. A. Martí, L. Cuadra, A. Luque, Quantum dot analysis of the space charge region of intermediate band solar cell, in *Proceedings of the 199th Electrochemical Society Meeting*. (The Electrochemical Society, Pennington, 2001), pp. 46–60
17. A. Luque, A. Martí, N. López, E. Antolín, E. Cánovas, C.R. Stanley, C. Farmer, P. Díaz, J. Appl. Phys. **99**(1), 094503 (2006)
18. A. Martí, L. Cuadra, A. Luque, IEEE Trans. Electron Dev. **49**(9), 1632 (2002). doi: 10.1109/TED.2002.802642
19. A. Martí, E. Antolín, E. Cánovas, P.G. Linares, A. Luque, MRS Proceedings, Spring Meeting, San Francisco, vol. 1101E, pp. KK06 (2008)
20. I. Tobías, A. Luque, A. Martí, J. Appl. Phys. **104**(3), 034502 (2008)
21. A. Mellor, I. Tobías, A. Luque, A. Martí, M.J. Mendes, Upper limits to adsorption enhancement in thick solar cells using diffraction gratings, Progress in Photovoltaics: Research and Applications, Vol. 19, pp. 676–687
22. A. Luque, A. Martí, M.J. Mendes, I. Tobías, J. Appl. Phys. **104**(11), 113118 (2008)
23. M.J. Mendes, A. Luque, I. Tobías, A. Martí, Appl. Phys. Lett. **95**(7), 071105 (2009)
24. A. Martí, L. Cuadra, and A. Luque, Quantum dot intermediate band solar cells in Photovoltaic Specialists Conference, 2000. Conference Record of the Twenty-Eighth IEEE, held in Anchorage, 2000 pp. 940–943.
25. J. Loehr, M. Manasreh, *Semiconductor Quantum Wells and Superlattices for Long-Wavelength Infrared Detectors* (Artech House, Boston, 1993)
26. A. Martí, L. Cuadra, A. Luque, Physica E **14**, 150 (2002)
27. A. Martí, N. López, E. Antolín, E. Cánovas, C. Stanley, C. Farmer, L. Cuadra, A. Luque, Thin Solid Films **511–512**, 638 (2006). doi: 10.1016/j.tsf.2005.12.122. http://www.sciencedirect.com/science/article/B6TW0-4J6W6YJ-3/2/ab8103ebdf162540c13ce872c3cc1879; EMSR 2005 – Proceedings of Symposium F on Thin Film and Nanostructured Materials for Photovoltaics – EMRS 2005 – Symposium F
28. A. Luque, A. Martí, N. López, E. Antolín, E. Cánovas, C. Stanley, C. Farmer, L.J. Caballero, L. Cuadra, J.L. Balenzategui, Appl. Phys. Lett. **87**(8), 083505 (2005)
29. A. Martí, E.A. n, C.R. Stanley, C.D. Farmer, N. López, P. Díaz, E. Cánovas, P.G. Linares, A. Luque, Phys. Rev. Lett. **97**(24), 247701 (2006)
30. V. Popescu, G. Bester, M.C. Hanna, A.G. Norman, A. Zunger, Phys. Rev. B **78**, 205321 (2008)
31. S.M. Hubbard, C.D. Cress, C.G. Bailey, R.P. Raffaelle, S.G. Bailey, D.M. Wilt, Appl. Phys. Lett. **92**(12), 123512 (2008)
32. R. Oshima, A. Takata, Y. Okada, Appl. Phys. Lett. **93**(8), 083111 (2008)
33. S. Blokhin, A. Sakharov, A. Nadtochy, A. Pauysov, M. Maximov, N. Ledentsov, A. Kovsh, S. Mikhrin, V. Lantratov, S. Mintairov, N. Kaluzhniy, M. Shvarts, Phys. Semiconduct. Tech. **43**(4), 537 (2009)
34. A. Luque, A. Martí, C. Stanley, N. López, L. Cuadra, D. Zhou, J.L. Pearson, A. McKee, J. Appl. Phys. **96**(1), 903 (2004)
35. D. Guimard, R. Morihara, D. Bordel, K. Tanabe, Y. Wakayama, M. Nishioka, Y. Arakawa, Appl. Phys. Lett. **96**(20), 203507 (2010). doi: 10.1063/1.3427392. http://link.aip.org/link/?APL/96/203507/1
36. A. Luque, A. Martí, The intermediate band solar cell: progress toward the realization of an attractive concept. Adv. Mater. **22**, 160–174 (2010)
37. A. Luque, P.G. Linares, E. Antolín, E. Cánovas, C.D. Farmer, C.R. Stanley, A. Martí, Appl. Phys. Lett. **96**(1), 013501 (2010)
38. A. Martí, N. López, E. Antolín, E. Cánovas, A. Luque, C.R. Stanley, C.D. Farmer, P. Díaz, Appl. Phys. Lett. **90**(23), 233510 (2007)
39. S.M. Hubbard, C.G. Bailey, C.D. Cress, S. Polly, J. Clark, D.V. Forbes, R.P. Raffaelle, S.G. Bailey, and D.M. Wilt, Short circuit current enhancement of GaAs solar cells using strain compensated InAs quantum dots in Conference Records of the 33rd Photovoltaic Specialists Conference held in San Diego, IEEE, 2008, pp. 1–6
40. A. Luque, A. Martí, L. Cuadra, IEEE Trans. Electron Dev. **48**(9), 2118 (2001)
41. A. Luque, A. Martí, L. Cuadra, IEEE Trans. Electron Dev. **50**(2), 447 (2003)

42. E. Antolín, A. Martí, P.G. Linares, I. Ramiro, E. Hernández, C.D. Farmer, C.R. Stanley, and A. Luque, Advances in quantum dot intermediate band solar cells, in *Conference Record of 35th IEEE Photovoltaic Specialists Conference (PVSC)*, 2010, Honolu, pp. 65–70
43. J. Wu, D. Shao, Z. Li, M.O. Manasreh, V.P. Kunets, Z.M. Wang, G.J. Salamo, Appl. Phys. Lett. **95**(7), 071908 (2009)
44. S. Tomic, T.S. Jones, N.M. Harrison, Appl. Phys. Lett. **93**(26), 263105 (2008)
45. P. Palacios, K. Sánchez, J.C. Conesa, J.J. Fernández, P. Wahnón, Thin Solid Films **515**(15), 6280 (2007)
46. B. Marsen, L. Steinkopf, I. Lauermann, M. Gorgoi, H. Wilhelm, T. Unold, R. Scheer, H.W. Schock, E-MRS Spring Meeting; Symp B, Strasbourg (2009)
47. A. Martí, C. Tablero, E. Antolín, A. Luque, R.P. Campion, S.V. Novikov, C.T. Foxon, Sol. Energ. Mater. Sol. cell. **93**(5), 641 (2009). doi: 10.1016/j.solmat.2008.12.031
48. K.M. Yu, W. Walukiewicz, J. Wu, W. Shan, J.W. Beeman, M.A. Scarpulla, O.D. Dubon, P. Becla, Phys. Rev. Lett. **91**(24), 246403 (2003)
49. W. Shan, W. Walukiewicz, J.W. Ager, E.E. Haller, J.F. Geisz, D.J. Friedman, J.M. Olson, S.R. Kurtz, Phys. Rev. Lett. **82**(6), 1221 (1999); Copyright (C) 2007 The American Physical Society Please report any problems to prola@aps.org PRL
50. K.M. Yu, W. Walukiewicz, J.W. Ager, D. Bour, R. Farshchi, O.D. Dubon, S.X. Li, I.D. Sharp, E.E. Haller, Appl. Phys. Lett. **88**(9), 092110 (2006)
51. N. López, L.A. Reichertz, K.M. Yu, K. Campman, W. Walukiewicz, Phys. Rev. Lett. **106**(2), 028701 (2011). doi: 10.1103/PhysRevLett.106.028701
52. E. Cánovas, A.Martí, A.Luque, W. Walukiewicz, Appl. Phys. Lett **93**, 174109 (2008)
53. W. Wang, A.S. Lin, J.D. Phillips, Appl. Phys. Lett. **95**(1), 011103 (2009)
54. W. Wang, A.S. Lin, J.D. Phillips, W.K. Metzger, Appl. Phys. Lett. **95**(26), 261107 (2009)
55. N.J. Ekins-Daukes, T.W. Schmidt, Appl. Phys. Lett. **93**(6), 063507 (2008)
56. M. Grätzel, J. Photochem. Photobiol. C Photochem. Rev. **4**, 145 (2003)
57. P. Palacios, I. Aguilera, K. Sánchez, J.C. Conesa, P. Wahnón, Phys. Rev. Lett. **101**(4), 046403 (2008)
58. R. Lucena, I. Aguilera, P. Palacios, P. Wahnón, J.C. Conesa, Chem. Maters **20**, 5125 (2008)
59. P. Palacios, J.J.F. ndez, K. Sánchez, J.C. Conesa, P. Wahnón, Phys. Rev. B (Condens. Matter Mater. Phys.) **73**(8), 085206 (2006)
60. P. Palacios, P. Wahnón, S. Pizzinato, J.C. Conesa, J. Chem. Phys. **124**(1), 14711 (2006)
61. A. Martí et al., Proceedings of 34th IEEE PVSC (2009)
62. P. Olsson, C. Domain, J.F. Guillemoles, Phys. Rev. Lett. **102**, 227204 (2009)
63. C. Tablero, Sol. Energ. Mater. Sol. Cell. **90**(5), 588 (2006)

Chapter 9
Modelling of Quantum Dots for Intermediate Band Solar Cells

Stanko Tomić

Abstract We present a theoretical model for design and analysis of semiconductor quantum dot (QD) array-based intermediate band solar cell (IBSC). The plane wave method with periodic boundary conditions is used in expansion of the **k·p** Hamiltonian for calculation of the electronic and optical structure of InAs/GaAs QD array. Taking into account realistic QD shape, QD periodicity in the array, as well as effects like band mixing between states in the conduction and valence band, strain and piezoelectric field, the model reveals the origin of the intermediate band formation inside forbidden energy gap of the barrier material. Having established the interrelation between QD periodicity and the electronic structure across the QD array Brillouin zone, conditions are identified for the appearance of pure zero density of states regions that separate intermediate band from the rest of the conduction band. For one realistic QD array, we have estimated all important absorption spectra in IBSC, and most important, radiative and nonradiative scattering times. Under radiative limit approximation, we have estimated efficiency of such IBSC to be 39%.

9.1 Introduction

Conventional single energy gap solar cells (SC) have an ultimate efficiency limit that was established by Shockley and Queisser[1] on the basis of detailed balance arguments. The "balance" in the model comes from the fact that it quantitatively accounts for two opposing fundamental processes that occur in any SC: absorption and emission. The later process occurs by virtue of the SC's finite temperature, T, since for $T > 0K$ the material behaves like a black body and emits photons into the environment. These photons come from free carriers recombination and thus

---

S. Tomić (✉)
Joule Physics Laboratory, University of Salford, Manchester M5 4WT, UK
e-mail: s.tomic@salford.ac.uk

reduce the output current. It was established that a single band gap SC device has an ultimate theoretical efficiency of 44% and a band gap of $E_g = 2.2\,\text{eV}$ if the SC is at $T = 0\,\text{K}$.[1] This figure is simply the ratio of the total power of the photons absorbed above E_g to the total power incident on the cell, under maximum concentration (provided by isotropic illumination of the cell with the radiance corresponding to the sun's photosphere). For SC at room temperature, the maximum efficiency is further reduced to 40.7% with $E_g = 1.1\,\text{eV}$ under maximum concentration condition, and to 31% with a $E_g = 1.3\,\text{eV}$, at one sun (i.e., when the solid angle subtended by the sun shining on a cell at normal incidence is taken into account). The main reason underlying those values is that only photons with an energy close to that of the semiconductor E_g are effectively converted. Photons with lower energy than E_g are simply lost (the semiconductor is transparent to them); and out of the photons with higher energy ($> E_g$), only a part of their energy, i.e., that equals the E_g energy is at best converted. The majority of high energy electrons generated by photons with $> E_g$, (hot carriers) decay thermally to the Fermi level of the conduction band before they can contribute to the current.

The principal aim here must be to make better use of the solar spectrum[2–4]. One such improvement is to take advantage of the incident photons with sub-band gap energy to be absorbed and contribute to increase photo-current, while in the same time the output voltage of device would ideally be preserved at its maximal value that is determined by the largest E_g that exists in the system (i.e., host material energy gap). A possible solution to that problem emerged in the form of *intermediate band solar cell* (IBSC) scheme[5–7]. The limiting efficiency of the IBSC concept for full concentration and at room temperature is 63.2%, with optimized absorption energies at $\sim 1.2\,\text{eV}$, $\sim 0.7\,\text{eV}$ and $\sim 1.9\,\text{eV}$, [5] significantly overcoming the SQ limit of 40.7% for a conventional single-gap SC under the same operating conditions.

Conceptually, an IBSC is manufactured by sandwiching an intermediate band (IB) material between two selective contacts, of p and of n type (Fig. 9.1). The IB material is characterized by the existence of an electronic energy band of allowed states within the conventional energy band gap E_g of the host material, splitting it into two sub-gaps, E_{gL} and E_{gH}. This band allows the creation of additional electron-hole pairs from the absorption of two sub-band gap energy photons. Under this assumption, first photon (2) pumps an electron from the valence band (VB) to the IB, and a second photon (3) pumps an electron from the IB to the conduction band (CB). To this end, it is necessary that the IB is half-filled with electrons so that it can supply electrons to the CB as well as receive them from the VB. This two-photon absorption process is also illustrated in Fig. 9.1 and has been experimentally detected in IBSCs based on quantum dots [8]. The electron-hole pairs generated in this way add up to the conventionally generated ones by the absorption of a single photon (1), the third one, pumping an electron from the VB to the CB. Therefore, the photocurrent of the solar cell, and ultimately its efficiency, are enhanced since this increment in photocurrent occurs without degradation of the output voltage of the cell. The output voltage is given by the split between electron

9 Modelling of Quantum Dots for Intermediate Band Solar Cells

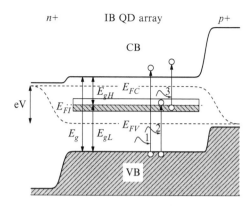

Fig. 9.1 Schematic view of the IBSC band diagram

and hole quasi-Fermi levels, E_{F_C} and E_{F_V}, that is still limited by the total band gap E_g [9–12].

In this chapter, we describe theoretical methods for design, and discuss possibility of using, semiconductor QD as an absorbing medium in IBSCs. In the QD implementation of IBSCs, the intermediate band is created inside host semiconductor (barrier) material by three-dimensional periodic array of QDs. In reality, it is rather a one-dimensional array of vertically aligned QD's[13–20]. For each electron bound energy level available inside a single QD, a periodic array of such QDs will produce a miniband. Those minibands will lie inside the forbidden energy gap of the barrier material. If there is only one bound energy level in the CB of QD, this may become the IB. With proper n-type doping, the quasi-Fermi level can be positioned within the IB. In this way, the IB can be half-full, providing coexistence of filled and empty stated in the IB, and enhancing the probability of creating photo-generated carriers.

The chapter is organized as follows: after Introduction, Sect. 9.1, in the second part, Sect. 9.2, we review theoretical methods for the calculation of electronic and optical properties of the QD array used for IBSCs. In Sect. 9.2.1, we briefly outline the $\mathbf{k} \cdot \mathbf{p}$ theory and its eight-band implementation, as a method used for quantum mechanical treatment of electronic structure of strained semiconductor QD arrays. In Sect. 9.2.2, we discuss advantages of the plane waves expansion method combined with periodic boundary conditions for the calculation of the electronic structure of QD arrays. In Sect. 9.2.3, we introduce the model QD array used throughout this chapter. In Sect. 9.2.4, we discuss the origin of intermediate band formation in QD arrays which is followed by the analysis of the electronic structure of QD arrays in Sect. 9.2.5. Results presented in Sect. 9.2.4 and Sect. 9.2.5 are used then to examine the absorption characteristics of QD arrays. In the third part of this chapter, Sect. 9.3, we discuss dominant radiative and non-radiative processes that might influence successful operation of QD array-based IBSC. It incudes theory of radiative processes in Sect. 9.3.1, theory of electron–phonon scattering processes in Sect. 9.3.2, and theory of the two most important non-radiative Auger processes:

electron cooling and bi-exitonic recombination discussed in Sect. 9.3.3. In Sect. 9.4, we give concluding remarks.

9.2 Theory of QD Array Electronic Structure

9.2.1 The k·p Theory

The quantum mechanical description of electrons in any material requires detailed knowledge of their wavefunctions, $\psi_{n,\mathbf{k}}(\mathbf{r})$, which are found by solving the Schrödinger equation (in the single-electron approximation):

$$H_0 \psi_{n,\mathbf{k}}(\mathbf{r}) = E_{n,\mathbf{k}} \psi_{n,\mathbf{k}}(\mathbf{r}). \tag{9.1}$$

The Hamiltonian in (9.1), $H_0 = p^2/2m_0 + V(\mathbf{r})$, is the function of the quantum mechanical momentum operator, $\mathbf{p} = -i\hbar\nabla$, and the crystal potential experienced by electrons, $V(\mathbf{r}) = V(\mathbf{r} + \mathbf{R})$, which is a periodic function, with the periodicity of the crystal lattice, $\mathbf{R}$. According to Bloch's theorem, the solutions to this Schrödinger equation can be written as: $\psi_{n,\mathbf{k}}(\mathbf{r}) = e^{i\mathbf{k}\cdot\mathbf{r}} u_{n,\mathbf{k}}(\mathbf{r})$ where $\mathbf{k}$ is electron wavevector, n is the band index, and $u_{n,\mathbf{k}}$ is the *cell-periodic* function, with the same periodicity as the crystal lattice. The cell-periodic function, $u_{n,\mathbf{k}}$, satisfies equation:

$$H_\mathbf{k} u_{n,\mathbf{k}} = E_{n,\mathbf{k}} u_{n,\mathbf{k}}, \tag{9.2}$$

where the Hamiltonian:

$$H_\mathbf{k} = H_0 + H'_\mathbf{k} = \frac{p^2}{2m_0} + V + \frac{\hbar^2 k^2}{2m_0} + \frac{\hbar \mathbf{k}\cdot\mathbf{p}}{m_0} \tag{9.3}$$

is given as a sum of two terms: the unperturbed, H_0, which in fact equals the exact Hamiltonian at $\mathbf{k}=0$ (i.e., at the Γ point in the Brillouin zone), and the "perturbation", $H'_\mathbf{k}$. Equation (9.3) is called the $\mathbf{k}\cdot\mathbf{p}$ Hamiltonian [21–23]. If the eigenvalues are non-degenerate, the first-order energy correction is given by $E'_{n,\mathbf{k}} \approx \langle u_{n,0}|H'_\mathbf{k}|u_{n,0}\rangle$ and there is no correction (to the first order, in the absence of non-diagonal matrix elements) in the eigenfunctions. The second order correction arises from non-diagonal terms; the energy correction is given by $E''_{n,\mathbf{k}} \approx \sum_{n'\neq n} |\langle u_{n,0}|H'_\mathbf{k}|u_{n',0}\rangle|^2/(E_{n,0} - E_{n',0})$. The "perturbation" term $H'_\mathbf{k}$ gets progressively smaller as $\mathbf{k}$ approaches zero. Therefore, the $\mathbf{k}\cdot\mathbf{p}$ perturbation theory is most accurate for small values of $\mathbf{k}$. However, if enough terms are included in the perturbative expansion, then the theory can in fact be reasonably accurate for any value of $\mathbf{k}$ in the entire Brillouin zone[24–28]. For a band n, with an extremum at $\mathbf{k} = 0$, and with no spin-orbit coupling, the result of $\mathbf{k}\cdot\mathbf{p}$ perturbation theory is (to the lowest nontrivial order):

9 Modelling of Quantum Dots for Intermediate Band Solar Cells

$$u_{n,\mathbf{k}} = u_{n,0} + \frac{\hbar}{m_0} \sum_{n' \neq n} \frac{\langle u_{n,0}|\mathbf{k} \cdot \mathbf{p}|u_{n',0}\rangle}{E_{n,0} - E_{n',0}} u_{n',0} \quad (9.4)$$

$$E_{n,\mathbf{k}} = E_{n,0} + \frac{\hbar^2 k^2}{2m_0} + \frac{\hbar^2}{m_0^2} \sum_{n' \neq n} \frac{|\langle u_{n,0}|\mathbf{k} \cdot \mathbf{p}|u_{n',0}\rangle|^2}{E_{n,0} - E_{n',0}}. \quad (9.5)$$

The parameters required to do these calculations, the band edge energies, $E_{n,0}$, and the optical matrix elements, $\langle u_{n,0}|\mathbf{p}|u_{n',0}\rangle$, are typically inferred from experimental data and detailed atomistic based theories.

A particular strength of the $\mathbf{k} \cdot \mathbf{p}$ theory is straightforward inclusion of the spin–orbit interaction and of the strain effects on the band structure via deformation potential theory[29].

Relativistic effects in the $\mathbf{k} \cdot \mathbf{p}$ method are included also perturbatively via the spin–orbit (SO) interaction Hamiltonian, $H_{SO} = (2\Delta/3\hbar^2)\mathbf{L} \cdot \mathbf{S}$, where $\mathbf{L}$ is the orbital angular momentum operator and $\mathbf{S}$ is the spin operator. Finally, the $\mathbf{k} \cdot \mathbf{p}$ Hamiltonian becomes:

$$H_{\mathbf{k}} = \frac{p^2}{2m_0} + V + \frac{\hbar^2 k^2}{2m_0} + \frac{\hbar \mathbf{k} \cdot \mathbf{p}}{m_0} + \frac{1}{4m_0^2 c^2}(\boldsymbol{\sigma} \times \nabla V) \cdot (\mathbf{k} + \mathbf{p}), \quad (9.6)$$

where $\boldsymbol{\sigma} = (\sigma_x, \sigma_y, \sigma_z)$ is a vector consisting of the three Pauli spin matrices.

In a strained system, the coordinate axes are stretched or compressed[29]. Therefore, the coordinates are transformed, assuming only effects up to the first order in strain tensor ϵ, as: $\mathbf{r}' = (1 + \epsilon)^{-1} \cdot \mathbf{r}$, where $\mathbf{r}$ is the unstrained coordinate. Consequently, the translational symmetry of the new, *strained cell-periodic functions*, $\tilde{u}_{n,\mathbf{k}'}(\mathbf{r}') = u_{n,\mathbf{k}}([1 + \epsilon]^{-1} \cdot \mathbf{r})$, is associated with the wavevector, $\mathbf{k}' = (1 + \epsilon) \cdot \mathbf{k}$. By solving the $\mathbf{k} \cdot \mathbf{p}$ equation in the new coordinate system defined by $(\mathbf{k}', \mathbf{r}')$, and using modified Bloch $\tilde{\psi}_{n,\mathbf{k}'}(\mathbf{r}')$, and cell-periodic $\tilde{u}_{n,\mathbf{k}'}(\mathbf{r}')$, functions, and neglecting all second-order effects due to strain tensor, ϵ, the strained $\mathbf{k} \cdot \mathbf{p}$ Hamiltonian becomes:

$$H_{\mathbf{k}} = \frac{p^2}{2m_0} + V + \frac{\hbar^2 k^2}{2m_0} + \frac{\hbar \mathbf{k} \cdot \mathbf{p}}{m_0} + \frac{1}{4m_0^2 c^2}(\boldsymbol{\sigma} \times \nabla V) \cdot (\mathbf{k} + \mathbf{p}) + \mathbf{D}^\epsilon \cdot \boldsymbol{\epsilon}, \quad (9.7)$$

where the $\mathbf{D}^\epsilon$ is the deformation potential operator. This Hamiltonian can be subjected to the same sort of perturbation-theory analysis as above[30].

The theoretical model of QD array electronic structure, presented in this chapter, is based on the 8-band $\mathbf{k} \cdot \mathbf{p}$ implementation[31] of the Hamiltonian (9.7) adapted for QD nanostructures[32, 33]. This Hamiltonian takes into account band mixing between the lower conduction band s-anti-bonding and top three valence band p-bonding states, all spin degenerate, in the underlying QD and barrier materials, and includes the strain and piezoelectric field.

9.2.2 Plane Waves Implementation of QD Array Electronic Structure Solver

The QD as a three-dimensional (3D) object breaks the translational symmetry of the bulk material along all three Cartesian directions implying operator replacement $k_\nu \to -i\partial/\partial\nu$ in (9.7), where $\nu = (x, y, z)$. To solve the multi-band system of Schrödinger equations, Eq.(9.7), the plane wave (PW) methodology is employed as an expansion method[34–38]. In the PW representation, the eigenvalues (E_n) and coefficients ($A_{n,\mathbf{k}}$) of the nth-eigenvector, [$\psi_n(\mathbf{r}) = \sum_\mathbf{k} A_{n,\mathbf{k}} e^{i\mathbf{k}\mathbf{r}}$], are linked by the relation

$$\sum_{m,\mathbf{k}'} h_{m,n}(\mathbf{k}', \mathbf{k}) A_{n,\mathbf{k}} = E_n \sum_\mathbf{k} A_{n,\mathbf{k}}, \tag{9.8}$$

where $h_{m,n}(\mathbf{k}', \mathbf{k})$ are the Fourier transform of the Hamiltonian matrix elements, and $m, n \in \{1, \ldots, 8\}$ are the band indexes of the 8-band $\mathbf{k} \cdot \mathbf{p}$ Hamiltonian. All the elements in the Hamiltonian matrix, (9.8), can be expressed as a linear combination of different kinetic and strain related terms and its convolution with the characteristic function of the actual QD shape, $\chi_{\text{qd}}(\mathbf{k})$[38, 39]. The whole $\mathbf{k}$-space is discretized by embedding the QD in a rectangular box of dimensions L_x, L_y, and L_z and volume $\Omega = L_x \times L_y \times L_z$ and choosing the $\mathbf{k}$-vectors in the form of $\mathbf{k} = 2\pi(n_x/L_x, n_y/L_y, n_z/L_z)$, where n_x, n_y, and n_z are integers whose change controls convergence of the method.

The PW-based $\mathbf{k} \cdot \mathbf{p}$ method inherently assumes *periodic Born-von Karman boundary conditions* and is particularly suited for analysis of the QD array structures. The electronic structure of such an array is characterized by a Brillouin zone (BZ) determined by the QD array dimensions[40–44]. To calculate the electronic structure, the only modification to the basis set is to replace the reciprocal lattice vectors in the PW expansion with those shifted due to the QD-superlattice (QD-SL):

$$k_\nu \to k_\nu + K_\nu, \tag{9.9}$$

where $0 \leq K_\nu \leq \pi/L_\nu$, and the L_ν are the super-lattice vectors in the $\nu = (x, y, z)$ directions. This allows sampling along the $\mathbf{K}$ points of a QD-SL Brillouin zone to be done at several points at the cost of the single QD calculation at each point. All the results presented in this chapter were obtained by using the kppw code[37, 38].

9.2.3 Model QD Arrays

The model QD arrays considered here consist of InAs/GaAs QDs with truncated pyramidal shape, Fig. 9.2. The size and shape of the QD is controlled by the pyramid base length, b, its height h, and truncation factor, t, defined as a ratio between

Fig. 9.2 Schematic of the model QD array (**a**), and QDs that makes array (**b**). QD truncation factor is defined as $t = a/b$, while all the other relevant parameters are defined in the main text

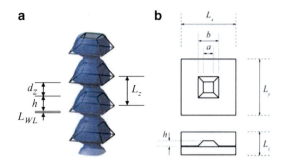

length of the pyramid side at h, and length of the pyramid base b. The QDs are embedded in the tetragonal-like unit cell, Ω. The vertical periodicity of the QD array is controlled by $L_z = d_z + h + L_{WL}$, where d_z is the variable vertical separation of the QDs in subsequent layers, i.e., the distance between the top of the QDs and the bottom of the WL in the following layer. The d_z is varied with $L_x = L_y$ chosen to be large enough to prevent lateral electronic coupling.

9.2.4 Wavefunction Delocalization and Intermediate Band Formation in QD Arrays

The origin of the intermediate band formation in the QD arrays lay in the electronic coupling of states in neighbouring QDs. When QDs are put close enough, the QD states eventually become delocalized over QDs in an array. By varying the QD size and vertical spacing between them, it is possible to control electronic properties of the QD array. In Fig. 9.3, the evolution of QD states delocalization, *"bonding"*, is shown. In Fig. 9.3a, the charge densities of the electron ground (e_0) and first excited (e_1) states in CB and the hole ground (h_0) and first excited (h_1) states in VB of a single isolated InAs/GaAs QD, with $b = 10$ nm, $h = 2.5$ nm and truncation factor $t = 0.5$, is shown. It can be seen that for such a QD all considered states are well localized inside the QD. Once QDs are put in an array structure, the wavefunctions start to delocalize. In Fig. 9.3b, c, the same size QD as in Fig. 9.3a are used, and put in an infinite periodic array with vertical spacing between the QDs of $d_z = 3$ nm and $d_z = 2$ nm, respectively. It is clear that closer packing of QDs enhances the wavefunction delocalization. This wavefunction delocalization is the main cause of *intermediate band formation* in QD array.

In analogy with solid-state physics, from Fig. 9.3b, one can quantify the character of bonding of QD states in an InAs/GaAs QD array as *covalent* rather than *metallic*. The wavefunctions are delocalized only to some degree, with charge density in the barrier region between QDs hardly exceeding 20% of its maximal value, and are far from being completely delocalized to result in nearly free electron motion in z

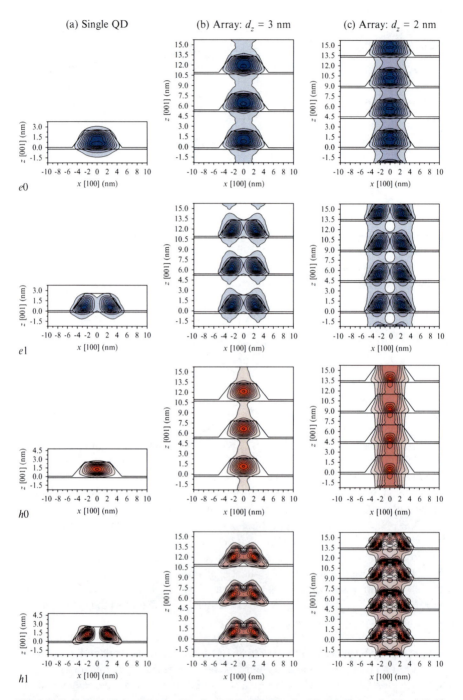

Fig. 9.3 Evolution of the QD wavefunctions delocalization in QD array as a function of the vertical proximity of QD in an array. The InAs/GaAs QD is with $b = 10$ nm, $h = 2.5$ nm and truncation factor is $t = 0.5$ (Reprinted with permission from [45]. Copyright 2010. American Physical Society)

direction, unlike the situation in quantum wires (QWR). This is mainly due to strong carrier confinement imposed by InAs/GaAs QD's potentials in CB and VB.

9.2.5 Electronic Structure of QD Arrays

When a large number of identical QD's are brought together to form a QD array, the number of wavefunctions becomes exceedingly large (infinite for an infinite array), and the difference in energy between them becomes very small (infinitesimal), so the levels may be considered to form continuous bands of energy rather than the discrete energy levels of the QD in isolation[46]. New normalized solution inside continuous bands of an QD array can be rewritten in the form of QD array Bloch states, $\propto e^{iK_z z}$, where the *QD array wavenumber*, K_z, is chosen within range $-\pi/L_z \leq K_z \leq \pi/L_z$. This range of the QD array wavenumbers form the QD array Brillouin Zone (BZ). However, some intervals of energy contain no wavefunctions, no matter how many QD are aggregated, forming energy gaps. Those energy gaps are between intermediate bands and not to be confused with energy gap of the underlying QD or barrier constituent materials. They will be referred later as a region of pure zero density of states inside CB that fundamentally distinguishes the properties of a QD array structure from those of QWRs, and are one of the key factors for successful operation of the IBSC. Those regions (gaps) of pure zero DOS can prevent rapid thermal depopulation of carriers from higher states in QD array's CB to IB originated from delocalized ground states (e_0).

To estimate the variation of the first few mini-bands (i.e. intermediate bands) with the vertical periodicity of the QD array, in Fig. 9.4, we have changed the vertical spacing, d_z between QD array layers in the range from 1 to 10 nm (note that this is the distance between bottom of the WL in $i + 1^{\text{th}}$ and top of the QDs in the ith growth layer). The lower and upper boundaries of an IB correspond to $K_z = 0$ and $K_z = \pi/L_z$, respectively. The width of the e_0 miniband at the close spacing of $d_z = 2$ nm is: 156 meV, at $d_z = 3$ nm is: 86 meV, at $d_z = 5$ nm is:

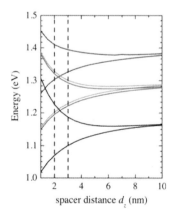

Fig. 9.4 Variation of the first four, optically strongest, mini-bands width in the CB with the QDs vertical spacing. The two vertical lines correspond to values of vertical spacing of $d_z = 2$ nm and $d_z = 3$ nm (Reprinted with permission from [45]. Copyright 2010. American Physical Society)

29 meV, and almost vanishes by $d_z = 10$ nm. While the energy gap between VB and CB states exists both in bulk and nanstructured semiconductor materials (like in quantum wells, quantum wires and quantum dots), one of the main reasons for using QD-arrays for IBSC is to open another energy gap, the gap between e_0-miniband induced by QD array, i.e. IB, and the rest of the CB spectra. In the QD arrays consisting of small size QDs, suggested as desirable combination for the IBSC [43], the e_1 and e_2 minibands are very close to or even overlap with the CB edge of the barrier material. In this case, the IB-CB energy gap can be quantified as the energy difference between the lower boundary of the e_1 miniband that corresponds to $K_z = 0$ and the upper boundary of e_0 miniband that corresponds to $K_z = \pi/L_z$. The IB-CB energy gap does not exist for the QDs closely spaced below some critical vertical distance $d_z^{(c)}$. This distance can be determined from the crossing point between the lower e_1 miniband and the upper e_0 miniband boundaries, i.e., from the condition $E_{e1}(K_z = 0) = E_{e0}(K_z = \pi/L_z)$. In our case, the region of zero IB-CB energy gap occurs for $d_z < 2.4$ nm. In this region, the QD-arrays effectively exhibit electronic properties of a QWR structure, with electron-free motion allowed in the z-direction. As can be seen in Fig. 9.3c at $d_z = 2$ nm, the wavefunction delocalization is substantial and form almost "metallic"-like bonds. For larger vertical distances, e.g. $d_z = 3$ nm, the IB-CB energy gap is 35 meV, while for $d_z = 5$ nm this gap increases to 88 meV. Although this value is much lower than optimal one, of ~700 meV, obtained under idealized assumptions,[5] it still proves the concept of realizing IB using QD arrays. The evolution of mini-bands broadening in the CB, together with profile of the CB edge of strained InAs QDs inside GaAs barrier material along z direction and for $(x, y) = (0, 0)$ is shown in Fig. 9.5. Due to the choice of QD (that has discrete δ-function like nature of the DOS as an artificial atom[47]) as a building block of the QD-array, the energy gaps between IBs in the conduction band are also characterized by pure zero DOS.

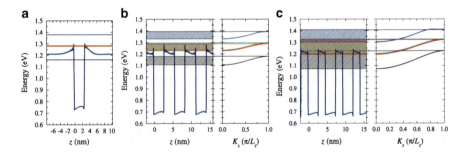

Fig. 9.5 Conduction band edge profile, along the z direction and $(x, y) = (0, 0)$, of the QD array, for: (**a**) single QD; (**b**) QD array made of the same QD as in (**a**) vertically spaced by $d_z = 3$ nm; and (**c**) QD array made of the same QD as in (**a**) vertically spaced by $d_z = 2$ nm. For QD arrays in (**b**) and (**c**), the dispersion of the first four, optically strongest, states in CB in the direction of the QD array is also shown. For QD array in (**b**), the energy gap between intermediate band e0 and intermediate band e1 can be identified, while this gap disappears for closely spaced QD in array (**c**) (Reprinted with permission from [45]. Copyright 2010. American Physical Society)

9 Modelling of Quantum Dots for Intermediate Band Solar Cells

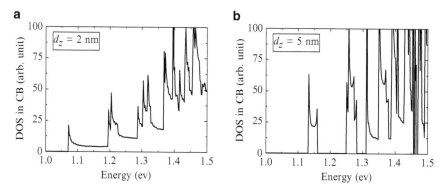

Fig. 9.6 Density of states in the conduction band for QD array with vertical periodicity (*left*) $d_z = 2$ nm and (*right*) $d_z = 5$ nm (Reprinted with permission from [45]. Copyright 2010. American Physical Society)

To distinguish the regions where QD array behaves effectively as a QWR from the one suitable for IBSC application, in Fig. 9.6 the DOS of QD-arrays is presented for two vertical spacer distances: (a) $d_z = 2$ nm and (b) $d_z = 5$ nm. In Fig. 9.6b, one can clearly recognize pure zero DOS regions between minibands, while such regions do not exist in Fig. 9.6a. Furthermore, sharp peaks at the edges of minibands correspond to van Hove singularities of the DOS at the critical points of the QD-array Brillouin zone, i.e., at Γ and X points.

An alternative way of designing the electronic structure of the QD arrays for IBSC is to change QD sizes in the arrays. It has been reported that for three characteristic QD sizes: (a) *small* $b = 6$ and $h = 3$ nm; (b) *medium* $b = 12$ and $h = 6$ nm; and (c) *big* $b = 20$ and $h = 10$ nm, the width of the e_0 IB at the closest spacing, $d_z = 1$ nm, is: 177 meV for (a), 86 meV for (b), 38 meV for (c).[43] It was estimated that for the *small* QD array (a) there is a gap between the e_0 and e_1 IBs of 106 meV at $d_z = 4$ nm. For this separation, the e_0 IB energy width is: 33 meV for (a), 14 meV for (b), and 6 meV for (c). The e_1 and e_2 IBs almost entirely overlap in structure (a) while in structures (b) and (c) a small gap appears between e_1 and e_2 for vertical spacing larger than 3 nm. By increasing the QD sizes in an array the optical gap between the electron and hole ground states (IB-VB gap) decreases. Due to much larger hole masses, the states in the VB are far more densely spaced. In all three structures considered, the spacing between states in the VB is much smaller than thermal energy at room temperature, $\sim$25 meV. At room temperature, these subbands in the VB are therefore essentially continuous.

9.2.6 Absorption Characteristics of QD Arrays

As shown schematically in Fig. 9.1, the rationale for using IBSCs as absorbing material is to create a partially occupied IB, thus affording sub-band-gap absorption

VB→IB into the empty states of the IB (process 2) and IB→CB from the occupied states of IB (process 3) in addition to the VB→CB of the host absorbing material (process 1). A few conditions have to be met for achieving good efficiency within such a concept: [48] (i) The VB→IB and IB→CB absorption spectra should ideally have no spectral overlap. As we have shown in Sect. 9.2.5, this condition can be achieved by designing the QD array with pure zero DOS between IB band and the rest of the CB electronic structure. This "photon-sorting" condition ensures maximum quantum efficiency for given positions of the CB-IB energy gap E_{gL} and IB-VB energy gap E_{gH} in Fig. 9.1. (ii) The VB→IB and IB→CB excitations must be optically allowed and strong. Thus, the quantum objects creating the IB must have significant concentration and oscillator strength.

The optical matrix element required for description of absorption and radiative related processes is defined as $|\hat{e} \cdot \mathbf{p}_{if}|^2$, where $\hat{e}$ is the light polarization vector and $\mathbf{p}_{if}(\mathbf{k}) = (m_0/\hbar)\langle i|\partial H_\mathbf{k}/\partial \mathbf{k}|f\rangle$ is the electron-hole momentum operator of the quantum structure, where $|i\rangle$ and $|f\rangle$ are initial and final states involved in the process. In the QD array structure, $\mathbf{k} \to \mathbf{k} + \mathbf{K}$. From the $\mathbf{K}$-dependent electronic structure, defined in Sect. 9.2.5, and optical dipole matrix element $\mathbf{p}_{ij}(\mathbf{K})$, the absorption characteristics of the QD array were calculated, in the dipole approximation:

$$\alpha(\hbar\omega) = \frac{\pi e^2}{c\epsilon_0 m_0^2 \bar{n}\omega b^2} \sum_{i,f,\mathbf{K}} |\hat{e} \cdot \mathbf{p}_{if}(\mathbf{K})|^2 (f_i - f_f)\delta[E_i(\mathbf{K}) - E_f(\mathbf{K}) - \hbar\omega], \quad (9.10)$$

where e is the electron charge, c is the speed of light in vacuum, m_0 is the rest electron mass, $\bar{n}$ is the refractive index of the GaAs, ϵ_0 is the vacuum permittivity, ω is the light frequency, b^2 is the area of the QD's base, and f_i and f_f are the Fermi distributions for the initial and final miniband respectively. The delta function, $\delta(x)$, is replaced with a Gaussian function $\exp[-(x/\sqrt{2}\sigma)^2]/(\sqrt{2\pi}\sigma)$, defined by the phenomenological broadening σ, to take into account random fluctuations in the structure of the QD array. Finally, the summation is replaced by integration over the wavevector K_z.

Assuming the IB formed by the QD-array minibad originated from the ground state e_0 separated by zero DOS from the rest of the CB spectra, the IB→CB and VB→IB involves e_0 as initial and final miniband, respectively. Due to a finite width of the e_0 miniband, in our IBSC absorption model, we can introduce an approximation that the final miniband states involved in the particular absorption process are always empty, i.e. $f_f = 0$. The corresponding absorption spectra $\alpha^{VB \to IB}(\hbar\omega)$, $\alpha^{IB \to CB}(\hbar\omega)$, and $\alpha^{VB \to CB}(\hbar\omega)$ are then calculated taking into account 1,000 hole minibands in the VB and e_0 miniband as the IB, the e_0 miniband as the IB and the next 100 minbands in the CB starting from e_1 miniband, and using 1,000 hole minibands in the VB and 100 minbands in the CB starting from e_1 miniband, respectively.

9 Modelling of Quantum Dots for Intermediate Band Solar Cells 241

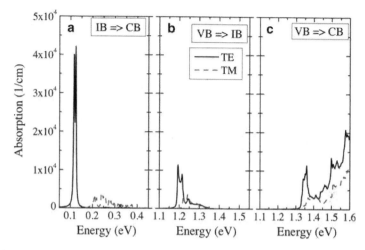

Fig. 9.7 Absorption spectra of QD array with vertical periodicity $d_z = 5$ nm. (**a**) IB→CB absorption, (**b**) VB→IB absorption and (**c**) VB→CB absorption (Reprinted with permission from [45]. Copyright 2010. American Physical Society)

Figure 9.7 shows all relevant absorption spectra of the QD array made of truncated pyramidal InAs/GaAs QDs with $b = 10$ nm, $h = 2.5$ nm, and $t = 0.5$, and with vertical spacing between the dots in the array of $d_z = 5$ nm.

First in Fig. 9.7a the transitions from QD array ground state e_0 miniband to all other electron minibands higher in energy are presented. A sharp double-peak that appears at 0.124 eV corresponds to all the dipole-allowed transitions between e_0 and e_1, e_2 minibands. This IB→CB absorption peak is completely TE polarized due to S symmetry of the QD states forming e_0 miniband and $P_{x,y}$ symmetry of the QD states forming e_1 and e_2 minibands. It should be emphasized that e_1 and e_2 minibands are already overlapped with the conduction band minima (CBM) of the barrier material. Existence of the pronounced peak at 0.124 eV suggests that the condition (ii) that IB→CB excitations must be optically allowed and strong can be fulfilled by relatively small $b = 10$ nm, InAs/GaAs QDs vertically stacked with $d_z = 5$ nm in the QD array. Further above the barrier CBM, in the energy range of 0.2–0.4 eV, a spectra of weak TE polarized transitions can be seen. This part of the IB→CB absorption spectra corresponds to transitions between e_0 miniband and largely delocalized states between QDs. The second absorption process relevant for IBSC operation is associated with the VB→IB transitions. This absorption spectra has been calculated again using (9.10), with the first electron miniband e_0 taken now as final and unoccupied. As initial minibands, we considered the first 1,000 hole minibands starting from h_0 miniband. The lowest energy of the 1,000th hole miniband goes well below the valence band maximum (VBM) in the barrier material. The calculated absorption spectra of VB→IB process is shown in Fig. 9.7b. The absorption peak is at ~ 1.2 eV and again of the TE polarization. This time the TE polarization originates from heavy hole QD's states that have no admixture of the orbital of the underlying bulk material with P_z symmetry. Due to highly strained InAs QD material in GaAs matrix, the biaxial component of the compressive strain splits the heavy from light hole states in the VB while hydrostatic component of the strain pushes heavy hole states upwards on the absolute energy scale making h_0 miniband of heavy hole character. The shoulder in the absorption spectra between 1.23 and 1.35 eV corresponds to transitions between deep VB minibands and e_0 miniband. The third absorption process, VB→CB, is shown in Fig. 9.7c. This absorption spectra were obtained using first 1,000 minibands in the VB as initial states and 100 empty minibands in the CB starting from e_1 miniband as final states. The absorption edge for this process is at 1.33 eV and is determined by energy distance between h_0 and e_1 minibands.

The intensity of the VB→IB and VB→CB absorption peaks is roughly the same, $\sim 10^4$ cm^{-1}, while IB→CB peak is 4 times larger, suggesting a strong, QD periodicity induced, *Bragg-like confinement* of states in e_1 and e_2 minibands. Those states partly overlap with CBM while the rest of those minibands are above CBM as can be seen in Fig. 9.5b.

It has been shown previously that using larger QDs ($b \sim 20$ nm) in the QD array the bulk-like absorption spectra can only be "red shifted" by almost continuum spectra of electron and hole minibands, while distinct absorption peak associated with IB cannot be identified [43].

9.3 Radiative and Non-Radiative Processes in QD Array

Apart from the requirements related to "photon sorting", strong absorption, and request to design suitable material that can provide $E_{gL} = 0.7$ eV, $E_{gH} = 1.2$ eV, and $E_g = 1.9$ eV, that will ultimately lead to maximal efficiency of the IBSC under concentrated light, there are a number of possible deleterious effects related to carrier lifetimes, that can affect carrier transport in IBSC, and need to be minimized. In the following sections, we will discuss radiative relaxation times between CB→IB and IB→VB in QD array and only those non-radiative times that perhaps might compete with radiative once: (1) electron longitudinal-optical phonon scattering between the CB and IB states, and (2) Auger-related non-radiative scattering times between states in CB and IB, *electron cooling* process, and Auger *bi-excitonic recombination* that occurs between carriers in IB and VB.

9.3.1 Radiative Processes in QD Arrays

The *radiative recombination* is an unavoidable process in direct gap semiconductor materials that causes loss of free electrons and holes from the transport process[49]. The radiative recombination is a spontaneous process defined as the transition probability of an electron from initial state $|i\rangle$ to the final state $|f\rangle$ that is followed

9 Modelling of Quantum Dots for Intermediate Band Solar Cells

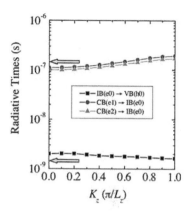

Fig. 9.8 Variation of the radiative transition times between CB($e2$) and IB($e0$); CB($e1$) and IB($e0$) that can be regarded as CB to IB radiative processes, and between IB($e0$) and VB($h0$) that can be regarded as IB to VB radiative process. *Horizontal arrows* mark the equivalent single QD radiative times (Reprinted with permission from [45]. Copyright 2010. American Physical Society)

by photon emission of frequency $\omega = (E_i - E_f)/\hbar$. The radiative recombination defines the ultimate efficiency of the SC, which is also referred to as *radiative limit efficiency*.

Starting from Fermi golden rule and assuming the wavelength of the sunlight much larger than the size of the QD array with which it interacts (i.e., in the dipole approximation), the expression for the radiative recombination in the QD array is:

$$\frac{1}{\tau_{if}^{\mathrm{rad}}(\mathbf{K})} = \frac{4}{3}\frac{\bar{n}[E_i(\mathbf{K}) - E_f(\mathbf{K})]}{\pi\hbar^2 c^3 \varepsilon_0}[|\hat{e}_x \cdot \mathbf{p}_{if}(\mathbf{K})|^2 + |\hat{e}_y \cdot \mathbf{p}_{if}(\mathbf{K})|^2 + |\hat{e}_z \cdot \mathbf{p}_{if}(\mathbf{K})|^2]. \tag{9.11}$$

Since the radiative lifetime, τ_{if}^{rad}, is determined mainly by the optical dipole matrix element $\mathbf{p}_{if}(\mathbf{K})$ and energy difference between initial and finale states, $E_i(\mathbf{K}) - E_f(\mathbf{K})$, the quantum nanostructures design can offer extra degree of freedom in modifying the value of this scattering time.

In Fig. 9.8, the variation of the radiative transition times across the BZ of QD array, between e_1, e_2 minibands (CB) and e_0 miniband (IB) and between e_0 miniband (IB) and h_0 miniband (VB) is shown. In a single InAs/GaAs QD, $(b, h) = (10, 2.5)$ nm, the radiative lifetime between electron and hole ground state is $\tau_{e0,h0}^{\mathrm{rad}} = 1.59$ ns. When the same size QD forms an array, at the $K_z = 0$ edge, the radiative time is $\tau_{\mathrm{IB}(e0),\mathrm{VB}(h0)}^{\mathrm{rad}}(K_z = 0) = 2.1$ ns. The increase of the radiative relaxation time of 32% in QD array when compared to single QD is explained in terms of e_0 and h_0 wavefunction delocalization in QD array due to electronic states coupling from neighboring QDs. Wavefunction delocalization reduces the value of the optical dipole matrix element which in turn increases the value of τ^{rad}.

In the single QD structure, the $\tau_{e1,e0}^{\mathrm{rad}} = 149$ ns and $\tau_{e2,e0}^{\mathrm{rad}} = 129$ ns. Although one might expect that e_0 miniband states, already shown to be delocalized by close proximity of QDs in the QD array, will further increase the values of radiative times $\tau_{\mathrm{CB}(e1),\mathrm{IB}(e0)}^{\mathrm{rad}}$ and $\tau_{\mathrm{CB}(e2),\mathrm{IB}(e0)}^{\mathrm{rad}}$, in the QD array structure the trend is indeed opposite. The radiative transition times between states in two minibands e_1 and e_2 that overlap with the CBM edge of barrier material and the e_0 miniband regarded as an IB, at the $K_z = 0$ edge, are $\tau^{\rm rad}_{{\rm CB}(e1),{\rm IB}(e0)}(K_z = 0) = 109$ ns and $\tau^{\rm rad}_{{\rm CB}(e1),{\rm IB}(e0)}(K_z = 0) = 98$ ns, and are smaller than for the equivalent radiative transitions in single QD. This trend emphasis the QD array induced Bragg like confinement of e_1 and e_2 miniband states that overlap or are only just above the CBM edge of the barrier material and not to be confused with strain induced confinement at the QD/barrier surfaces of the QD [50].

It should be mentioned that $\tau^{\rm rad}_{{\rm CB}(e1,e2),{\rm IB}(e0)}$ are intersubband transitions inside CB with energy difference in the range of ~ 100 meV, while $\tau^{\rm rad}_{{\rm IB}(e0),{\rm VB}(h0)}$ is intraband transition between states in CB and VB with energy difference of ~ 1 eV. This discrepancy in transition energies causes the big difference in radiative relaxation times, $\tau^{\rm rad}_{\rm CB,IB}/\tau^{\rm rad}_{\rm IB,VB} \sim 50$, between CB→IB and IB→VB transitions.

9.3.2 Electron–Phonon Interaction in QD Arrays

At finite temperatures, atoms in crystal can vibrate around their equilibrium positions. These lattice vibrations are quantized and are called *phonons*. Phonons create additional potential that perturbs otherwise stationary electronic states and causes transitions between them. Throughout this chapter, only polar coupling to optical phonons will be considered, because other types of phonon interactions are much weaker or irrelevant in the QD-based systems[51]. The Frölich interaction Hamiltonian describing polar coupling to optical phonons is given by[52]

$$\hat{H}_{\rm e-ph} = \sum_{i,f,\mathbf{q}} \alpha(\mathbf{q}) F_{if}(\mathbf{q}) \hat{a}_i^\dagger \hat{a}_f (\hat{b}_{\mathbf{q}} + \hat{b}^\dagger_{-\mathbf{q}}), \qquad (9.12)$$

where $\mathbf{q}$ is the phonon wavevector, $\hat{b}_{\mathbf{q}}$ and $\hat{b}^\dagger_{-\mathbf{q}}$ are phonon annihilation and creation operators, $\hat{a}_i^\dagger$ and $\hat{a}_f$ are annihilation and creation operators for electrons. The factor $\alpha(\mathbf{q}) \propto q^{-1}\sqrt{\hbar\omega_{\rm LO}}$, and the form factor $F_{if}(\mathbf{q}) = \langle i|e^{i\mathbf{q}\cdot\mathbf{r}}|f\rangle$. It is assumed that an electron in initial state is always at the band edge while momentum conservation is achieved with $\mathbf{q} = \mathbf{k} + K_z\hat{z}$, i.e., phonon wavevector equals electron momentum in the final state. Optical phonons are nearly dispersionless, and for simplicity a constant LO phonon energy $\hbar\omega_{\rm LO} = 36$ meV is assumed.

In a single QD structure, due to the discrete nature of energy levels together with very weak energy dispersion of the longitudinal-optical (LO) phonons, a simple consideration based on energy conservation only and Fermi golden rule predicted that scattering rates are zero unless the electron level spacing equals the LO-phonon energy[52]. This largely reduced relaxation rate from the inefficient phonon scattering in QDs is referred in literature to as *phonon bottleneck*. This condition can be relaxed in QD array structures due to finite widths of minibands. Even so, such an approach treats the electron and phonon systems separately with their interaction being only a perturbation. It is currently known that electrons and phonons in quantum dots form coupled entities, *polarons*, and that the polaron lifetime is determined by anharmonic decay of an LO phonon into two low energy bulk LA phonons[53–56].

To describe that scattering mechanism, we follow a Wigner–Weisskopf description for the carrier relaxation in QD through LO-phonon scattering. If the electron couples directly to the LO-phonon modes, quantum transition would result in a repeated energy exchange between the electron and phonon modes, known also as Rabi oscillation. However, due to the decay of the confined LO phonons, this oscillation will decay rapidly, thus the electron energy is dissipated away through the LO phonons. With such assumptions, the polaron lifetime is given by[54]:

$$\frac{1}{\tau_{if}^{LO}} = W_{ph} - \sqrt{2(R-X)}, \quad (9.13)$$

where $R = \sqrt{X^2 + Y^2}$ with $X = (g/\hbar)^2 + (\Delta_{if}^2 - W_{ph}^2)/4$ and $Y = W_{ph}\Delta_{if}/2$. Detuning from the phonon energy is $\Delta_{if} = E_i - E_f \mp \hbar\omega_{LO}$, while g is the coupling strength of an electron to all LO modes, and W_{ph} the phenomenological phonon decay rate due to LO phonon decay into two LA phonons[54,57]. The above theoretical considerations have been verified by experimental results on intraband carrier dynamics in QD[56,58].

In Fig. 9.9, the LO phonon absorption (open dots) and emission (solid dots) scattering rates are presented between the first 100 states in the CB. It is clear that both processes exhibit sharp peaks for electron transitions energy in the vicinity of ± 36 meV, which is the LO phonon energy. In this region, electron–phonon scattering is of the order 10^{-9}–10^{-12} s. This energy region is not particularly important for operation of the IBSC. Indeed, the fast phonon relaxation processes can help to prepare system in the state where electrons are in e_1 or e_2 miniband. At $K_z = 0$, the energy spacing between e_1 miniband and IB originated from e_0

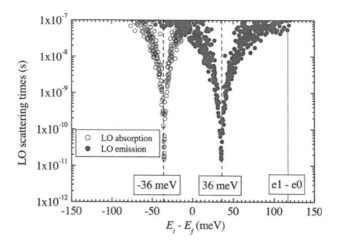

Fig. 9.9 LO absorption and emission scattering times (Reprinted with permission from [45]. Copyright 2010. American Physical Society)

states is ~117 meV. The LO phonon emission scattering time for this transition is $\tau_{CB(e1),IB(e0)}^{LO(e)} = 56$ ns, and it is increased for the QD array when compared to the same transition in the single QD, which is $\tau_{e1,e0}^{LO(e)} = 19$ ns. Increase of this scattering time is attributed to decrease of both $F(\mathbf{q})$ form factor and $\Delta_{e1,e0}$ energy detuning factor in QD array. Compared to the radiative time $\tau_{CB(e1),IB(e0)}^{rad} = 109$ ns, the $\tau_{CB(e1),IB(e0)}^{LO(e)}$ is somewhat smaller but still significantly larger than $\tau_{IB,VB}^{rad} = 2.1$ ns, suggesting that non-radiative phonon scattering-related processes between CB and IB might not be critical for operation of the IBSC based on QD array. Possible deteriorating effect of losing electrons from the IB, minband e_0, to the CB, minibands e_1 or e_2, is even less likely since phonon absorption process takes $\tau_{IB(e0),CB(e1)}^{LO(a)} = 650$ ns.

Acoustic phonon scattering is only significant when the states are closely spaced in energy (<10 meV). That is the case between higher miniband states in CB. For larger energy separation, like that between e_1 miniband and IB formed by e_0 state which is of the order ~100 meV, the interaction with LO-phonons is dominant[59].

9.3.3 Auger-Related Nonradiative Times in QD Arrays

Nonradiative Auger-related scattering processes play an important role in carrier dynamics in semiconductor nanostructures when both types of carriers (electrons and holes) are present. They become significant particularly in QDs, which have discrete electronic levels, which implies that the other nonradiative competing processes (like phonon scattering) could be strongly suppressed [60]. We consider two main Auger-related nonradiative processes: *electron cooling* and *bi-exciton recombination* – Fig. 9.10 – as they might compete on the time scale with its radiative recombination counterparts, $\tau_{CB,IB}^{rad}$ and $\tau_{IB,VB}^{rad}$ respectively. We adopt a phenomenological formula for the Auger rate derived under the standard time-dependent perturbation theory and using Fermi's golden rule[61, 62]:

$$\frac{1}{\tau_{if_n}^A} = \frac{2\pi}{\hbar} \sum_n |J(i,j;k,l)|^2 \delta(\Delta E + E_{f_n} - E_i), \quad (9.14)$$

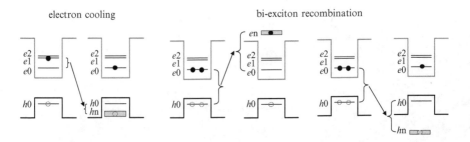

Fig. 9.10 Illustration of two different Auger processes: electron cooling (*left*) and bi-exciton recombination (*right*) (Reprinted with permission from [45]. Copyright 2010. American Physical Society)

w here i and f_n are initial and final electronic configuration involved in a particular Auger process, E_i and E_{f_n} are their energies, and ΔE is the energy transfer between initial and final configuration. In (9.14), we have used multiple final states f_n (where n includes spin as well), since each final state might have some contributions to the Auger rate. As in the case of absorption spectra, δ-function is replaced by a Gaussian function $\exp[-(x/\sqrt{2}\sigma)^2]/(\sqrt{2\pi}\sigma)$, defined by the phenomenological broadening σ, to take into account inhomogeneous line broadening due to size-distribution effects as well as homogeneous line broadening. To achieve convergence of our results, the number of final states used in particular Auger process is determined from the condition that $|\max\{E_{f_n}\} - \min\{E_{f_n}\}|/2 \geq 2\sigma$, i.e., to be inside two standard deviations, and $\sigma = 5$ meV.

In the electron cooling process, in which one electron is initially in e_1 state and hole is in its ground state h_0, the energy transfer occurs when electron relaxes to its ground state e_0 to transfer excess energy $\Delta E = E_{e1} - E_{e0}$ to the hole in order to excite it by ΔE deeper into VB. The Coulomb integral in (9.14) reads as $J(h_0, e_1; h_n, e_0)$. We have estimated that the Auger electron cooling in single InAs/GaAs QD is $\tau^A_{e-cool} = 1.37$ ps (excitonic gap 1.13 eV), which is in very good agreement with other theoretical and experimental results[62–64]. In QD array, the electron cooling time is increased to $\tau^A_{e-cool} = 2.05$ ps, despite further decrease in the excitonic gap to 1.09 eV. The main reason for this increase is delocalization of states involved in the process and long range Coulomb interaction between charges in neighbouring QDs in the array that was compensated in a single QD structure by Makov-Payne correction[38, 65].

In the bi-exiciton Auger relaxation, one exciton, composed of electron and hole in their ground states, recombines while the energy, $\Delta E = E_X$, released in this process is transferred to the other electron (hole) in the ground state (with opposite spin) to be excited to the states higher (lower) in CB (VB) by ΔE. These two processes are denoted as $\tau^A_{e,ex}$ and $\tau^A_{h,ex}$, while Coulomb integrals in (9.14) takes the form $J(e_{0,\alpha}, e_{0,\beta}; e_n, h_0) - J(e_{0,\beta}, e_{0,\alpha}; e_n, h_0)$ and $J(h_{0,\alpha}, h_{0,\beta}; h_n, e_0) - J(h_{0,\beta}, h_{0,\alpha}; h_n, e_0)$, respectively, where α and β stands for opposite spins. The bi-exciton recombination time is then: $1/\tau^A_{bx} = 2/\tau^A_{e,ex} + 2/\tau^A_{h,ex}$. In a single QD, we estimated: $\tau^A_{e,ex} = 209$ ns, $\tau^A_{h,ex} = 6.15$ ns and $\tau^A_{bx} = 3$ ns, while for QD array: $\tau^A_{e,ex} = 266$ ns, $\tau^A_{h,ex} = 18$ ns and $\tau^A_{bx} = 8.4$ ns. The trend that $\tau^A_{e,ex} \gg \tau^A_{h,ex}$ was already observed in semiconductor quantum well structures both experimentally and theoretically[66–68].

9.4 Conclusions

In this chapter, we have presented the comprehensive theoretical model for design and modelling of IBSC based on arrays of InAs/GaAs QDs. For one realistic QD array ($b = 10$ nm, $h = 2.5$ nm, $t = 0.5$, and $d_z = 5$ nm), we have estimated $E_{gH} = 1.2$ eV and $E_{gL} = 0.124$ eV. For these energies and absorption spectra found in

Table 9.1 Radiative, LO phonon emission/absorption, and Auger scattering times of InAs/GaAs QD array IBSC considered in the main text

Type	Minibands	Radiative (ns)	Phonons [e/a] (ns)	Auger (ns)
CB→IB	$e1 \to e0$	109	56/650	2×10^{-3}
IB→VB	$e0 \to h0$	2.1		8.4

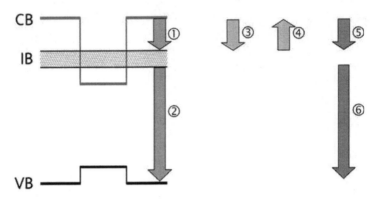

Fig. 9.11 Schematic representation of radiative and non-radiative relaxation times in IBSC: (1) radiative recombination between CB and IB, (2) radiative recombination between IB and VB, (3) phonon emission assisted CB to IB relaxation, (4) phonon absorption assisted IB to CB relaxation, (5) Auger electron cooling relaxation between CB and IB, and (6) Auger bi-exciton relaxation between IB and VB

Sect. 9.2.6, we have estimated ultimate efficiency of the IBSC in the radiative limit to be $\eta \approx 39\%$ for undoped structure for sunlight concentration factor of 1,000 [69]. This is increase of 56% compared to simple QD solar cell[70]. Our finding suggests that with appropriate design of the QD array structural parameters: (1) it is possible to achieve the regions of pure zero DOS between IB and the rest of the CB states, that is desirable for "photon sorting" and increased efficiency of the device, and (2) it is possible to achieve the strong optically allowed excitation between IB and CB. Analysis of various radiative and non-radiative times summarized in Table 9.1 and schematically presented in Fig. 9.11 indicates that: (1) the ratio between CB→IB and IB→VB radiative times is ∼50, (2) although phonon related relaxation time between CB and IB amounts to a half of the radiative relaxation time between CB and IB, it is still one order of magnitude larger than the IB→VB radiative time, (3) nonradiative phonon absorption process that promotes electrons from IB to CB is very slow and probably would not significantly affect the transport properties of the IBSC, (4) nonradiative Auger bi-exciton relaxation time is longer then radiative IB to VB relaxation time, indicating that this process will still be predominately radiative, (5) most detrimental effect on transport properties can originate from non-radiative Auger electron cooling process, that is in the ps timescale and is three orders of magnitude faster that any other relaxation process in the IBSC. Special attention needs to be paid in the design of the IBSC structures in order to suppress the effects of electron cooling and to provide an increased efficiency of the IBSCs.

Acknowledgements The author wishes to thank to A. R. Adams FRS, M. Blake, M. Califano, N. M. Harrison, Z. Ikonić, T. S. Jones, A. Luque, A. Marti and E. P. O'Reilly for many useful discussions and suggestions. He would also like to thank to N. Vukmirović for enlightening discussions regarding the methods for quantum dot electronic structure calculations and for his work on development of the kppw code. The author is grateful to STFC Energy Strategy Initiative for financial support.

References

1. W. Shockley, H.J. Queisser, J. Appl. Phys. **32**(3), 510 (1961)
2. A. Nozik, Physica E **14**(1–2), 115 (2002)
3. M. Green, Progr. Photovoltaics **9**(2), 123 (2001)
4. R.R. King, Nat. Photonics **2**(5), 284 (2008)
5. A. Luque, A. Martí, Phys. Rev. Lett. **78**(26), 5014 (1997)
6. A. Luque, A. Marti, A.J. Nozik, MRS Bull. **32**(3), 236 (2007)
7. A. Luque, A. Marti, Adv. Mater. **22**, 160 (2010)
8. A. Martí, E. Antolín, C.R. Stanley, C.D. Farmer, N. López, P. Díaz, E. Cánovas, P.G. Linares, A. Luque, Phys. Rev. Lett. **97**(24), 247701 (2006)
9. A. Luque, A. Marti, Prog. Photovoltaics **9**(2), 73 (2001)
10. L. Cuadra, A. Marti, A. Luque, IEEE Trans. Electron Dev. **51**(6), 1002 (2004)
11. A. Luque, A. Marti, N. Lopez, E. Antolin, E. Canovas, C. Stanley, C. Farmer, L. Caballero, L. Cuadra, J. Balenzategui, Appl. Phys. Lett. **87**(8) (2005)
12. A. Luque, A. Marti, E. Antolin, C. Tablero, Physica B **382**(1–2), 320 (2006)
13. Q. Xie, A. Madhukar, P. Chen, N.P. Kobayashi, Phys. Rev. Lett. **75**(13), 2542 (1995)
14. D.M. Bruls, P.M. Koenraad, H.W.M. Salemink, J.H. Wolter, M. Hopkinson, M.S. Skolnick, Appl. Phys. Lett. **82**(21), 3758 (2003)
15. A. Martí, N. López, E. Antolín, E. Cánovas, A. Luque, C.R. Stanley, C.D. Farmer, P. Díaz, Appl. Phys. Lett. **90**(23), 233510 (2007)
16. D. Alonso-Álvarez, A.G. Taboada, J.M. Ripalda, B. Alén, Y. González, L. González, J.M. García, F. Briones, A. Martí, A. Luque, A.M. Sánchez, S.I. Molina, Appl. Phys. Lett. **93**(12), 123114 (2008)
17. R. Oshima, A. Takata, Y. Okada, Appl. Phys. Lett. **93**(8), 083111 (2008)
18. K. Akahane, N. Yamamoto, T. Kawanishi, (2009), *Proceedings of 21st International Conference on Indium Phosphide and Related Materials*, pp. 73–74, Newport Beach, CA, 10–14 May 2009
19. K. Akahane, N. Yamamoto, T. Kawanishi, IEEE Photon. Technol. Lett. **22**(2), 103 (2010)
20. Z. Fan, H. Razavi, J.w. Do, A. Moriwaki, O. Ergen, Y.L. Chueh, P.W. Leu, J.C. Ho, T. Takahashi, L.A. Reichertz, S. Neale, K. Yu, M. Wu, J.W. Ager, A. Javey, Nat. Mater. **8**(8), 648 (2009)
21. J.M. Luttinger, W. Kohn, Phys. Rev. **97**(4), 869 (1955)
22. E.O. Kane, J. Phys. Chem. Solids **1**(4), 249 (1957)
23. E.O. Kane, Semiconduct. Semimetal. **1**(), 75 (1966)
24. M. Cardona, F.H. Pollak, Phys. Rev. **142**(2), 530 (1966)
25. S. Richard, F. Aniel, G. Fishman, Phys. Rev. B **70**(23), 235204 (2004)
26. D. Rideau, M. Feraille, L. Ciampolini, M. Minondo, C. Tavernier, H. Jaouen, A. Ghetti, Phys. Rev. B **74**(19), 195208 (2006)
27. B.A. Foreman, Phys. Rev. B **76**(4), 045327 (2007)
28. I. Saïdi, S.B. Radhia, K. Boujdaria, J. Appl. Phys. **107**(4), 043701 (2010)
29. G.L. Bir, G.E. Pikus, *Symmetry and Strain-Induced Effects in Semiconductors* (Wiley, New York, 1974)
30. J.P. Loehr, *Physics of Strained Quantum Well Lasers* (Kluwer, Dordrecht, 1998)

31. C.R. Pidgeon, R.N. Brown, Phys. Rev. **146**(2), 575 (1966)
32. C. Pryor, Phys. Rev. B **57**, 7190 (1998)
33. O. Stier, M. Grundmann, D. Bimberg, Phys. Rev. B **59**, 5688 (1999)
34. M.A. Cusack, P.R. Briddon, M. Jaros, Phys. Rev. B **54**(4), R2300 (1996)
35. A.D. Andreev, E.P. O'Reilly, Phys. Rev. B **62**, 15851 (2000)
36. N. Vukmirović, D. Indjin, V.D. Jovanović, Z. Ikonić, P. Harrison, Phys. Rev. B **72**, 075356 (2005)
37. S. Tomić, A.G. Sunderland, I.J. Bush, J. Mater. Chem. **16**, 1963 (2006)
38. N. Vukmirović, S. Tomić, J. Appl. Phys. **103**(10), 103718 (2008)
39. A.D. Andreev, J.R. Downes, D.A. Faux, E.P. O'Reilly, J. Appl. Phys. **86**, 297 (1999)
40. S.S. Li, J.B. Xia, Z.L. Yuan, Z.Y. Xu, W. Ge, X.R. Wang, Y. Wang, J. Wang, L.L. Chang, Phys. Rev. B **54**, 11575 (1996)
41. A.D. Andreev, (1998), *Proceedings of SPIE*, vol. 3284, pp. 151–161. Conference on In-Plane Semiconductor Lasers – From Ultraviolet to Mid-Infrared II, San Jose, CA, 26–28 Jan 1998
42. O.L. Lazarenkova, A.A. Balandin, J. Appl. Phys. **89**(10), 5509 (2001)
43. S. Tomić, T.S. Jones, N.M. Harrison, Appl. Phys. Lett. **93**(26), 263105 (2008)
44. J.W. Klos, M. Krawczyk, J. Appl. Phys. **106**(9), 093703 (2009)
45. S. Tomić, Phys. Rev. B **82**, 195321 (2010)
46. R.d.L. Kronig, W.G. Penney, Proc. Roy. Soc. (London) **A 130**, 499 (1931)
47. M. Grundmann, J. Christen, N.N. Ledentsov, J. Böhrer, D. Bimberg, S.S. Ruvimov, P. Werner, U. Richter, U. Gösele, J. Heydenreich, V.M. Ustinov, A.Y. Egorov, A.E. Zhukov, P.S. Kopev, Z.I. Alferov, Phys. Rev. Lett. **74**(20), 4043 (1995)
48. V. Popescu, G. Bester, M.C. Hanna, A.G. Norman, A. Zunger, Phys. Rev. B **78**(20) (2008)
49. J. Nelson, *The Physics of Solar Cells* (World Scientific, Singapore, 2003)
50. V. Popescu, G. Bester, A. Zunger, Appl. Phys. Lett. **95**(2), 023108 (2009)
51. G. Mahan, *Many-Particle Physics* (Kluwer, Dordrecht, 2000)
52. M.A. Stroscio, M. Dutta, *Phonons in Nanostructures* (Cambridge University Press, London, 2001)
53. S. Hameau, Y. Guldner, O. Verzelen, R. Ferreira, G. Bastard, J. Zeman, A. Lemaître, J.M. Gérard, Phys. Rev. Lett. **83**, 4152 (1999)
54. X.Q. Li, H. Nakayama, Y. Arakawa, Phys. Rev. B **59**, 5069 (1999)
55. O. Verzelen, R. Ferreira, G. Bastard, Phys. Rev. B **62**, R4809 (2000)
56. S. Sauvage, P. Boucaud, R.P.S.M. Lobo, F. Bras, G. Fishman, R. Prazeres, F. Glotin, J.M. Ortega, J.M. Gérard, Phys. Rev. Lett. **88**, 177402 (2002)
57. X. Li, Y. Arakawa, Phys. Rev. B **57**(19), 12285 (1998)
58. E.A. Zibik, L.R. Wilson, R.P. Green, G. Bastard, R. Ferreira, P.J. Phillips, D.A. Carder, J.P.R. Wells, J.W. Cockburn, M.S. Skolnick, M.J. Steer, M. Hopkinson, Phys. Rev. B **70**, 161305 (2004)
59. N. Vukmirović, Z. Ikonić, I. Savić, D. Indjin, P. Harrison, J. Appl. Phys. **100**, 074502 (2006)
60. L.W. Wang, Energ. Environ. Sci. **2**(9), 944 (2009)
61. L.W. Wang, M. Califano, A. Zunger, A. Franceschetti, Phys. Rev. Lett. **91**(5), 056404 (2003)
62. G.A. Narvaez, G. Bester, A. Zunger, Phys. Rev. B **74**(7), 075403 (2006)
63. T. Müller, F.F. Schrey, G. Strasser, K. Unterrainer, Appl. Phys. Lett. **83**(17), 3572 (2003)
64. T.B. Norris, K. Kim, J. Urayama, Z.K. Wu, J. Singh, P.K. Bhattacharya, J. Phys. D Appl. Phys. **38**(13), 2077 (2005)
65. G. Makov, M.C. Payne, Phys. Rev. B **51**(7), 4014 (1995)
66. S. Sweeney, T. Higashi, A. Andreev, A. Adams, T. Uchida, T. Fujii, Phys. Stat. Sol. b **223**(2), 573 (2001)
67. S. Sweeney, A. Adams, M. Silver, E. O'Reilly, J. Watling, A. Walker, P. Thijs, Phys. Stat. Sol. b **211**(1), 525 (1999)
68. A.D. Andreev, E.P. O'Reilly, Appl. Phys. Lett. **84**(11), 1826 (2004)
69. R. Strandberg, T.W. Reenaas, J. Appl. Phys. **105**(12), 124512 (2009)
70. V. Aroutiounian, S. Petrosyan, A. Khachatryan, K. Touryan, J. Appl. Phys. **89**(4), 2268 (2001)

Chapter 10
The Quantum Dot Intermediate Band Solar Cell

Colin R. Stanley, Corrie D. Farmer, Elisa Antolín, Antonio Martí, and Antonio Luque

Abstract This chapter concentrates on the development of the intermediate band solar cell (IBSC) using InAs quantum dots (QD) embedded within an (Al,Ga)As single p-n junction. The growth and optimisation of the QD structures using molecular beam epitaxy (MBE) are discussed, and issues related to the manufacture of solar cell die for concentrator photovoltaic system applications are addressed. Advanced characterisation techniques have been employed to compare the performance of QD-IBSCs against reference (un-modified) single junction (Al,Ga)As cells, and results from these studies are assessed. It is concluded that the main operating principles of the IBSC have been demonstrated, including the generation of photocurrent by the absorption of two sub-bandgap photons, and voltage preservation when thermal and/or tunnelling-assisted escape from the intermediate band are suppressed. However, efficiencies in excess of those for un-modified (Al,Ga)As reference cells have not yet been achieved.

10.1 Introduction

The intermediate band solar cell (IBSC) was proposed by Luque and Martí [1–3] as a radically different approach towards the realisation of solar cells with high conversion efficiencies. It consists of a single semiconductor p-n junction within which modified semiconductor is embedded such that an intermediate energy band (IB) is formed within its conventional bandgap. This IB section is electrically isolated from the n- and p- emitters of the solar cell. Its purpose is to provide a "route" whereby photons with sub-bandgap energies within the solar spectrum, which would otherwise be wasted, can be absorbed by the semiconductor to create

C.R. Stanley (✉)
School of Engineering, University of Glasgow, Glasgow G12 8QQ, UK
e-mail: Colin.Stanley@glasgow.ac.uk

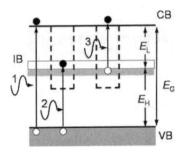

Fig. 10.1 Diagram illustrating the optical transitions in an intermediate band semiconductor, which lead to the creation of an electron–hole (e–h) pair by the absorption of sub-bandgap energy photons: (1) conventional e-h generation by absorption of a photon with energy $hf_1 > E_g$, the band gap of the host semiconductor; (2) absorption of a photon with sub-bandgap energy $hf_2 < E_g$ which promotes an electron from the VB to the IB; (3) absorption of a photon with sub-bandgap energy $hf_3 < E_g$, which promotes an electron from the IB to the CB

additional photo-current through a two-photon absorption process. This is illustrated schematically in Fig. 10.1. Photon 1 has an energy greater than the bandgap, E_g, and creates an electron-hole pair by the normal means through a valence band (VB) to conduction band (CB) transition. Under steady-state conditions, the absorption of both photon 2 and photon 3 is required to produce another electron-hole pair. One electron is promoted from the VB to the IB by photon 2, and photon 3 then excites another electron from the IB to the CB. The additional photo-current is achieved, in theory, without degradation of the output voltage of the solar cell. Efficiencies in excess of 60% under maximum concentration of sunlight (assumed as a black body at 6,000 K) have been predicted for the device when the values of the gaps at 300 K are optimum, namely; $E_g = 1.95$ eV, $E_L = 1.24$ eV and $E_H = 0.71$ eV.

Since the concept was first proposed, interest in IBSCs has grown dramatically. Prototype IBSCs have been reported by several groups using InAs quantum dots (QDs) embedded in a GaAs host semiconductor [4–8]. In this approach, the IB would arise from the confined states of the electrons in the QDs [9], although efficiencies comparable to those of good single junction GaAs cells have not been realised, thus far. A number of reasons for this degraded performance have been discussed [4]. A particular limitation of the QD approach is the relatively weak absorption coefficient of the VB→IB and the IB→CB transitions, especially the IB→CB transition [10]. A high volumetric concentration of QDs is required to enhance the overall absorption, which can lead to the generation of misfit and/or threading dislocations as strain builds up in multi-layer stacks [11]. One solution to this problem is to cap the QDs with a material which is in tensile strain, to offset the residual compressive stress associated with the QDs/wetting layers (WL). GaAsN [5] and GaP [6–8] have been explored for this purpose, while a similar strain-balancing technique has been applied to InAs QD stacks grown on InP substrates with InAlGaAs spacer layers [12]. Non-QD solutions to the manufacture of IBSCs have also been proposed and are discussed in more detail in other chapters in this book. They include dilute II–VI oxide semiconductors [13], e.g. oxygen-doped ZnTe [14]; Mn-doped InGaN [15]; highly mismatched alloys (HMAs) [16]; Ti-implanted Si [17, 18]; transition metal ions in III–V [19, 20] and chalcogenide semiconductors [21, 22], and Ti in copper chalcopyrite semiconductors [23].

The purpose of this chapter is to review the current status of the IBSC using InAs QDs embedded in a single junction (Al,Ga)As hetero-structure. Section 10.2 covers an introduction to the molecular beam epitaxial growth (MBE) of III–V semiconductors, while Sect. 10.3 addresses some of the fundamental issues of growing InAs QDs on and within GaAs. This is the material system most widely explored for QD-IBSCs and is discussed in more detail in Sect. 10.4. Some of the challenges posed by the processing of QD-IBSCs, particularly for use in concentrator photovoltaic (CPV) systems, are outlined in Sect. 10.5. Section 10.6 discusses advanced analytical techniques employed to demonstrate key operating principles of the IBSC, and overall conclusions are presented in Sect. 10.7.

10.2 Molecular Beam Epitaxy of III–V Semiconductors

10.2.1 Introduction

Molecular beam epitaxy (MBE) is carried out under clean, ultra-high vacuum (UHV) conditions and is used to grow a wide range of different semiconductors, metals and insulators. The term "molecular" refers to the regime where the atoms or molecules needed for the growth of the epitaxial structure traverse the vacuum between the sources and the substrate in straight lines, without undergoing intra-beam collisions. MBE has evolved from the three temperature deposition technique developed by Gunther in the late 1950s [24], where "three temperature" describes conditions under which a III–V compound is deposited on to a substrate (held at temperature T_1) using fluxes of the constituent elements, generated by heating these elements in separate Knudsen sources maintained at temperatures T_2 and T_3. MBE has gained acceptance in recent years as a production tool for the volume manufacture of III–V semiconductor devices including the laser diode used in CD players and the high electron mobility transistor (HEMT) found in mobile phones.

The growth of the III–V's involves one (or more) volatile element (the group V) and one (or more) relatively involatile element (the group III). A schematic diagram illustrating the main components of a MBE growth chamber is shown in Fig. 10.2. Eight or more sources, usually thermal, are arranged symmetrically on a large flange and directed towards a holder which can both heat and rotate the wafer attached to it. The atomic/molecular fluxes are set by the temperature of their respective sources, and are regulated by means of computer operated shutters, which are opened and closed in pre-programmed sequences. This allows complex, multi-layer structures to be deposited with precise control over the thickness, composition, interface flatness and doping concentration of each layer.

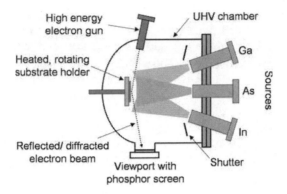

Fig. 10.2 Schematic diagram of a MBE growth chamber. Typically, between 8–12 cells are arranged symmetrically on a source flange (three are shown for clarity; Ga, In and As). The substrate is mounted so that it can be both heated and rotated. A key component is the reflected high energy electron diffraction (RHEED) system, comprising a 10–30 keV electron gun and phosphor screen. Programmed shutters are opened and closed in controlled sequences to change the thickness, composition and doping of the material being grown

10.2.2 MBE Growth Kinetics

Experiments based on modulated beam mass spectrometry (MBMS) [see [25] for a detailed discussion of the MBMS technique] have established that the group III elements vapourize to atoms on heating in vacuum, for example:

$$In(l) \xrightarrow{heat} In(g),$$

while the group V elements sublime as V_4 (tetrameric) molecules:

$$4As(s) \xrightarrow{heat} As_4(g).$$

It is often beneficial to use V_2 (dimeric) molecules since these can promote the growth of higher quality epitaxial layers. As_2, for example, is obtained by cracking As_4 in a secondary furnace held at a temperature $\geq 800°C$ and attached to the output of the tetramer source:

$$As_4(g) \xrightarrow{\geq 800°C} 2As_2(g).$$

Past MBMS research has also shed light on the surface kinetics of III–V MBE [26, 27]. Under normal growth conditions, group III atoms adhere to the wafer surface with unity sticking coefficient – s – i.e. once resident on the wafer surface, none of the atoms re-evaporates. In contrast, the sticking coefficient of the group V species is either 0.5 (As_4) or 1.0 (As_2) if there are free group III atoms on the substrate surface to which it can bond, but zero when there are none. Therefore, III–V MBE growth is effectively self-regulating, and it is sufficient to supply a small excess of the group V to ensure the deposition of stoichiometric layers. By way of illustration, gallium arsenide is grown from fluxes of Ga and either As_4 or As_2 (F atoms or molecules $cm^{-2} s^{-1}$) according to the following reactions:

$$4Ga(l) + 2As_4(g) \rightarrow 4GaAs(s) + As_4(g) \quad \text{with} \quad F_{As_4} > \frac{1}{2} F_{Ga} \quad [26]$$

$$2Ga(l) + As_2(g) \rightarrow 2GaAs(s) \quad \text{with} \quad F_{As_2} > \frac{1}{2} F_{Ga} \quad [27].$$

The MBE growth described later in this chapter was performed exclusively with As_2.

The unity sticking coefficient of group III atoms on a (100) III-V semiconductor surface at most temperatures of interest also means that the composition of an alloy layer such as $Al_xGa_{1-x}As$ is determined simply from a ratio of the incident group III beam fluxes: $x = F_{Al}/(F_{Ga}+F_{Al})$. Further, the growth rate of the epitaxial layer is governed by the total flux of group III atoms.

10.2.3 The Utility of Reflection High Energy Electron Diffraction in MBE

MBE growth takes place in an UHV environment, and in situ diagnostic equipment can be attached to the system to monitor and characterize epitaxial layer growth. The most important tool from the practical point of view is a reflection high energy electron diffraction (RHEED) facility, consisting of an electron gun with energy typically between 10 and 30 keV and a phosphor screen. These are mounted on opposite sides of the growth chamber so that the electron beam is made to interact with the wafer surface at a glancing angle of incidence before passing on to the phosphor screen where a diffraction pattern is revealed. In addition, the RHEED equipment allows the growth rate of the epitaxial layer to be determined with great precision by recording oscillations in the intensity of electrons specularly reflected from the layer surface when growth is commenced [28]. These oscillations are indicative of an essentially 2-D layer-by-layer or Frank-van der Merwe growth mode, with the completion of one cycle in the intensity of the specular spot corresponding to the deposition of one monolayer (ML) of epitaxial material. For the III-V semiconductors, 1 ML is equivalent to a thickness of $a_0/2$, where a_0 is the lattice constant. By recording growth over many oscillations on separate calibration substrates (GaAs for the Al and Ga fluxes, and InAs for the In flux), the individual group III fluxes can be set with high precision by adjusting the temperature of each cell.

The surface can be monitored by RHEED before, during and after growth. Both (Al,Ga)As and InAs exhibit an As-stable (2 × 4) diffraction pattern under an excess of arsenic when the electron beam interacts with the (100) surface of a wafer/epitaxial layer. The crystal directions and examples of the orthogonal ×2 and ×4 As-stable GaAs RHEED patterns are shown in Fig. 10.3a–c. The "streaky"

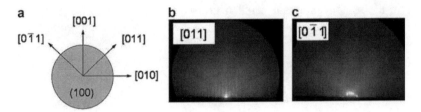

Fig. 10.3 (**a**) (100)-oriented III–V wafer with main crystal directions labelled. (**b**) Diffraction pattern generated by the (100)–(2 × 4) As-stable reconstructed surface of GaAs viewed by RHEED along the [011] direction; (**c**) the same surface viewed at 90°, along the [0$\bar{1}$1] direction. The whole surface structure would rotate 90° under Ga-stable conditions to become a (100)–(4 × 2) reconstruction. The spot associated with the specularly reflected electron beam is evident near the base of the 00 rod in (**b**)

nature of the diffraction pattern is further evidence for a surface which is nearly flat on the atomic scale. If there is an insufficient group V flux, the pattern flips through 90° to a (4 × 2) group III stabilized surface, which can deteriorate into a group III-rich surface, or one covered with liquid metal droplets.

10.3 MBE Growth of InAs Quantum Dots on (100)-GaAs

10.3.1 Introduction

The MBE of (Al,Ga)As on GaAs or InAs on InAs yields layers which are mirror-like in appearance after many microns of growth. Although the surface of the epitaxial layer acquires an average surface roughness with time which is manifest through a damping in the amplitude of the RHEED oscillations [28], near atomic flatness can be recovered by halting growth for a short period and maintaining the wafer at normal growth temperature in an arsenic flux. Such a growth interrupt is often employed to restore the smoothness of key interfaces during the deposition of a complex, multi-layer hetero-structure.

10.3.2 MBE Growth of InAs Quantum Dots

The MBE growth of InAs on GaAs proceeds initially layer-by layer (2-D) but changes quickly to 3-D once the thickness of InAs exceeds ∼1.7–2 ML. This is an example of Stranski–Krastanow (SK) growth and is illustrated schematically in Fig. 10.4. The spontaneous formation of coherent InAs islands or QDs on a thin InAs wetting layer supported by the GaAs substrate is a direct consequence of

10 The Quantum Dot Intermediate Band Solar Cell

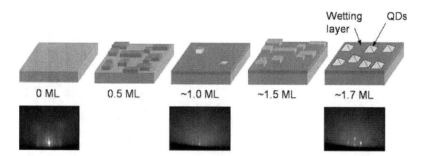

Fig. 10.4 A sequence of diagrams to illustrate the Stranski–Krastanow growth of InAs QDs on a GaAs substrate. Note the presence of the thin wetting layer. Both the QDs and the wetting layer will contain a proportion of Ga atoms due to intermixing. The streaky RHEED pattern for 0 ML corresponds to a (100)-GaAs surface with a c (4 × 4) reconstruction when viewed along the [011] direction. At 1 ML, a (2 × 4) As-stable (100)-InAs reconstruction is evident, and the specular spot intensity has recovered. After ∼1.7 ML, InAs QDs have formed, and these cause the RHEED pattern to change from streaky (2-D growth) to spotty (3-D growth). The exact thickness at which the self-assembled QDs form depends on the growth temperature and to a lesser extent the As flux. The RHEED images were recorded for an InAs deposition rate and the substrate temperature of ∼0.1 ML s^{-1} and ∼520°C, respectively

the strain generated by the ∼7% mismatch in lattice constants between the two materials. The precise thickness at which the self-assembled QDs form depends on the growth temperature and to a lesser extent the arsenic over-pressure [29]. Significantly, their nucleation can be detected through a change in the RHEED pattern, as illustrated by the images in Fig. 10.4. The streaky RHEED pattern for 0 ML corresponds to a (100)-GaAs surface with a c(4 × 4) reconstruction when viewed along the [011] direction. After 1 ML has been deposited, a (2 × 4) As-stable (100)-InAs reconstruction is evident, and the specular spot intensity has partially recovered. After ∼1.7 ML, InAs QDs have formed, which cause the RHEED pattern to change from streaky (2-D growth) to spotty (3-D growth). Pre-calibration of the In flux from RHEED intensity oscillations means that the critical thickness of InAs for the appearance of the QDs can be determined to fractions of a monolayer. The diffraction spots develop into chevrons or "arrow-heads" (Fig. 10.5) whose angle can provide insight into the crystal facets which define the shape of the InAs QDs [30]. Confirmation that 3-D InAs QDs are present also comes from post-growth atomic force microscopy (AFM) of the surface (see Fig. 10.6).

10.3.3 Capping of InAs QDs

An important step for technological reasons is the capping of the InAs QDs with (typically) GaAs. This is usually accomplished in two stages; first, a layer of GaAs is deposited at the QD growth temperature of ≤520°C to a thickness of 7–15 nm,

Fig. 10.5 RHEED pattern after the deposition of 2.3 ML of InAs. The diffraction spots seen immediately after the formation of the QDs (Fig. 10.4) have developed into chevrons

Fig. 10.6 AFM image, 1 μm × 1 μm in area, of 2.7 ML of InAs QDs formed by Stranski–Krastanow growth on (100)-GaAs at ∼520°C. The InAs deposition rate was 0.033 ML s^{-1}. The QD density is ∼2 × 10^{10} cm^{-2}, and the QD height and base length are ∼12 nm and ∼50 nm, respectively

depending on the initial height of the QDs. The GaAs covers the QDs and it displays a streaky c(4 × 4) RHEED reconstruction associated with the 2-D surface seen immediately prior to the deposition of the InAs QDs. Second, the wafer is heated to the normal GaAs growth temperature of ∼580°C where the As-stable (2 × 4) RHEED pattern is restored, and further GaAs is grown. Figure 10.7a shows a schematic of a dual-layer QD structure designed for photoluminescence (PL) spectroscopy of the InAs QDs capped with GaAs (buried layer), and for AFM on a surface layer of QDs to determine their approximate density and size distribution. The thickness of the GaAs cap is ≥100 nm to eliminate coupling between the two QD layers. However, the surface QDs will differ slightly from those in the buried layer for various reasons. These include how quickly the wafer temperature can be quenched following QD deposition to avoid Ostwald ripening, and the degree of alloying that takes place when the first layer of dots is capped with the GaAs. Alloying will increase their size and shift their bandgap [31]. It is found that the surface QDs make a negligible contribution to the total PL output.

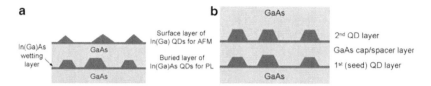

Fig. 10.7 (a) Diagram of a test structure designed to assess the optical properties of InAs QDs capped with GaAs (thickness ≥100 nm) by photoluminescence (PL) spectroscopy (buried layer), and to determine the approximate density and size distribution of the dots by AFM (surface layer). (b) Modified structure with two layers of QDs separated by ∼10 nm of GaAs. The dots in the second layer nucleate preferential over the dots in the "seed" layer due to residual strain at the surface of the GaAs cap/spacer layer

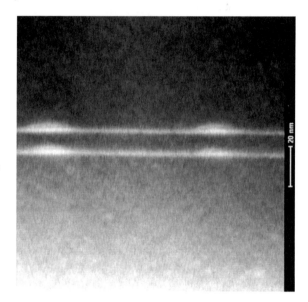

Fig. 10.8 Scanning transmission electron microscope (STEM) images using a High Annular Angle Dark Field (HAADF) detector. The contrast is due predominantly to Z-contrast, regions with high In-content showing up lighter than those composed of GaAs. Both the QDs and the wetting layer are seen clearly. Alignment of the dots in the second layer over the QDs in the "seed" layer is also evident

10.3.4 "Seeded" InAs QD Growth

When the GaAs cap thickness is restricted to ∼10 nm, a small strain field remains across its surface. This originates from the buried InAs QDs and results in a local reduction of the critical thickness of InAs needed to form a second layer of QDs, which therefore tend to nucleate directly over the underlying dots of the first (seed) layer (Fig. 10.7b). For the same total amount of InAs deposited, larger QDs with a tighter size distribution are created [32, 33]. The scanning transmission electron microscope (STEM) image shown in Fig. 10.8 using a High Annular Angle Dark Field (HAADF) detector illustrates clearly both the alignment of the dots in the second layer over the QDs in the "seed" layer and their larger size. The contrast in this image is due predominantly to Z-contrast, regions with high In-content showing up lighter than those composed of GaAs. Both the QDs and the wetting layer (WL) are well resolved.

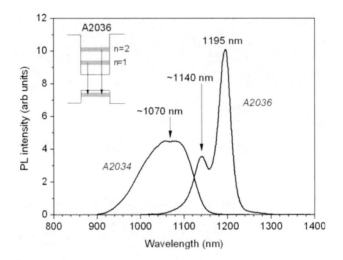

Fig. 10.9 11 K PL spectra for A2034 (single layer of InAs QDs, 2.5 ML thick) and A2036 (a bilayer consisting of two 2.5 ML layers of InAs QDs, separated by 7.5 nm of GaAs – see sketch in Fig. 10.7b). The broad peak of the PL spectrum for A2034 centred at ∼1,070 nm corresponds to smaller QDs with a much larger (and possibly bi-modal) size distribution compared with A2036. The inset shows an interpretation of the two main transitions seen for A2036. The $n = 1$ and $n = 2$ levels are shown slightly broadened to reflect the range of emission wavelengths generated by recombination in QDs with a small but finite distribution in sizes

11K PL spectroscopy also highlights the differences between "seeded" and "un-seeded" growth. The spectra shown in Fig. 10.9 are for a single 2.5 ML thick layer of InAs QD (A2034), whilst A2036 is a bi-layer consisting of two 2.5 ML thick InAs QD layers separated by a 7.5 nm GaAs cap/spacer layer. The PL emission from A2034 is broad and centred at ∼1,070 nm, indicating a wide distribution in dot sizes (the distribution may even be bi-modal). In contrast, the luminescence from A2036 is shifted to longer wavelengths and is narrower, confirming the growth of larger and more homogeneous QDs when a seeding layer is used. The spectrum of A2036 has (at least) two distinct peaks centred at 1,195 and 1,140 nm. The longer wavelength peak corresponds to transitions of electrons from the $n = 1$ bound conduction band state, while the shorter wavelength peak results from electrons recombining from the $n = 2$ bound state (see inset in Fig. 10.9). The intensity of the higher energy peak decreases rapidly as the PL pump intensity is reduced. Note that there is no evidence for PL emission from the seed layer of A2036.

10.3.5 Multi-Layer Stacks of InAs QDs

The cycle of growing InAs QDs followed by a GaAs capping layer can be repeated many times to produce a structure with stacked layers of QDs. The exact number that

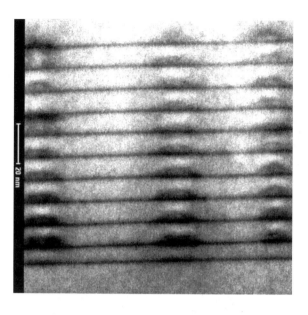

Fig. 10.10 Bright field STEM image of a cross section through ten InAs QD layers and a seed layer (bottom of the stack). The marker bar represents 20 nm. Each layer of QDs contains 2.5 ML of InAs. The GaAs spacer thickness over the seed layer is 7.5 nm, whilst the spacer thickness in the stack is 10 nm. The dots in the ten layers of the stack are larger than those in the seed layer, and are aligned vertically (Reprinted with permission from [11]. Copyright 2007. American Institute of Physics)

can be deposited before strain relaxation occurs and misfit/threading dislocations are generated is determined by the amount of InAs used to form each layer of QDs, and the thickness of the GaAs capping layers. A bright field scanning transmission electron microscope image of ordering of dots in a ten-layer stack is shown in Fig. 10.10. Vertical registration of the dots in successive layers along with the individual WLs are apparent, and there is no evidence for misfit/threading dislocations caused by strain relaxation. As mentioned in Sect. 10.1, the number of QD layers is essentially unlimited when a capping material in tensile strain is employed to off-set the residual compressive stress associated with the QDs/WLs [5–8, 12].

10.4 MBE Growth of InAs Quantum Dots for IBSCs

10.4.1 Introduction

The energy gaps associated with InAs QDs embedded in a GaAs matrix are not the optimum values required for an IBSC with the maximum conversion efficiency of 63.2% predicted by theory [1]. However, InAs-GaAs is a well researched materials system and thus provides a convenient vehicle for exploring both the physical concepts of the QD-IBSC and for assessing its performance. Before addressing the practical issues of growing suitable InAs QDs, it is worth summarising some QD-IBSC design parameters. These include (1) a host semiconductor with a band gap of ~1.95 eV (cf a 300 K GaAs band gap of 1.424 eV); (2) an IB positioned

~0.7 eV below the CB (large QDs); (3) a high areal density of QDs (to increase the absorptance) with a tight size distribution (to avoid significant overlap of absorption coefficients [34]); (4) multiple layers of QDs (again, to increase the absorptance); (5) doping to produce one electron per dot (half-filled IB), and (6) an IB electrically isolated from the *p*- and *n*-emitters of the cell.

10.4.2 Optimising the CB→IB Energy Separation

The CB→IB energy separation (EL) must be adjusted to maximise the IBSC conversion efficiency. Broadly, this requires the QDs to be as large as possible without losing their coherence, and their PL emission wavelength (QD CB→VB transition) to be as long as possible. The development of 1,300 nm QD laser diode poses similar growth issues [35]. In practice, ~3 ML of InAs can be deposited before dots begin to coalesce into larger dots [36]. There is very little increase in the QD emission wavelength beyond 3 ML and moreover, the integrated PL intensity from the QD layer decreases due to strain relaxation through the generation of misfit dislocations within the larger dots [37].

Capping the InAs QDs with GaAs causes erosion of material from the dots as well as alloying effects, both of which act to decrease E_L. Figure 10.11a shows a simplified energy band diagram for an InAs QD embedded in GaAs and includes the effect of the omni-present In(Ga)As WL. The WL level acts to lower the CB edge, and the band gap is further reduced by the many closely spaced QD confined hole states, which are likely to form a quasi-continuum with the VB [4]. The 300K band gap of the host semiconductor therefore shrinks from ~1.42 to

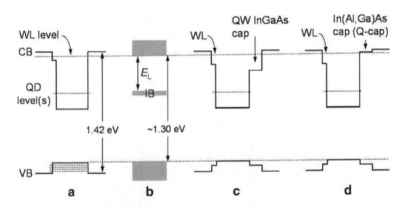

Fig. 10.11 (a) Energy band diagram for an InAs QD embedded in GaAs. Note the introduction of the level associated with the In(Ga)As wetting layer (WL). (b) Modified energy levels due to presence of the WL and the multiple confined QD hole states, resulting in band gap shrinkage. (c) Energy band diagram of an InAs QD capped with InGaAs and (d) In(Al,Ga)As. The conduction band discontinuity between the InGaAs and GaAs capping layers forms a quantum well, which can be virtually eliminated using In(Al,Ga)As or a Q-cap [38]

∼1.30 eV (Fig. 10.11b). Erosion of the QDs can be avoided by using a composite cap of InGaAs (a so-called strain-relief layer, or SRL) and GaAs, but this forms an unwanted quantum well (QW) due to the conduction band off-set between them (Fig. 10.11c). However, the conduction band discontinuity between the GaAs and the initial capping layer can be virtually eliminated by replacing InGaAs with InAlGaAs [38] (a quaternary- or Q-SRL [39]) (see Fig. 10.11d).

Ustinov et al. [37] have proposed an empirical formula for the total amount of InAs that can be deposited within the QDs and the SRL before strain relaxation sets in:

$$QD_\Sigma = QD_{InAs} + xd_{SRL} \sim 5ML,$$

where QD_Σ is the total amount of InAs in ML, QD_{InAs} is the amount of InAs deposited into the QDs, x is the composition fraction of the $In_x(Al_yGa_{1-y})_{1-x}As$ and d_{SRL} its thickness in ML. Given this constraint, it has been shown that the Q-SRL composition and thickness which produce the longest PL emission wavelength (large E_L), the narrowest PL line width (small QD size distribution) and the highest PL integrated intensity (low defect density) are $In_{0.2}Al_{0.2}Ga_{0.6}As$ and 2 nm, respectively. The two 11K PL spectra shown in Fig. 10.12 are for QDs formed from 2.7 ML of InAs deposited at ∼0.2 ML s^{-1} and capped with GaAs (measured PL intensity scaled by ×10) and a composite cap consisting of 2 nm of In(Al,Ga)As and 9 nm of GaAs (shown schematically in the inset of Fig. 10.12). Both cap layers were grown at the QD deposition temperature of ∼520°C. The composite cap produces a significant red-shift in the PL peak from ∼1.20 to 1.04 eV (1,192 nm)

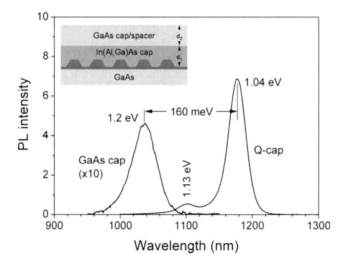

Fig. 10.12 Comparison of 11 K PL spectra for QDs formed from a 2.7 ML of InAs deposited at ∼0.2 ML s^{-1} with a GaAs cap (measured intensity increased by ×10) and a composite cap (Q-cap) consisting of 2 nm of In(Al,Ga)As and 9 nm of GaAs (see inset)

compared with the emission of the GaAs capped QDs. The peak with an energy of 1.13 eV corresponds to transitions from the $n = 2$ bound CB state of the QD, and decreases in intensity relative to the $n = 1$ transition as the PL pump power is reduced.

10.4.3 Multiple QD Layers

Ideally, an IBSC needs a large number of QD layers to maximise the absorption of sub-band gap photons and thus the additional short-circuit current it generates, whilst preserving the open-circuit voltage of the cell. Without the use of strain-compensating capping layers, the total number of QD layer + cap repeats before strain relaxation occurs is restricted by the amount of In incorporated in the QDs and the SRL (as discussed in Sect. 10.4.2) and the thickness of the In-free capping layer. There is a degree of choice here in the selection of the cap layer thickness. A thin capping layer will result in electronic coupling between QD layers and the formation of a mini-band, whereas a thick cap will suppress inter-layer electronic coupling. Figure 10.13 shows, for comparison, the 11 K PL spectra for a ×20 stack of 2.7 ML InAs QDs (capped with 3 nm of Q-SRL and 7 nm of GaAs – a thin cap) and a single 2.7 ML InAs QD layer, again with a composite Q-SRL plus GaAs cap.

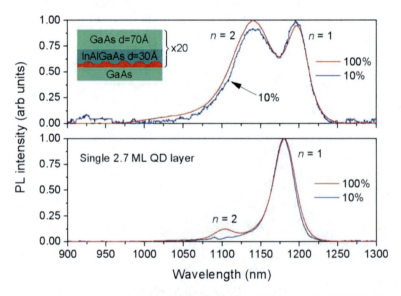

Fig. 10.13 *Upper trace*: normalised 11 K PL spectra for a 20-layer stack of 2.7 ML InAs QDs capped with a Q-SRL and GaAs. Lower trace: normalised 11 K PL spectra for a single 2.7 ML InAs QD layer also capped with a Q-SRL and GaAs. The full (100%) PL pump power and wavelength are ∼10 mW and 633 nm, respectively; the focussed beam diameter is ∼100 μm

Apart from the different peak output wavelengths, the variations of PL output as the pump intensity is varied of over 2–3 orders of magnitude or more, are distinctly different. For the case of the single QD layer (lower plot in Fig. 10.13), the intensity of the shorter wavelength $n = 2$ transition decreases quickly relative to the $n = 1$ transition. A stack with thick spacers exhibits similar behaviour (see Fig. 10.15). In contrast, the ratio of the $n = 2$ to $n = 1$ transition for the ×20 stack with thin spacers remains essentially constant (upper plot in Fig. 10.13), probably indicating that there is electronic coupling between layers and two partially overlapping bands have formed.

Inter-layer coupling may be desirable, for example, in order to tailor the IB→CB absorption coefficient so that a better overlap between wave-functions of the two bands exists [40]. In contrast to the QD-IBSC, the formation of an electronic band in bulk intermediate band materials is essential in order to inhibit non-radiative recombination [41].

10.4.4 Populating the IB with Electrons

One electron per QD is required to half-fill the IB, and this can be achieved by modulation doping. The QDs are deposited and capped with In(Al,Ga)As and GaAs, growth is interrupted and Si donor impurities are supplied to produce δ- or atomic plane doping. Growth is then recommenced and the GaAs cap layer completed, burying the Si atoms within the GaAs. The growth temperature for this series of steps is held constant at ~520°C, partly to avoid diffusion/surface segregation of the Si. It should be mentioned, however, that there is no obvious way of controlling the spatial distribution of Si atoms so that every QD captures just one electron.

Since the QD density is low compared with the atomic density of GaAs, the Si flux and the deposition time are critical to achieving accurate Si concentrations. The Si flux is determined from Hall measurements on a separate Si-GaAs sample grown at $1\,\mu\text{m}\,\text{hr}^{-1}$ with a target Si density around $4 \times 10^{16}\,\text{cm}^{-3}$. This leads to an acceptable Si deposition time of 36 s to produce an areal density of $4 \times 10^{10}\,\text{cm}^{-2}$ (the approximate QD density) with the same Si furnace temperature.

10.5 Manufacture of QD-IBSC Die

10.5.1 Introduction

The QD-IBSC and reference (Al,Ga)As PV structures have been grown on 3" diameter Si-GaAs(100) substrates in a solid source MBE system using As$_2$. Si has been used for n-type doping, and Be and C have been compared as p-type dopants. Two sets of QD-IBSCs have been produced; (1) devices incorporating 10-, 20- or 50-period QD stacks with either GaAs caps/spacer layers, or composite

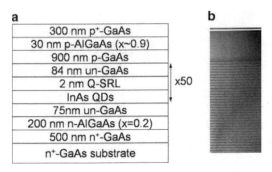

Fig. 10.14 (a) Schematic diagram of a full MBE-grown IBSC structure incorporating a ×50 period QD stack. The QDs are capped with 2 nm of In(Al,Ga)As and 9 nm of GaAs (Si δ-doping not shown), followed by a further 75 nm of GaAs. (b) Low magnification cross-sectional TEM image of the epitaxial structure taken under dark field 002 imaging conditions. The bright band at the top of the image is the 90% AlGaAs window layer. TEM by S.I. Molina, A. M. Sánchez and T. Ben, Universidad de Cádiz (Reprinted with permission from [42]. Copyright 2010. American Institute of Physics)

caps/spacer layers comprising In(Al,Ga)As+GaAs, both with total thicknesses ≤ 12 nm; (2) IBSCs with either 30 or 50 periods of QDs and a composite In(Al,Ga)As+GaAs cap/spacer layer with a total thickness of 84 nm. A diagram of the latter structure incorporating a ×50 period QD stack is shown in Fig. 10.14a (the Si δ-doping has been omitted). A low magnification cross-sectional TEM image of the corresponding epitaxial structure taken under dark field 002 imaging conditions (Fig. 10.14b) indicates that the structure is free of crystallographic defects. The upper $Al_{0.9}Ga_{0.1}As$ window layer is clearly visible as a bright band in the image. Examination of the QDs under higher magnification shows they have a lens shape with an average base length in the range 15–20 nm and a height between 5–7 nm.

A measure of the reproducibility of the MBE growth from one QD layer to the next, and between different growth runs can be gauged from the 300 K PL spectra recorded with a Fourier transform infra-red (FTIR) spectrometer and InGaAs detector, and shown in Fig. 10.15. The emission associated with the IB-VB transitions has a full width at half maximum (FWHM) line-width of only $\sim$30 meV, and the peak emission wavelengths are $\sim$1,272 nm and $\sim$1,279 nm for the ×30 and ×50 QD stacks, respectively. These wavelengths are comparable with the value for a single layer of QDs with a composite cap (see 11 K PL spectrum in Fig. 10.12).

The incorporation of (In,Ga)As QDs within the structure of an (Al,Ga)As solar cell does not present any special manufacturing issues, and QD-IBSCs are therefore produced with the same process techniques as used for the conventional (Al,Ga)As solar cells. In order to analyse the contribution of the QD-IB region to the solar cell output, reference III-V solar cells (without QDs, but otherwise nominally identical) are prepared along with the QD-IBSCs.

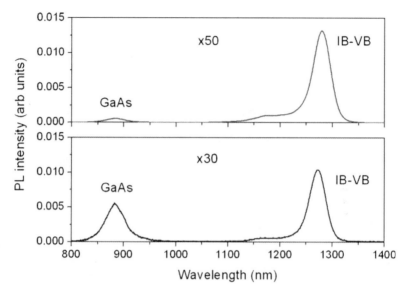

Fig. 10.15 300 K PL spectra from QD-IBSC *p-n* structures, recorded with an FTIR spectrometer and InGaAs detector. The spectrum for the ×30 QD layer device (*lower curve*) has been increased in amplitude by a factor of 10 compared with the ×50 layer one (*upper curve*). Note that the integrated intensities of the peaks at ∼ 883 nm associated with emission from the p$^+$-GaAs contact/cap layers are very similar

A summary of a typical fabrication process flow for making *p*-on-*n* concentrator solar cells is as follows, although the order of the individual process steps may vary:

1. Deposition of ohmic metal layers (e.g. AuGe) on the reverse side of the n^+-GaAs substrate.
2. Definition of busbar/gridline ohmic metal lift-off pattern using photoresist.
3. Deposition of p ohmic metal layers (e.g. TiPdAu) followed by lift-off.
4. Rapid thermal annealing of ohmic contacts.
5. Definition of busbar/gridline plating mold using photoresist.
6. Plating of busbar/gridline using gold electroplating solution.
7. Definition of isolation trench using photoresist.
8. Etching of isolation trench using chlorine-based plasma etch.
9. Deposition of passivation layer (e.g. SiN$_x$) on isolation trench sidewall.
10. Selective wet etching of p^+-GaAs contact layer over AlGaAs window.
11. Deposition of either a single layer SiN$_x$ or a dual layer ZnS/MgF$_2$ anti-reflection coating and lift-off.

The various levels of the fabricated die are depicted by the schematic diagram in Fig. 10.16, and a plan-view optical micrograph of a completed device is shown in Fig. 10.17. In the following sections, the main process modules are described briefly.

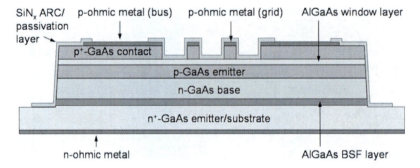

Fig. 10.16 Schematic diagram showing a cross-section through a processed GaAs solar cell. The QD-IBSC is similar except for the inclusion of the QD layers within the n-GaAs base region

Fig. 10.17 Plan-view optical image of a concentration solar cell (mounted and bonded on a custom-designed heat sink)

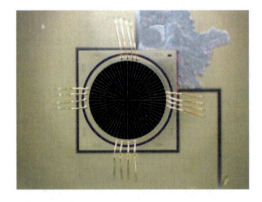

10.5.2 Low Resistance Ohmic Metallisations to n- and p-Type Emitters

The n-metallisation is a AuGeNi multi-layer, which is deposited over the whole of the back-side of the n^+-GaAs wafer. The TiPdAu p-metallisation layer on the light-receiving semiconductor surface takes the form of a grid whose precise geometry involves a trade-off between the requirements for low contact and in-plane resistances to current flow, and minimal shadowing in order to maximize the collection of the incident sunlight.

The QD-IBSC and benchmark GaAs solar cells share the same front metal grid design, often optimised for maximum efficiency of the benchmark solar cell under test conditions. This may be a reasonable comparison, in the first instance. However, given the promise of enhanced current output from a QD-IBSC, it follows that the front metal grid should be optimised separately in order to fully realise the efficiency gains of the QD-IBSC operated under concentration.

The peak efficiency of the solar cell is improved by the use of a grid design that is based upon lines with relatively high aspect ratio ($\sim$1:1) and conductivity (gold or silver). For example, Fig. 10.18 demonstrates 4 μm-high by 4 μm-wide gold metal

10 The Quantum Dot Intermediate Band Solar Cell

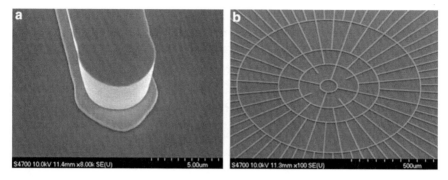

Fig. 10.18 (a) A scanning electron microscope (SEM) image of part of a Au-plated top surface grid designed for a solar cell operating under concentrations up to ×1,000 suns. The metal line is ~4 μm wide and ~4 μm high. The thin, 5 μm wide "seed" layer for the Au-plating process in direct contact with the p^+-GaAs is visible. (b) A lower magnification SEM image to show a larger area of the grid contact

lines formed by electro plating onto 5 μm-wide TiPdAu ohmic contact lines through a photoresist mold.

10.5.3 Definition of Diode Area

A non-selective wet-etched or a dry-etched mesa, for example chlorine-based reactive ion etching (RIE) or inductivity-coupled plasma etching (ICP), provides electrical confinement for photo-generated carriers. The etched sidewall must be passivated to suppress recombination currents associated with the perimeter. SiN_x can be used for this purpose.

10.5.4 Exposure of AlGaAs Window Layer

A highly doped p^+-GaAs contact layer, typically 300–600 nm thick, is included over the p-$Al_xGa_{1-x}As$ window layer ($x \sim 0.8$–0.9). This exhibits high absorption and recombination and must be removed between the gridlines. A self-aligned selective wet etch (e.g. $C_6H_8O_7$-H_2O_2 or NH_4OH-H_2O_2 for selective etching of GaAs over $Al_xGa_{1-x}As$) is generally employed for this purpose.

10.5.5 Broad Band Anti-Reflection Coating

A broadband anti-reflection coating (ARC) is essential to reduce Fresnel losses over the spectral range in which the product of the solar spectral irradiance and internal quantum efficiency of the solar cell is significant. A single layer of SiN_x or a ZnS/MgF_2 bilayer ARC used in conventional single junction solar cells may suffice for experimental comparisons. However, neither has sufficient bandwidth to cover the proposed operational range of the QD-IBSC. To realise the full potential of a QD-IBSC, ARCs must be developed that provide low reflection and absorption over the visible and near-IR regions (i.e. around $0.4 < \lambda < 3\,\mu m$) [43].

10.6 Characterisation of Quantum Dot Intermediate Band Solar Cells

Numerous semiconductor and electronic device characterization techniques can, in general, be applied to QD-IBSC devices. However, we will focus in this section on techniques that have been used to demonstrate basic but important principles of operation of the IBSC, namely, the production of sub-bandgap photo-current and the preservation of the open-circuit voltage.

Quantum efficiency measurements on very early QD-IBSC devices verified the production of photo-current for below bandgap energy photons (Fig. 10.19). However, some words of explanation are required. For the ideal IBSC to operate with maximum theoretical efficiency, there must be no overlap between the absorption coefficients associated with each of the three optical transitions involved. In other words, this photon selectivity implies that if, for example, a photon can be absorbed by means of an electronic transition from the IB to the CB, it cannot be absorbed through any other transition. Typical quantum efficiency measurements (without bias light) are taken by shining a known amount of monochromatic light on the

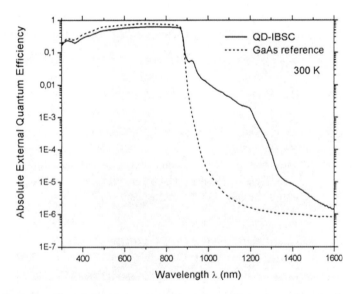

Fig. 10.19 Comparison of the absolute external quantum efficiencies versus wavelength for an InAs/GaAs QD solar cell and a single gap GaAs reference cell (Reprinted with permission from [44]. Copyright 2007. Elsevier B.V.)

10 The Quantum Dot Intermediate Band Solar Cell

cell and measuring the current produced under short-circuit conditions. Therefore, the fact that the quantum efficiency of an IBSC shows photo-response for below bandgap energy photons in this measurement reveals that the device is not ideal. In fact, the monochromatic light may be pumping electrons from the VB to the IB, but electrons subsequently escape from the IB to the CB thermally or by tunneling, that is, without a second photon being absorbed from the IB to the CB [42]. If this is the case, the IBSC theory predicts that the IB cell will enhance its photo-current when compared to the single gap host solar cell but will not increase its efficiency beyond that achievable by a mere bandgap tailoring.

Therefore, it is necessary to apply characterization techniques in order to demonstrate that the absorption of a second photon causing transitions from the IB to the CB actually takes place. A suitable method is based on a modified version of the monochromatic quantum efficiency measurement in which a second light

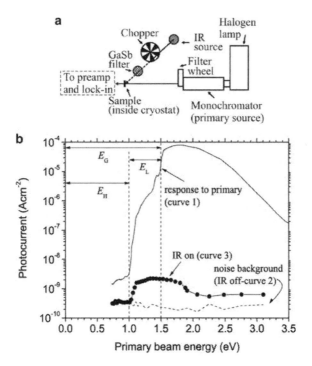

Fig. 10.20 (a) Modified quantum efficiency set-up suitable for demonstrating the production of photo-current from the absorption of two below bandgap energy photons. (b) Results of the measurement showing the photo-current produced in the QD-IBSC samples as a function of the energy of the photons of the primary light source. Curve 1 (response to primary) represents the photo-current produced when pumping with the chopped primary source only (cell biased at $-1.5\,\text{V}$, $T = 4.2\,\text{K}$). Curve 2 (noise background) is the photo-current measured when the IR source is off while the chopper, located in front of the IR source, is kept spinning (cell biased at $0\,\text{V}$, $T = 36\,\text{K}$). Curve 3 (IR on) is the photo-generated current when the IR source is turned on and chopped (cell biased at $0\,\text{V}$, $T = 36\,\text{K}$). (Reprinted with permission from [45]. Copyright 2006. American Physical Society)

source, with its wavelength tuned only to pump electrons from the IB to the CB (IR source), is used simultaneously with the monochromatic and scanning light source (Fig. 10.20a). If the transitions from the IB to the CB are not already saturated by thermal or tunnel escape, the photons from the second light source should increase the cell photo-current. At room temperature, when the bandgap from the IB to the CB is small (as is the case in a practical InAs/GaAs QD-IBSC), the IB to CB transition is usually saturated and it is necessary to perform the experiment at low temperatures to decrease the thermal escape component. Also, the experimental IB to CB absorption is so small that lock-in techniques have to be used to detect the associated increment in photo-current. Nevertheless, the technique has been successfully used to demonstrate that electron hole pairs are formed in InAs/GaAs QD-IBSCs from the absorption of two below bandgap energy photons [45].

Once the feasibility of the absorption of two below bandgap energy photons has been proved, the next step is to demonstrate that the IB does not limit the open-circuit voltage of the cell. This is known as "voltage preservation" and implies that the open-circuit voltage of the cell is limited only by the total bandgap. Because practical InAs/GaAs QD devices are still dominated at room temperature by thermal escape, voltage preservation can only take place at low temperatures. Achieving this effect at higher temperatures implies devices with a larger bandgap E_L (Fig. 10.11).

Figure 10.21 [46] shows the open-circuit voltage of some QD-IBSCs at different temperatures compared with a single gap GaAs reference cell (Fig. 10.21a). The remarkable result, connected with the arguments outlined above, is that, like the GaAs reference cell, the open-circuit voltage of a QD-IBSC is also capable of approaching the total bandgap of the cell (Fig. 10.21c) at reduced temperatures where the escape mechanism(s) from the IB to the CB has been suppressed. In contrast, a QD-IBSC in which these mechanisms have not been effectively suppressed is not capable of preserving the voltage even when the temperature is decreased (Fig. 10.21b).

10.7 Conclusions

The intermediate band solar cell has been the focus of increasing research since it was first proposed in 1997 [1]. Its attractiveness lies in its potential for very high conversion efficiencies, which are predicted to reach more than 60% under maximum concentration of sunlight, without the need for multi-junctions. Real devices, in particular ones based on InAs QDs embedded within a single (Al,Ga)As p–n junction, have been reviewed in this chapter. Other, bulk materials with IB characteristics have been identified and are discussed elsewhere in this book. A key principle of operation of the IBSC is the generation of electron-holes pairs by means of the absorption of two sub-bandgap energy photons which produce a transition between the VB→IB and second one between the IB→CB, resulting in additional photo-current in an external circuit. Our understanding of how the QD-IBSC

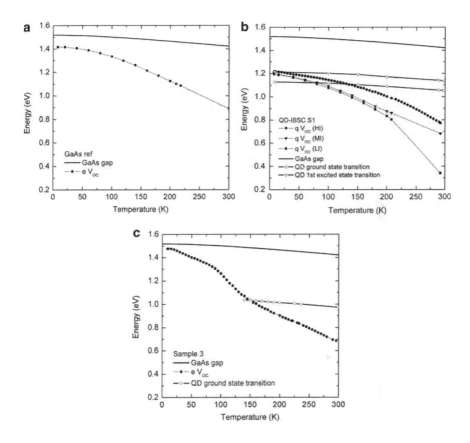

Fig. 10.21 Comparison between open-circuit voltage (filled symbols) and location of the intermediate band energy levels (open symbols) at different temperatures for: (**a**) a GaAs reference cell, (**b**) a QD-IBSC in which thermal and tunnel escape from the IB to the CB are *not* effectively suppressed (LI, MI, and HI stand for low, medium and high illumination, respectively) and (**c**) a QD-IBSC in which thermal and tunnel escape from the IB to the CB are effectively suppressed. The GaAs bandgap (solid line) is shown for reference in all three plots. (Reprinted with permission from [46]. Copyright 2010. IEEE)

works, as well as its limitations, has been helped by the development of advanced characterisation techniques which have been used to compare the performance of QD-IBSCs against reference (un-modified) single junction (Al,Ga)As cells. Results from these comparative measurements have informed refinements to QD-IBSC epitaxial designs. Thus far, the main operating principles of the IBSC have been demonstrated. These include the generation of photo-current by the absorption of two sub-bandgap photons, and voltage preservation when thermal and/or tunnelling-assisted escape from the IB are suppressed. However, efficiencies in excess of those for un-modified (Al,Ga)As reference cells have not yet been achieved. QD-IBSCs with improved efficiencies will require the manufacture of devices with bandgaps closer to the optimum values, and the application of "photon harvesting" to increase their absorptance if the material system itself is not capable of providing high sub-bandgap absorption coefficients [3].

Acknowledgements This work has been supported by the European Commission through grant number 211640, IBPOWER and NANOGEFFES (ENE2009-14481-C02-01) funded by the Spanish National Programmes. The authors are indebted to numerous colleagues at the University of Glasgow and the Universidad Politécnica de Madrid for their contributions to the work described in this chapter.

References

1. A. Luque, A. Martí, Phys. Rev. Lett. **78**(26), 5014 (1997)
2. A. Luque, F. Flores, M. A., J.C. Conesa, P. Wahnon, J. Ortega, C. Tablero, R. Pérez, L. Cuadra, Intermediate band semiconductor photovoltaic cell (2002). US Patent no. 6,444,897
3. A. Luque, A. Martí, Adv. Mater. **22**(2), 160 (2010)
4. A. Martí, E. Antolín, E. Cánovas, N. López, P.G. Linares, A. Luque, C.R. Stanley, C.D. Farmer, Thin Solid Films **516**(20), 6716 (2008)
5. R. Oshima, T. Hashimoto, H. Shigekawa, Y. Okada, J. Appl. Phys. **100**(8), 083110 (2006)
6. D. Alonso-Álvarez, A.G. Taboada, J.M. Ripalda, B. Alen, Y. González, L. González, J.M. García, F. Briones, A. Martí, A. Luque, A.M. Sánchez, S.I. Molina, Appl. Phys. Lett. **93**(12), 123114 (2008)
7. R.B. Laghumavarapu, M. El-Emawy, N. Nuntawong, A. Moscho, L.F. Lester, D.L. Huffaker, Appl. Phys. Lett. **91**(24), 243115 (2007)
8. S.M. Hubbard, C.D. Cress, C.G. Bailey, R.P. Raffaelle, S.G. Bailey, D.M. Wilt, Appl. Phys. Lett. **92**(12), 123512 (2008)
9. A. Martí, L. Cuadra, A. Luque, in *Conference Record of the Twenty-Eighth IEEE Photovoltaic Specialists Conference* (2000), pp. 940–943. Anchorage, USA, 15–22 Sep 2000
10. S. Tomić, T.S. Jones, N.M. Harrison, Appl. Phys. Lett. **93**(26), 263105 (2008)
11. A. Martí, N. López, E. Antolín, E. Cánovas, A. Luque, C.R. Stanley, C.D. Farmer, P. Díaz, Appl. Phys. Lett. **90**(23), 233510 (2007)
12. Y. Okada, N. Shiotsuka, H. Komiyama, K. Akahane, N. Ohtani, in *Conference Record of the 20th European Photovoltaic Solar Energy Conference and Exhibition* (2005), pp. 51–54. Barcelona, Spain, 6–10 Jun 2005
13. K.M. Yu, W. Walukiewicz, J. Wu, W. Shan, J.W. Beeman, M.A. Scarpulla, O.D. Dubon, P. Becla, Phys. Rev. Lett. **91**(24), 246403 (2003)
14. W.M. Wang, A.S. Lin, J.D. Phillips, W.K. Metzger, Appl. Phys. Lett. **95**(26), 011103 (2009)
15. A. Martí, C. Tablero, E. Antolín, A. Luque, R.P. Campion, S.V. Novikov, C.T. Foxon, Sol. Energ. Mater. Sol. Cell. **93**(5), 641 (2009)
16. V. Pacbutas, G. Aleksejenko, A. Krotkus, J.W. Ager, W. Walukiewicz, H. Lu, W.J. Schaff, Appl. Phys. Lett. **88**(19), 092110 (2006)
17. J. Olea, M. Toledano-Luque, D. Pastor, G. Gónzalez-Díaz, I. Mártil, J. Appl. Phys. **104**(1), 016105 (2008)
18. G. Gónzalez-Díaz, J. Olea, I. Mártil, D. Pastor, A. Martí, E. Antolín, A. Luque, Sol. Energ. Mater. Sol. Cell. **93**(9), 1668 (2009)
19. P. Wahnon, C. Tablero, Phys. Rev. B **65**(16), 165115 (2002)
20. C. Tablero, P. Wahnon, Appl. Phys. Lett. **82**(1), 151 (2003)
21. I. Aguilera, P. Palacios, P. Wahnon, Thin Solid Films **516**(20), 7055 (2008)
22. D. Fuertes Marrón, A. Martí, A. Luque, Thin Solid Films **517**(7), 2452 (2009)

23. B. Marsen, L. Steinkopf, A. Singh, H. Wilhelm, I. Lauermann, T. Unold, R. Scheer, H.W. Schock, Sol. Energ. Mater. Sol. Cell. **94**(10), 1730 (2010)
24. H. Freller, K.G. Gunther, Thin Solid Films **88**(4), 291 (1982)
25. C.T. Foxon, M.R. Boudry, B.A. Joyce, Surface Sci. **44**(1), 69 (1974)
26. C.T. Foxon, B.A. Joyce, Surface Sci. **50**(2), 434 (1975)
27. C.T. Foxon, B.A. Joyce, Surface Sci. **64**(1), 293 (1977)
28. J.H. Neave, B.A. Joyce, P.J. Dobson, N. Norton, Appl. Phys. A-Mater. Sci. Process. **31**(1), 1 (1983)
29. S.D. Bimberg, M. Grundmann, N.N. Ledentsov, *Quantum Dot Heterostructures* (Wiley, New York, 1999)
30. H. Lee, R. Lowe-Webb, W.D. Yang, P.C. Sercel, Appl. Phys. Lett. **72**(7), 812 (1998)
31. P.B. Joyce, T.J. Krzyzewski, G.R. Bell, B.A. Joyce, T.S. Jones, Phys. Rev. B **58**(24), 15981 (1998)
32. P.B. Joyce, E.C. Le Ru, T.J. Krzyzewski, G.R. Bell, R. Murray, T.S. Jones, Phys. Rev. B **66**(7), 075316 (2002)
33. P. Howe, E.C. Le Ru, E. Clarke, B. Abbey, R. Murray, T.S. Jones, J. Appl. Phys. **95**(6), 2998 (2004)
34. L. Cuadra, A. Martí, A. Luque, IEEE Trans. Electron Dev. **51**(6), 1002 (2004)
35. M. Grundmann, Phys. E-Low-Dimensional Syst. Nanostructures **5**(3), 167 (1999)
36. J.M. Moison, F. Houzay, F. Barthe, L. Leprince, E. Andre, O. Vatel, Appl. Phys. Lett. **64**(2), 196 (1994)
37. V.M. Ustinov, A.E. Zhukov, A.R. Kovsh, N.A. Maleev, S.S. Mikhrin, A.F. Tsatsuñikov, M.V. Maximov, B.V. Volovik, D.A. Bedarev, P.S. Kopév, Z.I. Alferov, L.E. Vorobév, D.A. Firsov, A.A. Suvorova, I.P. Soshnikov, P. Werner, N.N. Ledentsov, D. Bimberg, Microelectron. J. **31**(1), 1 (2000)
38. P.G. Linares, C.D. Farmer, E. Antolín, S. Chakrabarti, A.M. Sánchez, T. Ben, S.I. Molina, C.R. Stanley, A. Martí, A. Luque, Energy Procedia **2**(1), 133 (2010)
39. R. Heitz, N.N. Ledentsov, D. Bimberg, A.Y. Egorov, M.V. Maximov, V.M. Ustinov, A.E. Zhukov, Z.I. Alferov, G.E. Cirlin, I.P. Soshnikov, N.D. Zakharov, P. Werner, U. Gosele, Phys. E-Low-Dimensional Syst. Nanostructures **7**(3–4), 317 (2000)
40. A. Martí, E. Antolín, E. Cánovas, N. López, A. Luque, C.R. Stanley, C.D. Farmer, P. Díaz, C. Christofides, M. Burhan, in *Conference Record of the 21st European Photovoltaic Solar Energy Conference and Exhibition* (2006), pp. 99–102. Dresden, Germany, 4–8 Sep 2006
41. A. Luque, A. Martí, E. Antolín, C. Tablero, Phys. B-Condens. Matter **382**(1–2), 320 (2006)
42. E. Antolín, A. Martí, C.D. Farmer, P.G. Linares, E. Hernández, A.M. Sánchez, T. Ben, S.I. Molina, C.R. Stanley, A. Luque, J. Appl. Phys. **108**, 064513 (2010)
43. D.J. Aiken, Sol. Energ. Mater. Sol. Cell. **64**(4), 393 (2000)
44. E. Antolín, A. Martí, C.R. Stanley, C.D. Farmer, E. Cánovas, N. López, P.G. Linares, A. Luque, Thin Solid Films **516**(20), 6919 (2008)
45. A. Martí, E. Antolín, C.R. Stanley, C.D. Farmer, N. López, P. Díaz, E. Cánovas, P.G. Linares, A. Luque, Phys. Rev. Lett. **97**(24), 247701 (2006)
46. E. Antolín, A. Martí, P. García Linares, I. Ramiro, E. Hernández, C.D. Farmer, C.R. Stanley, A. Luque, in *Conference Record of the 35th IEEE Photovoltaic Specialists Conference* (2010), pp. 65–70. Honolulu, USA, 20–25 Jun 2010

Chapter 11
Thin-Film Technology in Intermediate Band Solar Cells: Advanced Concepts for Chalcopyrite Solar Cells

David Fuertes Marrón

Abstract Combining the two key factors of high performance and low cost into a single solar cell is the major challenge of research on photovoltaics. It is not easy to conceive a practical approach to such a device if not based on thin-film technology. Yet, it appears equally clear that current thin-film solar cells must upgrade their performance by some means in order to meet satisfactory energy conversion efficiencies. The incorporation of novel photovoltaic concepts, particularly the intermediate band solar cell, into thin-film technologies is expected to cross-fertilize both fields. In this chapter, we will outline the potential benefits of thin-film intermediate band solar cells (TF-IBSC) and describe two different approaches toward its practical implementation. Particular attention will be devoted to devices based on chalcopyrite absorbers, currently leading the efficiency records of thin-film solar cells, and characterized by material properties well suited for this purpose.

Thin-film photovoltaics is emerging as an important contributor to the development of solar electricity. For the first time in the history of photovoltaics, cost figures below 1 \$/Wp have been reported and, not surprisingly, the announcement was made by a thin-film solar cell manufacturer [1]. This milestone has confirmed the expectations on achieving low costs with reasonable efficiencies put on the technology. At the same time, some of the so-called novel photovoltaic concepts, so designed as to break the fundamental limitations imposed by thermodynamically consistent detailed balance considerations on conventional devices, are closer to reality than ever. Among such concepts, the intermediate band solar cell [2] is fairly well situated: key operational principles have been demonstrated and first prototypes are being fabricated and investigated [3].

D.Fuertes Marrón (✉)
Instituto de Energía Solar – ETSIT, Universidad Politécnica de Madrid
e-mail: dfuertes@ies-def.upm.es

Generally speaking thin-film technologies and novel concepts as a whole have been classified into two different fields [4], deserving separate symposia at international photovoltaic conferences and independent technical manuals.[1] However, some novel concepts, and particularly the intermediate band (perhaps also the hot carrier solar cell), could be implemented as a thin-film solar cell. We will discuss this idea further in some detail, highlighting the intrinsic advantages that thin-film technologies can offer to the realization of novel photovoltaic devices and pointing out the critical issues that will surely appear on the way.

11.1 Why Not Thin-Film Intermediate Band Solar Cells? Theoretical Aspects

First prototypes of intermediate band solar cells consist of nanostructured III–V compound semiconductors grown by molecular beam epitaxy, with careful control over growing conditions to guarantee structures with high crystal quality, as discussed by Stanley in a previous chapter. While this approach is highly attractive for investigation purposes (demonstrating operational principles), the question arises as to whether such a path is the proper way to go in terms of costs and throughput to meet the terawatt challenge that photovoltaics is to face in the coming years [5]. It could be argued that such expensive and technologically demanding approaches are still affordable when combined with optical concentration systems, similar to highly efficient multijunction devices. Furthermore, theoretical studies suggest that the use of sunlight concentration may help squeeze the full potential out of the intermediate band solar cell, alleviating voltage losses associated with recombination events via the intermediate band, even in the radiative limit, i.e., when all recombination events proceed radiatively [6]. The potential of nanostructured, wafer-based technology for the fabrication of intermediate band solar cells appears thus linked to the high-cost end of photovoltaic production technologies.

On the other hand, thin-film solar cells, which are designed to operate under one sun irradiation, have reduced the cost of the photovoltaic Wp more significantly than any technology based on wafer processing. This reduction stems from the small volume of active material required for fabricating operative cells. Indeed, the cost related to the physical substrate onto which the solar module is fabricated (typically glass, metal or plastic foils) is already becoming a major fraction of the overall cost figure [5]. Obviously, the challenge for any cheap thin-film production, once the production costs are already low, is to increase the efficiency of the devices. But how? Could the intermediate band concept be of any help for this purpose? Also, would it be possible to find a mass-production technology at the low-cost end capable to supply high-efficiency solar cells?

---

[1] See, for example, the programs of the past main photovoltaic conferences worldwide, http://www.photovoltaic-conference.com, http://www.ieee-pvsc.org.

11 Thin-Film Technology in Intermediate Band Solar Cells

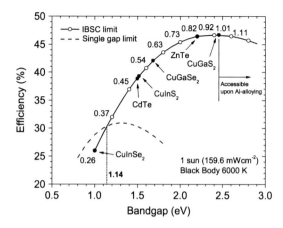

Fig. 11.1 Detailed balance calculation of the theoretical limiting efficiency of an intermediate band solar cell when operated at one sun (simulated by a black body at 6,000 K) as a function of the total semiconductor bandgap E_G. Figures in the plot indicate the optimum value (in eV) of the lowest sub-bandgap formed between the intermediate band and the closest main band edge. The detailed balance limiting efficiency of single gap solar cells is also shown for comparison (*dashed line*) (Reprinted with permission from [7]. Copyright 2008. American Institute of Physics)

Theoretical calculations have given what seems a positive answer to these questions. Figure 11.1 shows the calculated theoretical efficiency under ideal conditions (particularly, in the radiative limit) and one-sun irradiation (simulated as from a black body at 6,000 K) of an intermediate band solar cell as a function of its main bandgap (solid line) [7]. The curve shows a broad maximum above 46.7% efficiency for an optimal bandgap slightly above 2.4 eV. Numbers along the solid curve denote the optimal position within the bandgap (properly speaking, the relative energy in eV) of the intermediate band with respect to either the conduction or the valence band. The reason for this symmetry around midgap energy is that in the ideal case of infinite mobilities the sequential absorption of two photons leads to either a free electron or a hole behaving as a minority carrier in an equally efficient manner [2].

The dashed line in Fig. 11.1 represents the detailed balance calculation of maximum efficiency as calculated by Shockley and Queisser for a single gap absorber, i.e., the most that could be achieved by a conventional device (not just thin-film based) under one sun irradiation [8]. Three issues deserve attention when comparing the detailed balance calculations for the intermediate band solar cell (solid) and its single gap counterpart (dashed):

- The maximum theoretical efficiency achievable under one sun irradiation when incorporating an intermediate band in the device represents an improvement of more than 53% relative to that of the ideal single gap device.
- In the case of an intermediate band device, the optimal bandgap under the same irradiation conditions is shifted toward higher energies by about 1 eV. The design of the intermediate band requires a high main gap energy that transitions

involving the intermediate band may profit from sufficient spectral content of the sunlight. However, virtually no operating single gap solar cell with an absorber bandgap as high as 2.4 eV has ever been fabricated, as the useful part of the solar spectrum above such an energy is clearly deficient. The intermediate band solar cell therefore has no reference device with which to compare its performance directly. Nevertheless, devices operating in the reverse mode of a solar cell, i.e., light emitting diodes and lasers, have used wide gap semiconductors for such purposes. The eventual manufacture of intermediate band solar cells to operate at one sun irradiation may thus find some guidance and material candidates in the field of solid-state light sources.

- The curves corresponding to the ideal intermediate band and single gap solar cells cross one another at a bandgap energy of 1.14 eV. At energies below this value, there is no expectation of improving the performance of a solar cell by introducing an intermediate band within it, even when optimally positioned. Absorbers with bandgap energies greater than 1.14 eV may, however, improve their photovoltaic performance if provided with an intermediate band. The reason for this particular dependence expressed in simple words is related to the fact that a solar cell based on a low bandgap absorber already delivers a high current at a relatively low voltage. While the intermediate band seeks enhancement of the photocurrent with no net improvement in voltage,[2] the recombination rate introduced by the intermediate band in a low gap absorber, even when being exclusively radiative, outweighs the net gain in photocurrent. This stems from the reverse process of the two-photon absorption assisted by the intermediate band to which the photocurrent gain is owed.[3]

These three results bring some light to the potential and possibilities of intermediate band devices operating under one sun, the typical illumination conditions for thin-film solar cells. First, the predicted gain in maximum efficiency makes worthwhile nearly any attempt to implement an operating intermediate band in a thin-film absorber; and second, when looking for highest efficiencies one needs a wide gap absorber with $E_g > 2$ eV acting as a host for the intermediate band, which is to be situated at about 0.7 eV or more from any of the main band edges (we will come to this point later). Nevertheless, any absorber with a bandgap above 1.14 eV is, in principle, a candidate to host an intermediate band.

What material candidates are out there?. Thin-film technologies as we know them include Cu-containing chalcopyrite-based solar cells, CdTe, and a-Si:H, the

---

[2] In fact, the open-circuit voltage of the intermediate band solar cell is slightly degraded under one-sun operation with respect to that of an equivalent, single gap device. Sunlight concentration helps recovering the voltage as limited by the main bandgap, according to detailed balance calculations of Martí and Luque [6].

[3] Remind that the expression "two-photon absorption" in the context of the intermediate band solar cell does not correspond to the traditional understanding of the three-particle process involving the *simultaneous* absorption of two photons via a virtual electronic state, as used in spectroscopy and fluorescence microscopy. See [9].

latter in single and tandem structures,[4] with best efficiencies of 20.3%, 17.3%, and 12.1% under standard AM1.5G conditions, respectively [10]. Following criteria of maximum efficiency, the family of chalcopyrite compounds appears to be a conspicuously promising candidate to implement an intermediate band in it. Such ternary compounds of the type I–III–VI$_2$ also present some unique and amazing material properties, which most likely have not as yet been fully exploited in the engineering of related devices. We will mention just three of them and comment on some intrinsic advantages of the related processing[5]:

- The fact that the highest efficiencies of chalcopyrite-based solar cells have been achieved out of microcrystalline material, well ahead monocrystalline samples [12]. This remarkable experimental fact, also found in CdTe-based devices, has challenged the traditional understanding of electronic transport in polycrystalline semiconductors [13] and has driven scientific research to a profound understanding of the nature and role of grain boundaries,[6] surfaces [15], and the effects of compositional bandgap profiles across the structures [16].
- The benign character of the abundant intrinsic defects. Despite a high number of possible point defects (12 different types for a ternary compound) and a remarkable tolerance of the compounds to deviations from the nominal stoichiometry (up to 20 at% in some cases [17]), defect pairing into neutral complexes and DX-like centres have been proposed as a fundamental mechanism of electrical neutralization and stabilization of the compounds [18–20]. Of particular importance is the reaction $2V_{Cu}^{0} + In_{Cu}^{0} \rightarrow 2V_{Cu}^{-} + In_{Cu}^{2+}$, which is considered responsible for a number of intriguing experimental findings in selenide compounds [18], such as grain boundary electronics [21], preferential surface reconstruction into nominally polar facets [22], large off-stoichiometry and existence of the so-called ordered vacancy compounds [18], and self-regulatory control of the electronic properties of the compounds [23]. This point is illustrated in Fig. 11.2. The independent vacancy and antisite point defects result in a deep donor state and a shallow acceptor state in the bandgap of the semiconductor. V_{Cu} is considered to be largely responsible for the native p-type doping of Cu-containing chalcopyrites. The reaction into a defect pair, energetically favourable when growing the material under conditions employed for high efficiency devices, clears out the bandgap of deep states (at doping levels, acting as minority killers [24]), resulting in a shallow state close to the conduction band and a state resonant with valence band states [18].
- The possibility of alloying ternary compounds with additional elements to obtain multinary compounds of the type (Cu,Ag)(Al,Ga,In)(S,Se,Te)$_2$. Alloying

---

[4]a-Si and related concepts appear less suited for high efficiency concepts, mainly due to their intrinsic stability weaknesses under illumination and will not be considered further.

[5]For general references see, for example, [11].

[6]A good number of references on the issue of grain boundaries in thin film photovoltaic materials are collected in [14].

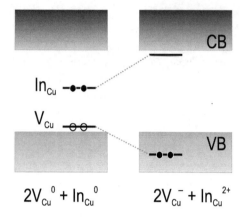

Fig. 11.2 Effect of point defect pairing on the removal of deep gap states in CuInSe$_2$. The formation of such pairs is spontaneous (exothermic) when growing the compound under conditions as used for high-efficiency devices [18]

permits tailoring of the optical and electronic properties continuously as desired, particularly the absorption onset (with energy bandgaps ranging from 0.96 up to 3.5 eV), but also crystal properties such as lattice parameters, of importance for lattice-matched epitaxial structures. Particularly the incorporation of Al into the group-III sub-lattice permits widening of the main bandgap above the values for In,Ga-alloys, as indicated in Fig. 11.1.

The microcrystalline nature and the benign character of certain grain boundaries of the films used for devices, together with the reduced thickness (typically below 5 μm for a complete device), the high absorption coefficients of the absorbers, and the self-doping of the semiconductors within optimal operating values via native defects and self-compensation are common to chalcopyrites and CdTe absorbers. However, the advantage of alloying opens the possibility of tailoring the bandgap profile within the thin-film structure so as to optimize the carrier transport across the device. This is an exclusive advantage of ternary and multinary compounds [25, 26].

Labels in Fig. 11.1 indicate the bandgaps of selected ternary chalcopyrite compounds and the corresponding maximum theoretical efficiencies. CuGaS$_2$ appears as an excellent candidate to act as a host for an intermediate band, provided that the intermediate band lies at its optimal position within the bandgap about 1 eV away from one of the main band edges. Interestingly, CuGaS$_2$ has been studied for light-emitting applications [27]. Two more ternaries, CuGaSe$_2$ and CuInS$_2$, are also included in the figure. The former has been studied as an absorber for a potential thin-film, high-voltage solar cell and as a top cell for all-chalcopyrite tandem devices together with low gap counterparts. The latter has recently entered into commercial production as a mature thin-film photovoltaic technology. As stated previously, any desired bandgap value along the solid line of Fig. 11.1 is achievable upon alloying different elements, e.g., a Cu(In,Ga)(S,Se)$_2$ multinary compound. The three compounds included may serve as a reference to estimate the potential of implementing an operating intermediate band in typical chalcopyrite devices. Indeed, the forecast for eventual CdTe-based intermediate band devices goes along with that of CuInS$_2$, due to their similar bandgap values (1.50 and

1.53 eV, respectively). All four materials, the three chalcopyrites and CdTe, are expected to lead to improved figures of efficiency in related devices under favourable intermediate band operation with respect to current state of the art and ideal single gap operation (dashed line).

An interesting candidate for the realization of thin-film intermediate band solar cells is ZnTe. Its main bandgap (2.3 eV) lies in the range of high theoretical efficiency, with figures above 45% according to Fig. 11.1. The introduction of light and very electronegative elements, such as O or N, in the semiconductor matrix leads to the formation of the so-called highly mismatched alloys. Such alloys show a singular electronic structure, characterized by the presence of an intermediate band within the bandgap of the host. The width and position of the intermediate band in these compounds has been successfully described by the *band-anticrossing model* and can be tailored by a careful control of the impurity concentration [28–33]. First prototypes of intermediate band solar cells based on ZnTe:O epitaxially grown onto GaAs substrates have been recently reported [34]. Yet with low overall conversion efficiencies, below 1%, ZnTe:O devices have demonstrated the two-photon excitation process predicted by the theory and have exhibited an enhanced photoresponse in comparison with ZnTe reference structures.

We have referred so far to maximum efficiencies of intermediate band solar cells whenever the intermediate band occupied its optimal position within the band gap.[7] But, how tolerant is such a device towards deviations in the relative position of the intermediate band?. Figure 11.3 shows the calculated maximum theoretical efficiencies of intermediate band solar cells in the radiative limit based on $CuInSe_2$, $CuInS_2$, $CuGaSe_2$, and $CuGaS_2$ as a function of the relative position of the intermediate band with respect to one of its main band edges. Horizontal dotted lines denote absolute maxima of single gap and intermediate band devices according to detailed balance calculations. It can be observed that for $CuInS_2$, $CuGaSe_2$, and $CuGaS_2$ the curves present a single maximum, corresponding to the optimal energies, and that the width of the maxima above the line denoting the single gap maximum efficiency ranges from about 0.35 to 0.55 eV. These values give an estimation of the tolerance of the device with respect to misplacing of the intermediate band. As the curves are not symmetric about their apexes, the device will tolerate better slight deviations in position of the intermediate band toward the closest of the main band edges, rather than toward midgap.

Plentiful experimental evidence leads to the conclusion that most thin-film-based solar cells do not operate in the radiative limit. By way of example, not all the absorbers on which thin-film devices are based show luminescence at room temperature, implying the dominance of alternative, non-radiative minority carrier recombination mechanisms. Achieving better performance from such devices requires that the radiative limit be approached. However, current high-efficiency thin-film solar cells show diode ideality factors, as obtained from I–V curves fitting,

---

[7]Additionally, the width of the intermediate band within the bandgap of the host material is zero in the ideal case, see [35].

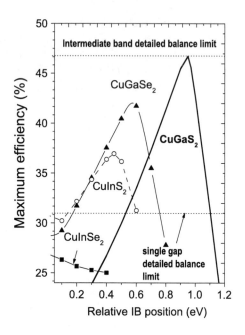

Fig. 11.3 Maximum theoretical efficiency of intermediate band solar cells based on four representative chalcopyrite absorbers, as a function of the relative position of the intermediate band with respect to the edge of one of the main bands. *Dotted lines* denote the absolute upper limits as determined from detailed balance calculations for single gap and intermediate band solar cells. Reprinted with permission from [36]. Copyright 2009. Elsevier

well above unity, indicating that the electronic transport is affected by non-radiative recombination, taking place either within the space-charge region (mostly the case for lowgap-based, highly effcient devices) or at the interface with the buffer layer (most commonly found in widegap-based cells). Regardless of the actual nature of the dominating non-radiative recombination mechanism, the associated electronic losses are the limiting factor of the current technology. We have tried to transfer these limitations to the calculation of the theoretical performance of intermediate band devices, albeit approximately, so as to estimate the impact of non-idealities on the predicted efficiencies. Starting from the ideal case, the photocurrent of the cell can be described by [2]:

$$J_{\text{tot}} = [\dot{N}(E_L, E_H, 0, T_{\text{sun}}) + \dot{N}(E_G, \infty, 0, T_{\text{sun}}]$$
$$-[\dot{N}(E_L, E_H, \mu_e, T_{\text{cell}}) + \dot{N}(E_G, \infty, V, T_{\text{cell}}], \quad (11.1)$$

where E_G, E_H, E_L denote the three gaps of the cell (main, high sub-bandgap, and low sub-bandgap, respectively), $\dot{N}$ the photon flux entering the absorber, J_{tot} the total current circulating in the device, T_{sun}, T_{cell} the sun and cell temperatures, μ_e the electrochemical potential of the electrons (assuming electrons as minority carriers) and V the output voltage of the cell under bias. This equation, in its simplest form, basically states that the maximum total photocurrent delivered by the cell is given by the number of photons emitted from a thermal source at $T = T_{\text{sun}}$ ($\mu = 0$) entering the device in the energy range between E_L and E_H and above E_G minus the number of photons leaving it ($T = T_{\text{cell}}$, $\mu = \mu_e, qV$). Once the current is known, the voltage can be determined from the single diode equation of Shockley. Thereby, one can obtain the expected efficiency.

Including non-idealities in the model requires an expression a bit more elaborated than (11.1). The new expression must include those experimental observations that are known to limit the performance of current conventional thin-film devices, particularly (1) current losses, (2) the impact of non-radiative recombination, and (3) the diode contribution, via saturation current and ideality factor. We arrive at the following expression:

$$\begin{aligned} J_{\text{tot}} = &[\dot{N}(E_\text{L}, E_\text{H}, 0, T_{\text{sun}}) + \dot{N}(E_\text{G}, \infty, 0, T_{\text{sun}})](1 - \textbf{\textit{CL}}) \\ &- \textbf{\textit{NRF}}[\dot{N}(E_\text{L}, E_\text{H}, \mu_e, T_{\text{cell}}) + \dot{N}(E_\text{G}, \infty, V, T_{\text{cell}})] \\ &- \textbf{\textit{J}}_0(\exp(qV)/(\textbf{\textit{n}}kT) - 1). \end{aligned} \qquad (11.2)$$

The photocurrent then consists of carriers photogenerated by optical transitions between the three bands ending with a free electron at the conduction band, as before, corrected by the factor CL, which accounts for incomplete collection (current losses), minus the fraction of carriers lost by radiative recombination, this term scaled now by a factor NRF accounting for non-radiative events, minus an additional term accounting for non-idealities proper of the diode junction, where J_0 stands for the saturation current, and n for the corresponding ideality factor. Factors typed in bold related to non-idealities are described in some detail as follows:

- The current losses considered account for the difference between the maximum obtainable short-circuit current (a function of the bandgap of the absorber material and the spectrum considered) and the experimental records. The losses stem from incomplete collection of photogenerated carriers and optical losses due to reflection. In the case of a CuGaSe$_2$-based solar cell ($E_g = 1.67$ eV), the current losses are as high as 35% [36], whereas for high-efficiency Cu(In,Ga)Se$_2$ ($E_g = 1.14$ eV), the losses are reduced to 23% [7].
- The impact of non-radiative recombination is critical in the physics of thin-film materials. The quantitative estimation of the impact, and therefore its incorporation into pseudo-analytical models, of non-radiative processes is extremely difficult. A simple and rather rough model has been proposed in the frame of detailed balance calculations by assuming the contribution due to non-radiative recombination to be simply a numerical factor (NRF in the equation) to be added to the fraction of radiative processes. This crude oversimplification (basically *counting non-radiative recombination processes as if they were radiative*, and adding them up to the net recombination rate) is obviously incorrect, as the physical process describing non-radiative recombination (a three- or multiple-particle process) differs from that of radiative recombination (two-particle process). Nevertheless, the quasi-analytical model expressed in the equation gives a qualitatively correct picture of the impact of excess recombination ($NRF > 1$) over the purely radiative case ($NRF = 1$). Figure 11.4 shows the case of a typical device based on CuGaSe$_2$ considering different values of

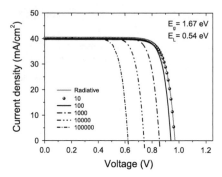

Fig. 11.4 Calculated I–V curves for a TF-IBSC based on CuGaSe$_2$ ($E_g = 1.67$ eV) under one sun of black-body radiation at 6,000 K (1,595 W m^{-2}) in the radiative limit and for values of the non-radiative factor between 10 and 10^5, accounting for an increasing contribution of non-radiative recombination following (11.2). Reprinted with permission from [36]. Copyright 2009. Elsevier

NRF and otherwise realistic values of current losses and diode contribution (see next paragraph) [36, 37]. It can be observed that increasing the non-radiative factor NRF severely affects the open-circuit voltage of the device, as expected. Only for NRF values well above 10^6 is the photocurrent measurably reduced.

- The diode contribution originates from the fact that thin-film solar cells typically show a non-negligible inverse saturation current and ideality factors different from unity, in agreement with the non-radiative character of the dominant recombination mechanisms controlling the electronic transport. The diode term in the equation has been included and experimental values of saturation current and ideality factor can be fed into the model, assuming that the implementation of the intermediate band will improve none of these aspects as compared to current devices. No effect of parasitic resistances has been included. It is considered that the optimization of their values is purely an engineering issue not related to fundamental aspects of the device operation.

A further source of non-idealities is the incomplete selectivity of absorption coefficients. The ideal model assumes that each absorbed photon can only contribute to one of the three absorption onsets expected in the intermediate band absorber, a very restrictive condition hard to fulfil in practice. A particular case of the relaxation of such constraint has been considered in [7] for the case of a thin-film intermediate band solar cell. The incomplete selectivity of the absorption coefficients is not limited to the case of thin-film devices, but rather affects all intermediate band concepts. For this reason, we will not treat it further here, referring the reader to [38], where this problem has been studied in some detail.

Having stated how far we can go theoretically with a thin-film intermediate band solar cell, we will turn in the following sections to the question of how to implement such a device in practice. Two different approaches will be presented and issues related to their eventual realization will be discussed.

11.2 The Impurity Approach

One way to incorporate an intermediate band into a given semiconductor host is to introduce sufficient amounts of the right impurity. It follows that there are two important issues surrounding this so-called *impurity approach*: (1) to find out what is meant by the "right impurity"; and (2) to find out what is meant by "in sufficient amounts". Starting from the second point, it is conventionally understood that the formation of a collection of electronic states with band-like character implies the delocalization of the electronic wave functions associated with the states. Such delocalization has in fact different grades and profound consequences on the electronic transport properties of solids, and has been a matter of study since the works of Anderson and Mott [39, 40]. It has been used to describe transport phenomena like hopping processes[8] [42–45] and the occurrence of characteristic signatures in Hall-effect and magnetoresistance experiments attributed to impurity bands [46–50]. In the frame of intermediate band theory, where electronic transport within the impurity band is of no relevance, and consequently has not been studied so far, the delocalization is assumed to occur following the model proposed by Luque et al., in which it is concluded that impurity concentrations in the range of 6×10^{19} cm^{-3} may suffice [24]. Concerning the question of the "right impurity", the appropriate answer is, obviously, the impurity which introduces a collection of states in the bandgap of the semiconductor host at the appropriate distance in energy from the main band edges, fulfilling the predictions stated above in terms of efficiency of the related devices as a function of the position of the intermediate band.

There is no a priori knowledge of which impurities these are for a given host compound, or if they even exist. Remember that we are talking of rather large amounts of impurities and not simply of concentrations in the doping range [51]. The consequence is that the massive introduction of impurities significantly modifies the electronic structure of the host compounds, not merely introducing localised states at certain energy levels. We are considering impurity concentrations at composition levels, rather than doping levels, and therefore the electronic structure of the compounds is largely modified, as in alloys. In fact, the appearance of collection of states identified as an intermediate band following the *impurity approach* results in most cases from the interaction of the electronic states of the host and of the impurity (hybridization), with restrictions imposed by the crystalline structure of the compound and symmetry considerations. Only in particular cases, like those involving rare earth metals and impurities with non- or low-interacting states, is the intermediate band expected to stem entirely from the impurity species. Ab-initio calculations are probably the most reliable way in order to predict the properties modified compounds may exhibit, bearing in mind that different numerical methods lead to different degrees of accuracy. These methods have helped and continue helping in screening the imaginable number of impurity species that

---

[8]For the case of hopping transport in chalcopyrites, see [41].

can be incorporated into a given semiconductor host to find potential candidates for intermediate band materials (see [52] and references therein). The use of numerical methods to calculate electronic structures has become widespread, and has been extended to the case of modified chalcopyrite hosts by the introduction of selected metal impurities Ti, V, Cr, and Mn in the quest for intermediate band materials [53–57]. Although interesting cases have been identified, eventual concerns about the chemical stability of the modified compounds with these particular impurities have been pointed out in light of total energy calculations.

Due to the demanding computational load of ab-initio calculations, a previous analysis of the eventual competing phases that may appear when attempting to introduce specific impurities into a semiconductor can simplify the search and save computational efforts. The problem can be stated as follows: a given number of atomic species, comprising the elements forming the host and the impurity, are expected to form a compound. Are there other possible atomic arrangements leading to the formation of different compounds that are more favourable energetically?

In the case of ternary chalcopyrites as host compounds, the impurity candidates studied up until now refer to substitutional impurities introduced in the cation sublattice of the host. Due to the high reactivity of the chalcogen species sulphur, selenium and tellurium, eventual competing phases are expected in the form of binary chalcogenides with the impurity atoms as cations. Indeed, the role played by binary phases in technological processes of relevance during the growth of bare chalcopyrite thin films, particularly in sequential and multistage processes in which the chemical composition of the film undergoes changes, is known to be crucial to the final electronic quality of the material. Furthermore, the reactivity of particular elements to be introduced as impurities with the chalcogens, typically introduced in excess during the fabrication process, may inhibit the incorporation of the impurity into the chalcopyrite on behalf of more stable binary chalcogenides.

The stability of the compound modified by the introduction of the selected impurity can be foreseen by comparing its enthalpy of formation to that of any possible competing phase. The enthalpy is the free energy under conditions of constant pressure, as is the case in most film-growth technological processes) and can be described as a function of the chemical potential of the respective atomic constituents. This procedure has been used by Zunger and co-workers [18, 19, 58] to determine the solubility limits of selected impurities into the chalcopyrite matrix. It is briefly explained in the following lines.

The thermodynamic stability of ternary chalcopyrite compounds of the type $CuGaX_2$, where X stands for S or Se ($CuGaS_2$ and $CuGaSe_2$ being the most promising chalcopyrite candidates for intermediate band solar cells, according to Fig. 11.1) requires that:

$$\Delta\mu_{Cu} + \Delta\mu_{Ga} + 2\Delta\mu_X = \Delta H_f(CuGaX_2), \quad (11.3)$$

where $\Delta\mu$ represents the change in the chemical potential of the constituents with respect to the isolated elements in equilibrium ($\Delta\mu = 0$ for the elemental solids) and ΔH_f the enthalpy of formation of the compound ($\Delta H_f < 0$ if stable). Reported

11 Thin-Film Technology in Intermediate Band Solar Cells

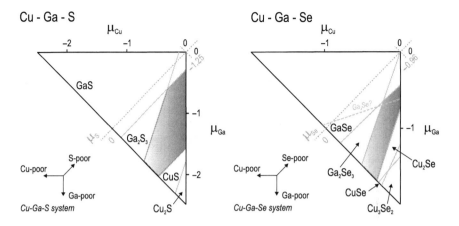

Fig. 11.5 Stability plots of the systems Cu–Ga–S (*left*) and Cu–GaSe (*right*) as a function of the chemical potential of their constituents. *Shaded areas* indicate the range of stability of the ternary chalcopyrite compounds. The *grey* scale reflects the variation of the chemical potential of the chalcogen species (light/dark corresponding to rich/poor chalcogen content). Some discrepancies have been found in the literature for the case of the selenide, see [58–63, 67]

values of ΔH_f for $CuGaS_2$ and $CuGaSe_2$ are -2.49 eV and -1.93 eV, respectively, as calculated from first principles [58–63]. Additionally, thermochemical tables provide tabulated data of enthalpies of formation, particularly of binary compounds [64–66]. Equation 11.3 can be represented graphically as a stability plot of the compound. This is shown in Fig. 11.5 for the cases of $CuGaS_2$ and $CuGaSe_2$ as right triangles. Each pair of catheti represents the chemical potential of the metals Cu and Ga, whose length is set in each case by the corresponding magnitude of ΔH_f. The chemical potential of the chalcogen is bound to those of the metals by (11.3) and is represented in each case as the corresponding hypotenuse of the right triangle. The bisector of the right angle in each plot denotes the scale of $\Delta \mu_X$, opposite to those of the metals (maximum metal-enrichment implies a chalcogen-poor material). Notice that the minimum value of $\Delta \mu_X$ is half the minimum value of $\Delta \mu_{Cu}$ and $\Delta \mu_{Ga}$ due to its stoichiometric coefficient. The area enclosed by the triangle represents the range of stability of the compound and the triangle itself denotes the situation of equilibrium between the compound and its constituents in the form of elemental solids.

As mentioned above, however, competing phases appear in the form of binary chalcogenides (like Cu_2S, CuS, Cu_2Se, Ga_2S_3, GaS, Ga_2Se_3 and $GaSe$; other ternaries of the type $CuGa_5S_8$, the so-called defect compounds, have not been considered). A set of conditions of the type:

$$c\Delta \mu_{Cu} + x\Delta \mu_X < \Delta H_f(Cu_c X_x), \tag{11.4}$$

$$g\Delta \mu_{Ga} + x\Delta \mu_X < \Delta H_f(Ga_g X_x), \tag{11.5}$$

for Cu- and Ga- compounds, respectively, state that the domain of existence of the chalcopyrite compound is bound by its constituents being in thermodynamical equilibrium with the respective binary chalcogenides, rather than with the elemental solids. Restrictions imposed by binary chalcogenides on the stability of the ternary compound are represented by solid lines in Fig. 11.5. Shaded areas represent the actual range of stability of the ternary compounds.

Now, the addition of foreign species, labelled as A, may impose further restrictions on the domain of existence of the chalcopyrite, should they form stable compounds upon reaction with one or more elements of the host compound. A new set of inequalities of the type of (11.4) and (11.5), with the impurity substituting the cation site, must hold simultaneously with (11.3)–(11.5). The chemical potentials of the impurity and of the constituents are, thus, interrelated, and ultimately depend on the growth process of choice. As long as we are using thermodynamic arguments, we are implicitly considering processes operating close to equilibrium. Within this idealization, we can further assume that changes in μ_A and μ_X in a compound such as $A_a X_x$ stem from events of atomic transfer between the compound and elemental atomic reservoirs. In this way, we can increase the content of species A in the compound by transferring A-atoms from its atomic reservoir into the compound. This implies that the chemical potential of A in the compound increases (indeed, at the expense of the chemical potential of X) and we can keep increasing it as long as the compound remains stable, maintaining the same $\Delta H_f(A_a X_x)$. The limiting case $\Delta \mu_A \to 0$, or equivalently $\Delta \mu_X \to \Delta H_f(A_a X_x)/x$, represents the situation of maximum enrichment of element A in the compound $A_a X_x$ (since the value of $\Delta \mu_A$ is negative in the stable compound) and states that the constituent A has reached equilibrium with its elemental solid. This might not be realistic for every case but, nevertheless, the associated mathematical expression imposes the most stringent condition possible to the accessible values of μ_X, under which element X would be expected as a constituent of the chalcopyrite compound (11.3) and would not form the spurious secondary phase. The stability of the ternary compound will be determined by comparing $\Delta \mu_X \to \Delta H_f(A_a X_x)/x$ with the allowed range of $\Delta \mu_X$ in the pure chalcopyrite: should $\Delta \mu_X \to \Delta H_f(A_a X_x)/x$ be higher than the minimum μ_X of the shaded area in Fig. 11.5, the chalcopyrite phase will be stable. It would then be, in principle, possible to select the appropriate chemical potentials for the constituents that allow the growth of single-phase material; on the other hand, should $\Delta \mu_X \to \Delta H_f(A_a X_x)/x$ be lower than the minimum μ_X of the shaded area in Fig. 11.5, no stability range for the chalcopyrite would be accessible for any combination of μ_{Cu}, μ_{Ga}, and μ_X, meaning that the chalcopyrite will decompose into more stable binary phases. Tables 11.2 and 11.3 in the Appendix show the results of a fairly complete elemental screening carried out for the cases of sulphide and selenide compounds. Elements, corresponding binary chalcogenides, and enthalpies of formation are included, together with values of $\mu_X^{\min}$.

Figure 11.6 shows some examples of the reduction of the stability range of the ternary compound $CuGaS_2$ upon incorporation of Fe, Co and Ni impurities, as a function of the expected competing binary phases that could eventually form during the growth of the material. The set of hypotenuses and the colour scale denote

11 Thin-Film Technology in Intermediate Band Solar Cells 291

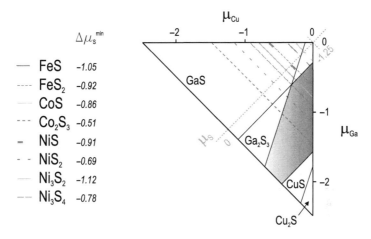

Fig. 11.6 Stability plot of the system Cu–Ga–S upon introduction of Fe, Co and Ni impurities and eventual formation of related competing binary chalcogenides. Each hypotenuse denotes the value μ_S^{min} for the corresponding binary compound. All secondary phases listed, except Ni_3S_2, satisfy $\mu_S^{min} > \mu_S^{chalco}$, allowing in principle the growth of single-phase material

values of μ_S^{min} for conditions of maximum impurity enrichment, as discussed before. Each dotted/dashed line restricts the stability range of the semiconductor host to the upper right part of the shaded area, i.e., only the range for which $\mu_S^{chalco} < \mu_S^{min}$ of the binary ensures accessible values of μ_{Cu} and μ_{Ga}. In the case of $CuGaS_2$, $\mu_S^{chalco} = -1.08$ eV [58–63]; whereas in the case of $CuGaSe_2$, μ_{Se}^{chalco} can range between -0.66 and -0.75 eV (depending on the stability of the Ga_2Se phase, see Fig. 11.5 [58–63, 67]). All secondary phases considered in Fig. 11.6, except Ni_3S_2, leave a (reduced) range of existence of the ternary still accessible. Elements for which the corresponding secondary phases do not annul the stability range of the ternary compound completely are typed bold in Tables 11.2 and 11.3.

The screening of impurity candidates proceeds then by classifying the foreign species according to the expected difficulty in incorporating them into the selected host. Table 11.1 summarizes the results of the thermochemical screening for $CuGaS_2$ and $CuGaSe_2$ hosts. Columns have been labelled as *no competing binary* for those cases in which no competing binary phase is expected to inhibit the formation of the modified chalcopyrite, *competing binary* for those impurities forming very stable phases with the chalcogen and thus inhibiting the formation of the desired compound, and *no data* for those impurities for which no reliable thermochemical data could be found in the literature. The latter is especially relevant for the case of selenide compounds.

Electronic structure calculations can then be performed on those candidates in columns *No comp. bin.* of Table 11.1 for each chalcopyrite host. Allowing substitutional sites at the cation sublattice of the host to locate the impurity, there are still two possibilities to consider, as the impurity can sit at either Cu or Ga sites. Normally, the valence of the impurity considered will imply a preferential

Table 11.1 Classification of impurity candidates according to the stability of the chalcopyrite host with respect to competing binary chalcogenides. The limiting case $\Delta\mu_A = 0$ representing maximum A-enrichment (A denoting the foreign element) has been considered

CuGaS$_2$			CuGaSe$_2$		
No comp. bin.	Comp. bin.	No data	No comp. bin.	Comp. bin.	No data
Ag	Cd	Au	Ag	Cd	C
Co	Ce	C	Au	La	Ce
Fe	Cr	Er	Fe	Li	Co
Ge	Eu	Gd	Ge	Mn	Cr
Ir	K	Hf	Mo	Nd	Er
Ni	La	Hg	Ni	Pb	Eu
Pb	Li	Ho	Pt	Zn	Gd
Pd	Mg	Lu	Ru		Hf
Pt	Mn	Nb	Sn		Hg
Rh	Mo	Os			Ho
Ru	Na	Re			Ir
Si	Nd	Sm			K
Sn	Pr	Ta			Lu
	Sc	Y			Mg
	Ti	V			Na
	W	Yb			Nb
	Zn				Os
	Zr				Pd
					Pr
					Re
					Rh
					Sc
					Si
					Sm
					Ta
					Ti
					V
					W
					Y
					Yb
					Zr

substitution, at least over a certain range of concentration. However, deviations in the stoichiometry of the host compound from nominal values can have a significant influence on the final destination of the impurity [68].

Two examples of electronic structure calculations are given in Figs. 11.7 and 11.8, where the band structure of modified CuGaS$_2$ incorporating Ir$_{Cu}$ and Ni$_{Cu}$ at a concentration of 12.5 %at, respectively, are shown. These calculations are based on the density functional theory (DFT) in the local spin density approximation (LSDA), using the supercell approach with periodic boundary conditions and allowing relaxation of the structure. Further numerical details and a complete discussion of

Fig. 11.7 Calculated electronic structure ($E_f = 0$) of IrCuGa$_2$S$_4$ along $X - \Gamma$ and $\Gamma - M$ directions (*left*) and projected total density of states (*right*). Courtesy of C. Tablero

Fig. 11.8 (Colour online) Left, calculated electronic structure ($E_f = 0$) of NiCuGa$_2$S$_4$ along $X - \Gamma$ and $\Gamma - M$ directions for spin up (*black*) and down (*gray*). Right, projected total density of states for each spin population. Courtesy of C. Tablero

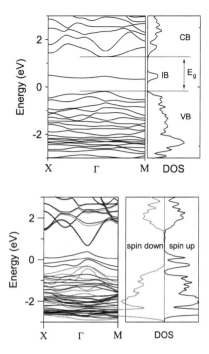

the results for the case of the CuGaS$_2$ matrix can be found in [52]. Calculations refer to ideal crystalline structures in which the only point defect considered corresponds to the substitutional impurity. Eventual defect pairing has not been considered. The particular cases of Figs. 11.7 and 11.8 illustrate interesting situations:

- For particular substitutions, a well-resolved intermediate band is found, separated in energy from the main band edges by a region of zero-density of states. The case of Ir$_{Cu}$ in Fig. 11.7 illustrates such a situation. Intermediate bands have been found for the cases of substitutional impurities of group IVa at either cation sublattice in the sulphide host [52]. It should be mentioned that, although no competing binary phase does in principle threaten the stability of the chalcopyrite structure, there exist other ternary and quaternary phases in the system Cu–Ga–(Si,Ge)–S that could induce phase segregation and even inhibit the incorporation of the impurity into the chalcopyrite matrix above a certain solubility limit. Some of these multinary competing phases are listed in Table 11.4. Further studies are necessary in this respect.
- Cases in which an intermediate band has been identified may still require doping to fulfill the requirements of partial filling. As an example, in Fig. 11.7 the intermediate band appears above the Fermi level ($E = 0$), being consequently empty of carriers, and thus requiring *n*-type doping. It is well known that doping chalcopyrites, and particularly doping wide gap compounds as *n*-type semiconductors, is far from trivial [19]. It is thus anticipated that those situations for which *p*-type doping might be required appear more favourable.

- Most of the transition metals considered as substitutional impurities induce magnetic properties in the modified hosts [52]. In Fig. 11.8, this is readily seen from the spin-selective electronic structure and the corresponding densities of states for the case of Ni$_{Cu}$ substitutions in CuGaS$_2$. Generally, such compounds can be described as half-metals, either as *spintronic* materials, showing, within the gap, zero density of states for one spin component and a finite value for the other, and as *possibly spintronic*, showing zero density of states for both spin components within the bandgap but with spin-selective bands around the Fermi energy.
- For magnetic impurities, two cases can be differentiated, corresponding to ferro- and antiferromagnetic ordering in the lattice. In some cases, like Fe$_{Ga}$ and Ir$_{Ga}$ in CuGaS$_2$, antiferromagnetic ordering leads to the appearance of intermediate bands of different complexity, partially occupied in the latter case and empty of carriers in the former. For ferromagnetic ordering, spintronic behaviour is predicted in both cases [52]. These last two points, referring to magnetic properties of modified compounds though not directly related to the operation of intermediate band solar cells, open new perspectives for the use of chalcopyrite-based materials in different applications.

The identification of modified chalcopyrites as possible intermediate band materials at a theoretical level is expected to drive the experimental research toward its practical realization. First attempts are currently on the way and preliminary results have been recently reported [69].

11.3 The Nanostructuring Approach

An alternative to introducing impurities in the host in search of the intermediate band in thin films is the so-called "nanostructuring approach" [70]. It follows the strategy of the quantum-dot solar cell, in which ideally zero-dimensional nanostructures of low-bandgap material are embedded into a matrix compound of larger bandgap. If a sufficient density of nanostructures are introduced, their confined states will develop minibands within the main gap of the matrix material. This approach may be advantageous over the "impurity approach" with regard to the limited solubility of certain impurities, as discussed in the previous section.

In recent years, there has been an increasing number of contributions related to the growth and characterisation of chalcopyrite compounds in the form of nanocrystals [71–113]. A large fraction of the literature describes approaches based on solvothermal methods [71–111], producing colloidal suspensions and inks of great potential interest for non-vacuum growth processes of conventional devices, but also of interest for the development of low-dimensional structures. The incorporation of nanostructures into solid-state matrices, as required in the design of intermediate band solar cells, is more difficult and the related literature scarce [113–115]. Nevertheless, the nature of chalcopyrite compounds makes them

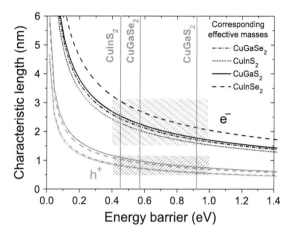

Fig. 11.9 Minimum diameter of nanostructures leading to carrier confinement, calculated for holes and electrons as a function of the energy barrier. *Vertical lines* indicate optimal values of energy barriers for the corresponding ternary compounds, according to Fig. 11.1 (Reprinted with permission from [70]. Copyright 2009. Wiley-VCH)

relatively simple to incorporate into binary compounds sharing atomic species. Such an idea has been demonstrated experimentally, using nanoscopic metallic precursors during the growth of binary chalcogenides [116] resulting in ternary compounds embedded in the binary matrix. This approach is considered a feasible route to the nanostructured thin-film intermediate band solar cell.

Independently of the chosen process for the practical realization of the nanostructures and their incorporation into a semiconducting matrix, some general considerations must be followed with respect to the dimensions of the nanocrystals to be used as precursors of the intermediate band. Simple calculations based on the effective mass approach, assuming the case of spherical particles, provide minimum values of their diameter D_{min} as to ensure the existence of at least one bound state when embedded in a known matrix [70, 117], according to:

$$D_{min} = \frac{\pi \hbar}{\sqrt{2m^* \Delta E_b}}, \tag{11.6}$$

where m^* is the effective mass of the carrier to be confined and ΔE_b is the energy barrier between the nanostructure and the matrix material. Figure 11.9 shows the dependency of the minimum diameter of spherical quantum dots on the energy barrier. Vertical lines indicate optimal values of the energy separation between the intermediate band and the nearest main band edge in the respective ternary hosts, as indicated in Fig. 11.1. Due to the different effective masses of electrons and holes in different chalcopyrite compounds [118], the minimum size of the quantum dot leading to confinement depends on the type of carrier considered. It can be concluded from Fig. 11.9 that, in the range of energy barriers of interest, $D_{min} = 1.5–3$ nm for electrons and $D_{min} = 0.5–1.2$ nm for holes. These ranges are indicated by shaded areas in the figure.

Upper limits to the characteristic dimensions of the nanostructures follow from the requirement of sufficient separation in energy between confined states, as to ensure that the thermal population at excited states of the nanostructures be small compared to that of the ground state. For a population ratio of some 5% between the first excited state E_1 and the ground state E_0, this means that [117]:

$$kT \leq \frac{1}{3}(E_1 - E_0). \tag{11.7}$$

Considering the confining potential of a 3D-harmonic oscillator, $E_1 - E_0 = \frac{2}{3}E_0$, with $E_0 = \frac{\pi^2 \hbar^2}{2m^*}$. Using the effective masses of [118], calculated values of D_{max} lie in the range of $D_{max} = 9.5\text{--}12.7$ nm for electrons and $D_{max} = 3.3\text{--}4.6$ nm for holes, depending on the ternary chalcopyrite compound of choice to be used in the form of nanocrystals [70]. Similar calculations have been carried out within the so-called finite-depth-well effective mass approximation, specifically designed to evaluate the confinement potential in colloidal suspensions of chalcopyrite nanocrystals and showing good agreement with a few experimental results [119]. Experimental evidence of size-dependent light emission in chalcopyrite nanocrystals in the form of colloidal suspensions has also been reported [120].

One of the main difficulties foreseen in the growth of embedded nanostructures of ternary compounds in, say, related binary host compounds is the possibility of interdiffusion of elemental constituents between confined and confining materials. As a result, no sharp edge of the nanostructure will be found, but rather a blurred interfacial region characterised by compositional gradients over a limited volume of the structure. The importance of interdiffusion phenomena in nanostructured ternary compounds has been studied on simple quantum well model structures of $CuGaSe_2/CuInSe_2/CuGaSe_2$ grown epitaxially onto GaAs substrates [121]. The left panel of Fig. 11.10 shows photoreflectance spectra and the corresponding theoretical fittings obtained from the structure schematically shown on the right panel, where both the nominal and the actual structure have been superimposed. Although interdifussion phenomena and the absence of monolayer-wide interfaces between nanostructures and barrier materials in the volume of the material may not necessarily prevent carrier confinement, their consequences could affect the design and dimensioning of the intended structures and should be taken into account.

The range of nanostructures that are potentially interesting for the realization of intermediate band solar cells also includes one-dimensional structures. Nanowires offer unique optoelectronic properties due to the singularity in their joint density of states [122]. The fabrication of heterostructures within quasi-1D semiconductors can be accomplished in two general ways: either as a core-shell structure, in which two different materials are grown sequentially along the radial direction of the wire, or as a 1D superlattice, in which the different compounds are grown sequentially along the growing direction of the wire. In the first case, the resulting radial heterostructures are a one-dimensional analogue to a planar configuration,

11 Thin-Film Technology in Intermediate Band Solar Cells

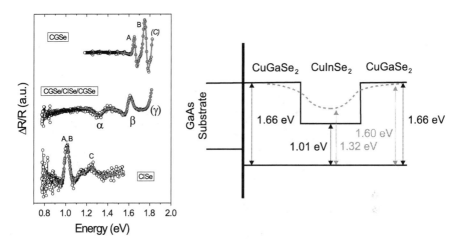

Fig. 11.10 *Left*: photoreflectance spectra (*open dots*) of epitaxial selenide layers on GaAs(001); CuGaSe$_2$ (*top*), CuInSe$_2$ (*bottom*), and CuGaSe$_2$/CuInSe$_2$/CuGaSe$_2$ stack. *Solid lines* represent fittings of the experimental data following the third-derivative functional form. *Right*: schematic band diagram of the multilayer stack. The *dashed line* represents the actual conduction band profile, according to the interpretation of photoreflectance results. Reprinted from Solar Energy Materials and Solar Cells, **94**, 1912 (2010), Copyright 2010, with permission from Elsevier [121]

subject to the singularities associated with the confinement in two dimensions [123]. In the second case, the situation of quantum-dot embedding in a barrier material can be recovered, this time constrained to 1D, by reducing the length of the segment of the low gap material to virtually zero [122]. Several works have reported on the growth of chalcopyrite-based nanowires [124–131]. In most cases, the wires have been obtained by the vapor–liquid–solid method, using a noble metal, typically Au, as a catalyst. The left image of Fig. 11.11 shows a CuGaSe$_2$ whisker grown by chemical vapor deposition on an Au-coated glass substrate. The gold nanoparticle driving the directional growth is readily seen at the apex of the whisker. The right image of Fig. 11.11 illustrates how the same whisker would have formed if an In-source, for example, were operated intermittently during the process, leading to the formation of quantum dots of a low gap material (shaded, representing CuInSe$_2$) in a 1D-matrix of a wide gap barrier compound, like CuGaSe$_2$. On the other hand, the core-shell approach has been followed by Cui and co-workers, who have reported on structures of CuInSe$_2$ 1D-nanowires coated with an ultrathin CdS layer, thus close to operational devices [131–135].

It is concluded that the nanostructuring approach to thin-film intermediate band solar cells is highly appealing, notwithstanding the foreseen technical difficulties and challenges related to the accurate control of the growth conditions leading to reproducible patterns and structures. In this sense, the embedding process of the nanostructures, either in the bulk of a film or within the nanostructure itself, is a

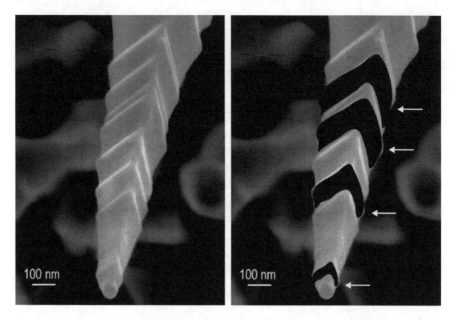

Fig. 11.11 *Left*: CuGaSe$_2$ whisker grown on glass by CVD using Au as catalyst. *Right*: recreation of an identical structure that would be formed if an In-source were used intermittently (*shaded areas*) during the growth of the structure. The resulting whisker would include nearly-zero-dimensional, low gap structures surrounded by nearly-one-dimensional wide gap barrier materials

key issue, fundamentally limited in the particular case of chalcopyrite compounds by kinetic aspects of elemental interdiffusion.

11.4 Conclusions and Outlook

The incorporation of intermediate bands in thin-film solar cells is an appealing approach with the potential to improve the efficiency of current state-of-the-art devices. Cu-containing chalcopyrites and binary chalcogenides, like CdTe and ZnTe, appear as interesting candidates to host intermediate bands. The theoretical frame of the ideal operation of intermediate band devices has been widened to incorporate non-idealities present in thin-film devices that are known to limit their performance. In this way, more realistic predictions of device operation can now be developed including the effects of misplacing the intermediate band from its ideal position, finite saturation currents, ideality factors differing from unity, significant non-radiative recombination or current losses. Regarding the practical realization of thin-film intermediate band devices, two approaches have been discussed. Following the *impurity approach*, an elemental screening has been carried out for two exemplary chalcopyrite compounds, in search of the appropriate substitutional impurities that may lead to operative intermediate bands. The screening has included thermochemical and ab-initio calculations and the results have been summarized in Tables 11.1–11.4. A deeper understanding of the defect chemistry of the semiconductor hosts including large amounts of foreign impurities is still required in order to make precise predictions regarding the existence and operation of intermediate bands. Alternatively, the *nanostructuring approach*, requiring the growth of nanostructures and their embedding into a barrier material, appears a feasible method alleviating eventual limitations imposed by a reduced solubility of impurities. Different schemes for the realization of nanostructured thin-film intermediate band materials have been discussed, together with difficulties foreseen in practical architectures. Both approaches are currently in development and it is expected that the experimental realization of prototypes will be achieved in the short term.

Acknowledgements The author is grateful to Profs. Luque and Martí for the invitation to participate in the workshop. Financial support from the Spanish Ministry of Science and Innovation within the program Ramón y Cajal is acknowledged.

Appendix

Table 11.2 Classification of selected elements and their potential competing binary sulphides; corresponding standard enthalpies of formation and minimum values for the chemical potential of sulphur. Elements typed bold present binary compounds for which $\Delta\mu_S^{min} > \Delta\mu_S^{min}(CuGaS_2)$. All values from Knacke et al., except where indicated: 1: Lide; 2: Chase (see [64–66])

Element	Binary	$\Delta_f H^0$ (kJ/mol)	$\Delta\mu_S^{min}$ (eV/at)
Ba	BaS (1)	−460.0	−4.77
Be	BeS	−234.299	−2.43
	BeS (1)	−234.3	−2.43
Ca	CaS	−473.202	−4.90
	CaS (1)	−482.4	−5.00
Cd	CdS	−154.565	−1.60
	CdS (1)	−161.9	−1.68
Ce	CeS	−456.474	−4.73
	CeS (1)	−459.4	−4.76
	Ce_3S_4	−1,652.680	−4.28
	Ce_2S_3	−1,188.256	−4.10
	CeS_2	−618.202	−3.20
Co	$CoS_{0.89}$	−94.558	−1.10
	CoS (1)	−82.8	−0.86
	Co_3S_4	−358.987	−0.93
	Co_2S_3 (1)	−147.3	−0.51
	CoS_2	−153.134	−0.79
Cr	CrS	−155.644	−1.61

(continued)

Table 11.2 (continued)

Element	Binary	$\Delta_f H^0$ (kJ/mol)	$\Delta\mu_S^{min}$ (eV/at)
Cs	Cs$_2$S (1)	−359.8	−3.72
Eu	EuS	−418.400	−4.33
Fe	FeS	−101.671	−1.05
	FeS (1)	−100.0	−1.04
	FeS (troilite) (2)	−101.67	−1.05
	FeS$_2$ (pyrite)	−171.544	−0.89
	FeS$_2$ (pyrite) (1)	−178.2	−0.92
	FeS$_2$ (pyrite) (2)	−171.54	−0.89
	FeS$_2$ (marcasite) (2)	−167.36	−0.87
Ge	GeS	−76.868	−0.80
	GeS$_2$	−156.900	−0.81
	GeS (1)	−69.0	−0.71
Hg	HgS	−53.346	−0.55
	HgS (1)	−58.2	−0.60
Ir	Ir$_2$S$_3$	−208.840	−0.72
	Ir$_2$S$_3$ (1)	−234.0	−0.81
	IrS$_2$	−133.001	−0.69
	IrS$_2$ (1)	−138.0	−0.71
K	K$_2$S	−376.560	−3.90
	K$_2$S (1)	−380.7	−3.95
La	LaS	−464.424	−4.81
	LaS (1)	−456.0	−4.73
	La$_2$S$_3$	−1204.992	−4.16
Li	Li$_2$S	−441.399	−4.57
	Li$_2$S (1)	−441.4	−4.57
Mg	MgS	−345.719	−3.58
	MgS (1)	−346.0	−3.58
	MgS (2)	−345.72	−3.58
Mn	MnS	−213.384	−2.21
	MnS (1)	−214.2	−2.22
	MnS (3)	−158.26	−1.64
	MnS$_2$	−225.710	−1.17
	MnS$_2$	−225.58	−1.17
Mo	Mo$_2$S$_3$	−406.868	−1.40
	MoS$_2$	−276.010	−1.43
	MoS$_2$ (1)	−235.1	−1.22
Na	Na$_2$S	−366.100	−3.79
	Na$_2$S (1)	−364.8	−3.78
Nd	NdS	−464.842	−4.82
	Nd$_2$S$_3$	−1,125.469	−3.89
Ni	Ni$_3$S$_2$	−216.312	−1.12
	Ni$_3$S$_2$ (2)	−216.31	−1.12

(continued)

11 Thin-Film Technology in Intermediate Band Solar Cells

Table 11.2 (continued)

Element	Binary	$\Delta_f H^0$ (kJ/mol)	$\Delta\mu_S^{min}$ (eV/at)
	NiS	−87.864	−0.91
	NiS (1)	−82.0	−0.85
	NiS (2)	−87.86	−0.91
	Ni_3S_4	−301.109	−0.78
	Ni_3S_4 (2)	−301.11	−0.78
	NiS_2	−133.888	−0.69
	NiS_2 (2)	−131.38	−0.68
Pb	PbS	−98.324	−1.02
	PbS (1)	−100.4	−1.04
Pd	Pd_4S	−69.036	−0.72
	PdS	−70.709	−0.73
	PdS (1)	−75.0	−0.78
	PdS_2	−78.211	−0.40
Pr	PrS	−451.872	−4.68
	Pr_3S_4	−1,554.356	−4.03
Pt	PtS	−81.671	−0.85
	PtS (1)	−81.6	−0.85
	PtS_2	−109.164	−0.56
	PtS_2 (1)	−108.8	−0.56
Rh	Rh_3S_4	−357.732	−0.93
	Rh_2S_3	−262.470	−0.91
Ru	RuS_2	−217.965	−1.13
Sc	Sc_2S_3	−1,171.520	−4.05
Si	SiS_2	−213.384	−1.10
Sn	SnS	−107.947	−1.12
	SnS (1)	−100.0	−1.04
	Sn_3S_4	−370.284	−0.96
	Sn_2S_3	−263.592	−0.91
	SnS_2	−153.552	−0.79
Sr	SrS (1)	−472.4	−4.89
Ti	TiS	−271.960	−2.82
	TiS_2	−407.103	−2.11
W	WS_2	−259.408	−1.34
Zn	ZnS	−205.016	−2.12
	ZnS (wurtzite) (1)	−192.6	−1.99
	ZnS (sphalerite)(1)	−206.0	−2.13
Zr	ZrS_2	−577.392	−2.99

Table 11.3 Classification of selected elements and their potential competing binary selenides; corresponding standard enthalpies of formation and minimum values for the chemical potential of selenium. Elements typed bold present binary compounds for which $\Delta\mu_{Se}^{min} > \Delta\mu_{Se}^{min}(CuGaSe_2)$. All values from Knacke et al., except where indicated: 1: Lide (see [64–66]); 3: Zhao et al. (see [58–63])

Element	Binary	$\Delta_f H^0$ (kJ/mol)	$\Delta\mu_{Se}^{min}$ (eV/at)
Au	AuSe	−9.008	−0.09
Ca	CaSe	−368.192	−3.81
Cd	CdSe	−145.615	−1.51
Fe	FeSe$_{0.961}$	−66.944	−0.72
Ge	GeSe	−69.036	−0.72
	GeSe$_2$	−112.968	−0.59
Hg	HgSe	−54.023	−0.56
La	LaSe	−359.824	−3.73
	La$_2$Se$_3$	−933.032	−3.22
Li	Li$_2$Se	−420.073	−4.35
Mn	MnSe	−154.808	−1.60
	MnSe (1)	−106.7	−1.10
	MnSe (3)	−122.55	−1.27
Mo	Mo$_3$Se$_4$	−336.368	−0.87
	MoSe$_2$	−153.469	−0.80
Nd	NdSe	−359.824	−3.73
	Nd$_2$Se$_3$	−941.400	−3.25
Ni	NiSe	−74.893	−0.78
	NiSe$_2$	−108.784	−0.56
Pb	PbSe	−99.997	−1.04
	PbSe (1)	−102.9	−1.07
Pt	Pt$_5$Se$_4$	−241.672	−0.63
Ru	RuSe$_2$	−170.548	−0.88
Sn	SnSe	−88.700	−0.92
	SnSe$_2$	−121.336	−0.63
Sr	SrSe (1)	−385.8	−4.00
Zn	ZnSe	−170.297	−1.76
	ZnSe (1)	−163.0	−1.69

Table 11.4 Potential competing phases in the form of ternary and multinary chalcogenides and reported experimental work on the incorporation of impurities into chalcopyrite compounds

Element	Other potentially competing phases	Comments
Au		Reported as dopant in $CuInSe_2$ [136]
Cd	$CdGa_2S_4$	Reported as dopant in $CuInSe_2$ [137]
Ce	$(CuCeSe_2)$	
Co	$(CuCo_2InSe_4)$	Reported as dopant in $Cu(Ga_{1-x}Al_x)(S,Se)_2$, $AgGaS_2$ [68, 138, 139]
Cr	Cu_xCrS_2, $Cr_2Ga_{0.67}S_4$, $Cr_{0.89}Ga_{1.78}S_4$ $(AgCrS_2)$	
Eu	$CuEu_2S_3$	
Fe	$CuFeS_2$	Reported as dopant in $Cu(Ga,Al)S_2$, $Cu(In,Ga)Se_2$ [68, 140–142]
Ge	$CuGaGeS_4$, $(Cu_2GeSe_3, CuGaGeSe_4)$	
Hg	$HgGa_2S_4$, $(HgInGaS_4)$	
Ir	$CuIr_2S_4$	
K	$(KGaSe_2, KGaTe_2)$	
La	$CuLaS_2$, $(CuLaSe_2)$	
Li	$LiGaS_2$, $(LiInS_2, LiGaSe_2, LiGaTe_2)$	
Mg	$MgGa_2S_4$, $(Mg_{0.5}Ga_2InS_5, MgInGaS_4, MgIn_{3.67}S_8)$	Reported as dopant in $CuInS_2$ [143]
Mn	$MnGa_2S_4$, $(MnGaInS_4)$	
Mo	$Cu_xMo_6S_8$	
Nd	$CuNdS_2$, $(CuNdSe_2)$	
Ni		Reported as dopant in $CuGa(S,Se)_2$, $AgGaS_2$, $CuAlS_2$ [68, 142]
Pr	$CuPrS_2$, $(CuPrSe_2)$	
Sc	$CuScS_2$	
Si	$CuGaSiS_4$, $(CuGaSiSe_4)$	
Sn	Cu_4SnS_4, $CuSn_{3.75}S_8$, $CuGaSnS_4$, $(CuGaSnSe_4)$	Reported as dopant in $CuInS_2$
Ta	Cu_3TaS_4, $(AgTaS_3)$	
Ti	Cu_xTiS_2, $(AgTiS_2, Cu_2TiTe_3)$	Reported as dopant in $CuAlS_2$ [144, 145]
V	$Cu_{0.75}VS_2$, $Cu_{0.65}VS_2$, CuV_2S_4, Cu_3VS_4, $CuVS_2$	Reported as dopant in $CuAlS_2$ [144]
Y	$CuYS_2$	
Zn		Reported as dopant in $CuInS_2$, $CuAlS_2$ [143, 146]
Zr	(Cu_2ZrTe_3)	

References

1. Press release: *First Solar Passes $1 Per Watt Industry Milestone*. (First Solar, Feb 2009) http://investor.firstsolar.com/phoenix.zhtml?c=201491&p=irol-newsArticle_print&ID=1259614&highlight=.Cited12Jan2010
2. A. Luque, A. Martí, Phys. Rev. Lett. **78**, 5014 (1997)
3. See previous chapters on IBSCs in this book
4. M.A. Green, *Third Generation Photovoltaics. Advanced Solar Energy Conversion* (Springer, Heidelberg, 2003), pp. 2–4
5. K. Zweibel, in *The Terawatt challenge for thin-film PV*. ed. by J. Poortmans, V. Arkhipov Thin-Film Solar Cells. Fabrication, Characterization and Applications (Wiley, Chichester, 2006), pp. 427–462; available as NREL technical report TP-520-38350, August 2005. http://nrelpubs.nrel.gov/Webtop/ws/nich/www/public/SearchForm.Cited12Jan2010
6. A. Martí, E. Antolín, E. Cánovas, N. López, P.G. Linares, A. Luque, C.R. Stanley, C.D. Farmer, Thin Solid Films **516**, 6716 (2008)
7. A. Martí, D. Fuertes Marrón, A. Luque, J. Appl. Phys. **103**, 073706 (2008)
8. W. Shockley, H.J. Queisser, J. Appl. Phys. **32**, 510 (1961)
9. M. Göppert-Mayer, Ann. Phys. **9**, 273 (1931)
10. M.A. Green, K. Emery, Y. Hishikawa, W. Warta, E. Dunlop, Prog. Photovolt Res. Appl. **19**, 565 (2011)
11. R. Klenk, M.Ch. Lux-Steiner, in *Thin-Film Solar Cells. Fabrication, Characterization and Applications*, ed. by J. Poortmans, V. Arkhipov (Wiley, Chichester, 2006), pp. 237–275, and refs. therein
12. C.H. Champness, H. Du, I. Shih, *Proceedings of the 29th IEEE PV Spec. Conf.*, p. 732 (IEEE, Piscataway, 2002)
13. J.Y.W. Seto, J. Appl. Phys. **46**, 5247 (1975)
14. U. Rau et al., Appl. Phys. A **96**, 221 (2009)
15. R. Scheer, Trends Vac. Sci. Techol. **2**, 77 (1997)
16. A. Gabor, J. Tuttle, M. Bode, A. Franz, A. Tennant, M.A. Contreras, R. Noufi, D. Jensen, A. Herman, Sol. Energy Mater. Sol. Cells **41–42**, 247 (1996)
17. S. Fiechter, Y. Tomm, K. Diesner, T. Weiss, Jpn. J. Appl. Phys. **39**, Suppl. 39-1, p. 123 (2000)
18. S.B. Zhang, S.-H. Wei, A. Zunger, H. Katayama-Yoshida, Phys. Rev. B **57**, 9642 (1998)
19. A. Zunger, Appl. Phys. Lett. **83**, 57 (2003)
20. S. Lany, A. Zunger, Phys. Rev. Lett. **100**, 016401 (2008)
21. C. Persson, A. Zunger, Phys. Rev. Lett. **91**, 266401 (2003)
22. J.E. Jaffe, A. Zunger, Phys. Rev. B **64**, 241304R (2001)
23. S. Schuler, S. Siebentritt, S. Nishiwaki, N. Rega, J. Beckmann, S. Brehme, M.Ch. Lux-Steiner, Phys. Rev. B **69**, 045210 (2004)
24. A. Luque, A. Martí, E. Antolín, C. Tablero, Physica B **382**, 320 (2006)
25. A. Morales-Acevedo, Sol. Energy Mater. Sol. Cells **93**, 41 (2009)
26. M.A. Green, Prog. Photovolt Res. Appl. **17**, 57 (2009)
27. J.L. Shay, B. Tell, H.M. Kasper, Appl. Phys. Lett. **19**, 366 (1971)
28. W. Shan, W. Walukiewicz, J.W. Ager III, E.E. Haller, J.F. Geisz, D.J. Friedman, J.M. Olson, S.R. Kurtz, Phys. Rev. Lett. **82**, 1221 (1999)
29. W. Walukiewicz, W. Shan, K.M. Yu, J.W. Ager III, E.E. Haller, I. Miotlowski, M.J. Seong, H. Alawadhi, A.K. Ramdas, Phys. Rev. Lett. **85**, 1552 (2000)
30. J. Wu, W. Shan, W. Walukiewicz, Semicond. Sci. Technol. **17**, 860 (2002)
31. K.M. Yu, W. Walukiewicz, J. Wu, W. Shan, J.W. Beeman, M.A. Scarpulla, O.D. Dubon, P. Becla, Phys. Rev. Lett. **91**, 246203 (2003)
32. W. Walukiewicz, Mater. Res. Soc. Symp. Proc. Vol. **865**, F5.7.1 (2005)
33. B. Lee, L.W. Wang, Appl. Phys. Lett. **96**, 071903 (2010)
34. W. Wang, A.S. Lin, J.D. Phillips, Appl. Phys. Lett. **95**, 011103 (2009)
35. M.Y. Levy, C. Honsberg, Phys. Rev. B **78**, 165122 (2008)

36. D. Fuertes Marrón, A. Martí, A. Luque, Thin Solid Films **517**, 2452 (2009)
37. D. Fuertes Marrón, A. Martí, C. Tablero, E. Antolín, E. Cánovas, P.G. Linares, A. Luque, in *Proceedings of the 23rd European Photovoltaic Solar Energy Conference and Exhibition*, Valencia, Spain, p. 33 (WIP 2008)
38. L. Cuadra, A. Martí, A. Luque, IEEE Trans. Electron Dev. **51**, 1002 (2004)
39. P.W. Anderson, Phys. Rev. **109**, 1492 (1958)
40. N.F. Mott, W.D. Twose, Adv. Phys. **10**, 107 (1961)
41. E. Arushanov, S. Siebentritt, T. Schedel-Niedrig, M.Ch. Lux-Steiner, J. Appl. Phys. **100**, 063715 (2006) and refs. therein
42. N.F. Mott, E.A. Davies *Electronic Processes in Non-Crystalline Materials* (Oxford Clarendon Press, Oxford, 1979)
43. N.F. Mott, *Metal-Insulator Transitions* (Taylor and Francis, London, 1990)
44. B.I. Shklovskii, A.L. Efros, *Electronic Properties of Doped Semiconductors* (Springer, Berlin, 1984)
45. M. Pollak, B.I. Shklovskii (eds.), *Hopping Transport in Solids* (North-Holland, Amsterdam, 1991)
46. C.S. Hung, J.R. Gliessman, Phys. Rev. **96**, 1226 (1954)
47. J.F. Woods, C.Y. Chen, Phys. Rev. **135**, A1462 (1964)
48. A.C. Beer, in *The Hall Effect and Its Applications*, ed. by C.L. Chien, C.R. Westgate (Plenum Press, NY, 1980), pp. 299–338
49. E. Arushanov, S. Siebentritt, T. Schedel-Niedrig, M.Ch. Lux-Steiner, J. Phys. Condens. Matter **17**, 2699 (2005)
50. J.T. Asubar, Y. Jinbo, N. Uchitomi, Phys. Status Solidi C **6**, 1158 (2009)
51. H.-W. Schock, Mat. Res. Soc. Symp. Proc. Vol. **763**, B1.6.1 (2003)
52. C. Tablero, D. Fuertes Marrón, J. Phys. Chem. C **114**, 2759 (2010)
53. P. Palacios et al., Phys. Stat. Sol A **203**, 1395 (2006)
54. P. Palacios et al., Thin Solid Films **515**, 6280 (2007)
55. P. Palacios et al., J. Sol. Energy Eng. **129**, 314 (2007)
56. P. Palacios et al., J. Phys. Chem. C **112**, 9525 (2008)
57. I. Aguilera et al., Thin Solid Films **516**, 7055 (2008)
58. S.B. Zhang, S.-H. Wei, A. Zunger, Phys. Rev. Lett. **78**, 459 (1997)
59. Y.-J. Zhao, A. Zunger, Phys. Rev. B **69**, 075208 (2004)
60. Y.-J. Zhao, C. Persson, S. Lany, A. Zunger, Appl. Phys. Lett. **85**, 586 (2004)
61. S. Lany, Y.-J. Zhao, C. Persson, A. Zunger, *Proc. 31st IEEE PV Spec. Conf.*, New York, 2005, p. 343
62. S. Lany, Y.-J. Zhao, C. Persson, A. Zunger, Appl. Phys. Lett. **86**, 042109 (2005)
63. C. Persson, Y.-J. Zhao, S. Lany, A. Zunger, Phys. Rev. B **72**, 03521 (2005)
64. See, for example, O. Knacke, O. Kubaschewski, K. Hesselmann, *Thermochemical Properties of Inorganic Substances* (Springer, New York, 1991)
65. M.W. Chase Jr., J. Phys. Chem. Ref. Data Monograph **9**, 1 (1998)
66. D.R. Lide (ed.), *CRC Handbook of Physics and Chemistry* (Taylor and Francis, Boca Raton, 2007)
67. D. Cahen, R. Noufi, J. Phys. Chem. Solids **53**, 991 (1992)
68. U. Kaufmann, A. Rauber, J. Schneider, Solid State Commun. **15**, 1881 (1974)
69. B. Marsen, L. Steinkopf, A. Singh, H. Wilhelm, I. Lauermann, T. Unold, R. Scheer, H.-W. Schock, *Effects of Ti-incorporation in CuInS$_2$ solar cells*. Solar Energy Materials & Solar Cells 94, 1730 (2010)
70. D. Fuertes Marrón, A. Martí, A. Luque, Phys. Status Solidi (a) **206**, 1021 (2009)
71. S.L. Castro, S.G. Bailey, R.P. Raffaelle, K.K. Banger, A.F. Hepp, Chem. Mater. **15**, 3142 (2003)
72. R.P. Raffaelle, S.L. Castro, A.F. Hepp, S.G. Bailey, Progr. Photovolt. Res. Appl. **10**, 433 (2001)
73. S. Ghoshal, L.B. Kumbhare, V.K. Jain, G.K. Dey, Bull. Mater. Sci. **30**, 173 (2007)
74. H. Junqing, D. Bin, T. Kaibin, L. Qingyi, J. Rongrong, Q. Yitai, Solid State Sci. **3**, 275 (2001)

75. L. Tain, J.J. Vittal, New J. Chem. **31**, 2083 (2007)
76. Q. Lu, J. Hu, K. Tang, Y. Qian, G. Zhou, X. Liu, Inorg. Chem. **39**, 1606 (2000)
77. D. Pan, X. Wang, Z.H. Zhou, W. Chen, C. Xu, Y. Lu, Chem. Mater. **21**, 2489 (2009)
78. J. Tang, S. Hinds, S.O. Kelley, E.H. Sargent, Chem. Mater. **20**, 6906 (2008)
79. J. Olejnicek, C.A. Kamler, A. Mirasano, A.L. Martinez-Skinner, M.A. Ingersoll, C.L. Exstrom, S.A. Darveau, J.L. Huguenin-Love, M. Diaz, N.J. Ianno, R.J. Soukup, Sol. Energy Mater. Sol. Cells, **94**, 8 (2010)
80. Y.-H.A. Wang, C. Pan, N. Bao, A. Gupta, Solid State Sci. **11**, 1961 (2009)
81. S. Ahn, C. Kim, J. Yun, J. Lee, K. Yoon, Sol. Energy Mater. Sol. Cells, **91**, 1836 (2007)
82. M.G. Panthani, V. Akhavan, B. Goodfellow, J.P. Schmidtke, L. Dunn, A. Dodabalapur, P.F. Barbara, B.A. Korgel, J. Am. Chem. Soc. **130**, 16770 (2008)
83. M. Uehara, K. Watanabe, Y. Tajiri, H. Nakamura, H. Maeda, J. Chem. Phys. **129**, 134709 (2008)
84. T. Nyari, P. Barvinschi, R. Baies, P. Vlazan, F. Barvinschi, I. Dekany, J. Crystal Growth **275**, e2383 (2005)
85. K.-T. Kuo, S.-Y. Chen, B.-M. Cheng, C.-C. Lin, Thin Solid Films **517**, 1257 (2008)
86. C. Wen, X. Weidong, W. Juanjuan, W. Xiaoming, Z. Jiasong, L. Lijun, Mater. Lett. **63**, 2495 (2009)
87. L. Zheng, Y. Xu, Y. Song, C. Wu, Y. Xie, Inorg. Chem **48**, 4003 (2009)
88. D. Pan, L. An, Z. Sun, W. Hou, Y. Yang, Z. Yang, Y. Lu, J. Am. Chem. Soc. **130**, 5620 (2008)
89. S. Han, M. Kong, Y. Guo, M. Wang, Mater. Lett. **63**, 1192 (2009)
90. K.V. Yumashev, A.M. Malyarevich, P.V. Prokoshin, N.N. Posnov, V.P. Mikhailov, V.S. Gurin, M.V. Artemyev, Appl. Phys. B **65**, 545 (1997)
91. J.S. Gardner, E. Shurdha, C. Wang, L.D. Lau, R.G. Rodriguez, J.J. Pak, J. Nanopart. Res. **10**, 633 (2008)
92. C. Yu, J.C. Yu, H. Wen, C. Zhang, Mater. Lett. **63**, 1984 (2009)
93. J.J. Nairn, P.J. Shapiro, B. Twamley, T. Pounds, R. von Wandruszka, T.R. Fletcher, M. Williams, C. Wang, M.G. Norton, Nano Lett. **6**, 1218 (2006)
94. Y. Qi, Q. Liu, K. Tang, Z. Liang, Z. Ren, X. Liu, J. Phys. Chem. C **113**, 3939 (2009)
95. B. Koo, R.N. Patel, B.A. Korgel, Chem. Mater. **21**, 1962 (2009)
96. Q. Guo, S.J. Kim, M. Kar, W.N. Shafarman, R.W. Birkmire, E.A. Stach, R. Agrawal, H.W. Hillhouse, Nano Lett. **8**, 2982 (2008)
97. M. Azad Malik, P. O'Brien, N. Revaprasadu, Adv. Mater. **11**, 1441 (1999)
98. H. Zhong, Y. Li, M. Ye, Z. Zhu, Y. Zhou, C. Yang, Y. Li, Nanotechnology **18**, 025602 (2007)
99. C.-H. Chang, J.-M. Ting, Thin Solid Films **517**, 4174 (2009)
100. H. Chen, S.-M. Yu, D.-W. Shin, J.-B. Yoo, Nanoscale Res. Lett. doi: 10.1007/s11671-009-9468-6 (2009)
101. B. Koo, R.N. Patel, B.A. Korgel, J. Am. Chem. Soc. **131**, 3134 (2009)
102. F. Bensebaa, C. Durand, A. Aouadou, L. Scoles, X. Du, D. Wang, Y. Le Page, J. Nanopart. Res. doi: 10.1007/s11051-009-9752-5 (2009)
103. J. Xiao, Y. Xie, Y. Xiong, R. Tang, Y. Qian, J. Mater. Chem. **11**, 1417 (2001)
104. M. Nanu, J. Schoonman, A. Goossens, Nano Lett. **5**, 1716 (2005)
105. H. Grisaru, O. Palchik, A. Gedanken, V. Palchik, M.A. Slifkin, A.M. Weiss, Inorg. Chem. **42**, 7148 (2003)
106. Y.-H.A. Wang, N. Bao, A. Gupta, Solid State Sci. doi: 10.1016/j.solidstatesciences.2009.11.019 (2009)
107. C. Wang, S. Xue, J. Hu, K. Tang, Jpn. J. Appl. Phys. **48**, 023003 (2009)
108. V.S. Gurin, Colloids Surf. A **142**, 35 (1998)
109. R. Xie, M. Rutherford, X. Peng, J. Am. Chem. Soc. **131**, 5691 (2009)
110. S.K. Batabyal, L. Tian, N. Venkatram, W. Ji, J.J. Vittal, J. Phys. Chem. C **113**, 15037 (2009)
111. J. Tang, S. Hinds, S.O. Kelley, E.H. Sargent, Chem. Mater. **20**, 6906 (2008)
112. N. Benslim, S. Mehdaoui, O. Aissaoui, M. Benabdeslem, A. Bousala, L. Bechiri, A. Otmani, X. Portier, J. Alloys Comp. **489**, 437 (2010)

113. I.V. Bodnar, V.S. Gurin, A.P. Molochko, N.P. Solovei, P.V. Prokoshin, K.V. Yumashev, Semiconductors **36**, 298 (2002)
114. I.V. Bodnar, V.S. Gurin, A.P. Molochko, N.P. Solovei, Inorg. Mater. **40**, 797 (2004)
115. E. Arici, D. Meissner, F. Schäffler, N.S. Sariciftci, Intern. J. Photoenergy **5**, 199 (2003)
116. D. Fuertes Marrón, S. Lehmann, M.Ch. Lux-Steiner, Phys. Rev. B **77**, 085315 (2008)
117. D. Bimberg, M. Grundmann, N.N. Ledentsov, *Quantum Dot Heterostructures* (Wiley, Chichester, 1999)
118. C. Rincón, J. González, Phys. Rev. B **40**, 8552 (1989)
119. T. Omata, K. Nose, S. Otsuka-Yao-Matsuo, J. Appl. Phys. **105**, 073106 (2009)
120. H. Nakamura, W. Kato, M. Uehara, K. Nose, T. Omata, S. Otsuka-Yao-Matsuo, M. Miyazaki, H. Maeda, Chem. Mater. **18**, 3330 (2006)
121. D. Fuertes Marón, E. Cánovas, M.Y. Levy, A. Martí, A. Luque, M. Afshar, J. Albert, S. Lehmann, D. Abou-Ras, S. Sadewasser, N. Barreau, *Optoelectronic Evaluation of the Nanostructuring Approach to Chalcopyrite-Based Intermediate Band Solar Cells*. Solar Energ. Mater. Sol. Cells **94**, 1912 (2010)
122. M. Dresselhaus, Y.-M. Lin, O. Rabin, M.R. Black, J. Kong, G. Dresselhaus, in *Springer Handbook of Nanotechnology*, ed. by B. Bhushan (Springer, Heidelberg, 2004), pp. 113–160, and refs. therein
123. M.-E. Pistol, C.E. Pryor, Phys. Rev. B **78**, 115319 (2008)
124. D. Fuertes Marrón, S. Lehmann, J. Kosk, S. Sadewasser, M.Ch. Lux-Steiner, Mater. Res. Soc. Symp. Proc. **1012**, 1012-Y02-07 (2007)
125. G. Harichandran, N.P. Lalla, Mater. Lett. **62**, 1267 (2008)
126. Y. Jiang, Y. Wu, S. Yuan, B. Xie, S. Zhang, Y. Qian, J. Mater. Res. **16**, 2805 (2001)
127. B. Li, Y. Xie, J. Huang, Y. Qian, Adv. Mater. **11**, 1456 (1999)
128. Y. Jiang, Y.Wu, X. Mo, W. Yu, Y. Xie, Y. Qian, Inorg. Chem. **39**, 2964 (2000)
129. J. Hu, Q. Lu, B. Deng, K. Tang, Y. Qian, Y. Li, G. Zhou, X. Liu, Inorg. Chem. Commun. **2**, 569 (1999)
130. M.X. Wang, L.S. Wang, G.H. Yue, X. Wang, P.X. Yan, D.L. Peng, Mater. Chem. Phys. **115**, 147 (2009)
131. S.T. Connor, C.-M. Hsu, B.D. Weil, S. Aloni, Y. Cui, J. Am. Chem. Soc. **131**, 4962 (2009)
132. D.T. Schoen, H. Peng, Y. Cui, J. Am. Chem. Soc. **131**, 7973 (2009)
133. H. Peng, C. Xie, D.T. Schoen, K. McIlwrath, X.F. Zhang, Y. Cui, Nano Lett. **7**, 3734 (2007)
134. H. Peng, D.T. Schoen, S. Meister, X.F. Zhang, Y. Cui, J. Am. Chem. Soc. **129**, 34 (2007)
135. H. Peng, S. Meister, C.K. Chan, X.F. Zhang, Y. Cui, Nano Lett. **7**, 199 (2007)
136. E. Hernández, J.G. Albornoz, J. González, S.M. Wasim, J.C. Sánchez, F. Sánchez Pérez, Jpn. J. Appl. Phys. **32**, 92 (1993)
137. P. Migliorato, J.L. Shay, H.M. Kasper, S. Wagner, J. Appl. Phys. **46**, 1777 (1975)
138. W.-C. Zheng, S.-Y. Wu, B.-J. Zhao, S.-F. Zhu, J. Phys. Chem. Solids **63**, 895 (2002)
139. M.-S. Jin, W.-T. Kim, C.-S. Yoon, J. Phys. Chem. Solids **54**, 1509 (1993)
140. K. Sakurai, H. Shibata, S. Nakamura, M. Yonemura, S. Kuwamori, Y. Kimura, S. Ishizuka, A. Yamada, K. Matsubara, H. Nakanishi, S. Niki, Mater. Res. Soc. Symp. Proc. Vol. **865**, F14.12.1 (2005)
141. S. Nomura, T. Takizawa, A. Goltzene, C. Schwab, Jpn. J. Appl. Phys. Suppl. **32–33**, 196 (1993)
142. M. Birkholz, P. Kanschat, T. Weiss, M. Czerwensky, K. Lips, Phys. Rev. B **59**, 12268 (1999)
143. T. Enzenhofer, T. Unold, R. Scheer, H.-W. Schock, Mater. Res. Soc. Symp. Proc. Vol. **865**, F11.3 (2005)
144. I. Aksenov, K. Sato, Jpn. J. Appl. Phys. **31**, L527 (1992)
145. U. Kaufmann, A Räuber, J. Schneider, J. Phys. C Solid State Phys. **8**, L381 (1975)
146. I. Aksenov, T. Yasuda, Y. Segawa, K. Sato, J. Appl. Phys. **74**, 2106 (1993)

Chapter 12
InGaN Technology for IBSC Applications

C. Thomas Foxon, Sergei V. Novikov, and Richard P. Campion

Abstract In this chapter, we will discuss the possible application of InGaN:Mn layers for Intermediate Band Solar Cell devices. We will consider the reasons for choosing to study this promising system; why plasma-assisted molecular beam epitaxy is the method of choice to produce such structures; the specific problems associated with InGaN growth and the progress made so far in achieving this objective. We also discuss briefly other possible dopants for IBSCs based on InGaN.

12.1 Introduction

In progress towards a next generation of photovoltaics, devices based on InGaN may play a significant role, due to the unique properties of the nitride-based materials. In particular, because the band gap covers the entire range from 0.64 eV for InN [1] to 3.4 eV for GaN [2], devices based on InGaN are now being extensively studied for various applications. InGaN-based opto-electronic devices already form the basis for industrial applications including, blue light emitting diodes (LEDs) [3] and blue/violet laser diodes [4] and for electronic devices including high-power RF transistors [5]. The same material system will form the basis for future solid state lighting, by combining efficient nitride based LEDs with appropriate phosphors to obtain the correct colour balance [6]. In many cases, a key issue is the quality and availability of suitable substrates, since unlike other III–V compounds, no bulk substrates are available for group III-Nitride semiconductors.

For photovoltaic applications, the ability to cover the whole of the spectral range in a single materials system could have significant implications for cost savings; however, there is a considerable mis-match in lattice parameter between InN and

C.T. Foxon (✉)
School of Physics and Astronomy, University of Nottingham, Nottingham NG7 2RD, UK
e-mail: C.Thomas.Foxon@nottingham.ac.uk

GaN, which would make conventional multi-junction solar cells difficult to produce without introducing a high defect density. An alternative approach to high-efficiency solar cells is to use the Intermediate Band Solar Cell (IBSC) approach pioneered by Luque and Martí [7], using various possible intermediate band materials in the active part of the InGaN device. The general concept behind this approach is mentioned in the earlier chapter of this monograph by Martí et al. (see Chap. 8) and will not be discussed here in detail. However, as noted earlier, InGaN is an attractive material to study because in that system we have both the ability to select different materials to form the intermediate band and the possibility to tune both the band gap and the band offsets.

In this chapter, we will discuss the growth method needed to produce suitable InGaN-based structures, in particular the use of Mn in InGaN to form the intermediate band within an InGaN p–i–n heterojunction. One of the unique features of the nitrides is the strong internal electric fields, which may be used perhaps to enhance separation of the electron holes in p–i–n devices; however, it is also well known that there are significant difficulties associated with p-doping for In-rich InGaN, which will also be discussed. Finally, we will discuss the progress made to date in reaching the overall goal of a working IBSC device based on InGaN, together with other options for the IB material in InGaN.

12.2 Motivation and Theoretical Consideration for InGaN:Mn in IBSC Applications

Detailed theoretical calculations show that Mn-doped InGaN can be a potential material for IBSC devices [8]. The $Mn^{3+/2+}$ level for Mn is approximately mid gap in GaN and close to the conduction band edge in InN, as shown in Fig. 12.1. Therefore, by choosing an appropriate In composition for the InGaN film, the $Mn^{3+/2+}$ level can be positioned approximately 1/3rd of the gap below

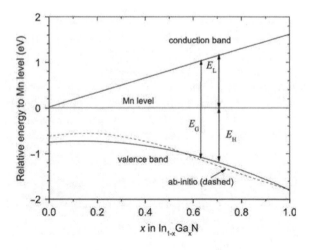

Fig. 12.1 Positions of the conduction band and valence band relative to the Mn level in $In_{1-x}Ga_xN$, assuming the Mn level is constant when referred to the vacuum level. *Dashed lines* are the result of ab-initio calculations (Reprinted with permission from [8]. Copyright 2009. Elsevier B.V)

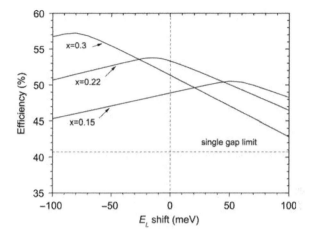

Fig. 12.2 The dependence of the limiting efficiency on a shift in the predicted position of the $Mn^{3+/2+}$ energy level. A positive shift means that the energy level approaches the center of the bandgap. A negative shift means that the energy level approaches the conduction band edge. The calculations assume a solar spectrum corresponding to the sun corresponds to a BB at 6,000 K (46,050 suns) (Reprinted with permission from [8]. Copyright 2009. Elsevier B.V)

the conduction band edge of the alloy, which is the optimum position to fulfil the conditions for IBSC operation (see Chap. 8). Detailed calculations show that the maximum efficiency under concentrator conditions can reach as high as 57% as shown in Fig. 12.2 [8]. Whilst this may not be achievable in practise, it clearly shows the potential for IBSC based on In-rich InGaN. This suggests that initial studies should concentrate on the composition range with a Ga content of ~20–30%.

However, to achieve this goal, we need to be able to incorporate a large concentration (~1–2%) of Mn in In-rich InGaN, which in turn must be otherwise lightly doped. Both are non- trivial tasks, requiring significant materials development. To date, using molecular beam epitaxy (MBE) we have demonstrated that it is possible to incorporate high levels of Mn (>10%) in both GaMnAs, by growth at low temperature (<200°C) [9] and in GaMnN (~4%) at higher temperatures (600–700°C) [10]. In both cases, growth occurs at a temperature well below the melting point, which is typically required for incorporating dopants at higher concentrations than the equilibrium solubility limit. MBE is the most suitable method for growth under extreme non-equilibrium conditions, since it is possible using this method to obtain good quality films at very low growth temperatures.

The general principles of growth by MBE are already discussed in the article in this monograph by Martí et al. (see Chap. 8) and will not be repeated here; however, the special requirements for the growth of In-rich InGaN and for the growth of the complete structure will be discussed below. The calculations shown in Figs. 12.1 and 12.2 indicate that the successful IBSC device will require an Indium composition of ~70–80%, which poses special problems for composition control. In addition, it is well known that *p*-doping for high In content InGaN is difficult to achieve, both will be discussed in detail below.

12.3 Growth of Mn Doped In-Rich InGaN

As discussed in the previous article by Martí et al. (see Chap. 8), in MBE, most material can be supplied by evaporation or sublimation from the elements using suitable "Knudsen cells". However, nitrogen is a special case since no suitable nitrogen containing liquid or solid materials are available. One option is to supply nitrogen from ammonia, which is available in high purity form and is used extensively in the metal-organic vapour phase growth of group III-Nitride semiconductors. However, under vacuum conditions at low pressure ($<10^{-5}$ Torr), ammonia does not decompose efficiently below $\sim 600°C$ [11]. Since lower growth temperatures are needed for In-rich InGaN (see below), this is not a viable option for growth of the IBSC device structures.

Active nitrogen is therefore typically supplied by decomposing nitrogen gas in an RF or ECR plasma source. In our case, the former is used and in this source an RF coil surrounds a boron nitride crucible from which active nitrogen gas emerges through a series of small apertures using RF excitation. Using optical spectroscopy, we have previously demonstrated that the plasma consists of both excited molecular nitrogen and atomic nitrogen [12] and by supplying the stable nitrogen isotope $^{15}N_2$ [13], we have established that within the plasma all the nitrogen gas is decomposed, but most recombines before leaving the source. Depending on the particular conditions, approximately 1–4% of active nitrogen in the form of atomic nitrogen is obtained in practise and due to the high binding energy of nitrogen, atomic nitrogen is an extremely active source, which readily combines with the group III elements to form III-Nitride semiconductors [2]. This method is commonly known as plasma-assisted MBE (PA-MBE) or sometimes referred to as plasma-enhanced MBE (PE-MBE).

Nitride semiconductors naturally form in the stable wurtzite (hexagonal) structure and are typically grown by hetero-epitaxy on substrates such as sapphire, SiC, GaAs since bulk nitride substrates are not readily available. Films typically grow in the [0001] Ga-polar or [000-1] N-polar direction depending on the nucleation conditions. The properties of films grown in the Ga or N-polar directions differ markedly both in structural quality and in incorporation of unwanted impurities. Films can also be grown in the metastable zinc-blende (cubic) form using arsenic as a surfactant (see below).

For MBE growth of most III–V materials, the group V species has a much higher vapour pressure than the group III elements, so growth takes place with excess group V flux, the excess re-evaporating to leave stoichiometric films. However, for group III-Nitrides grown by PA-MBE, this does not occur due to the extreme reactivity of the active nitrogen species. Films grown under N-rich conditions tend to be columnar as shown in Fig. 12.3a and under extreme conditions, this can lead to the growth of almost defect-free nano-columns as shown in Fig. 12.3b [14], the growth mechanism involved being partly due to geometric factors [15].

For GaN, growth under Ga-rich conditions results in compact films. At high growth temperatures and low III:V flux ratios, excess Ga will desorb, but with

Fig. 12.3 (a) An SEM photograph of the surface and edge of a GaN film grown under N-rich conditions by PA-MBE, showing a columnar growth mode. (b) An SEM micrograph of GaN nanocolumns grown under very nitrogen-rich conditions at a high substrate temperature

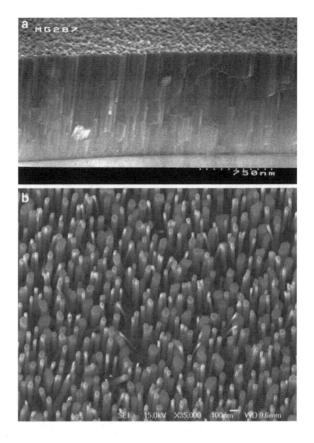

increasing III–V ratio eventually Ga droplets form on the surface. It is well established that the best properties are obtained for films grown under excess III-V flux, before the formation of droplets [16]. Growth under such conditions is therefore akin to growth by liquid phase epitaxy, the Ga coverage on the surface enabling active nitrogen to migrate on the surface beneath the Ga overlayer. However, even under optimum conditions, because there are no low-cost, defect-free lattice matched substrates available, the resulting films have extremely high defect density as shown in, which in turn can result in poor surface morphology as shown in Fig. 12.4a. Nevertheless, strong room temperature PL can be observed from such films, due to the localisation of carriers preventing recombination at non-radiative defects. Growth on low-defect density substrates supplied by UNIPRESS does provide material with extremely smooth surfaces as shown in Fig. 12.4b. Using such substrates results in structural properties similar to those obtained in other III–V materials, enabling high quality AlGaN/GaN low-defect density, low-dimensional structures to the grown using PA-MBE method [17].

The group III-Nitrides normally crystallise in the hexagonal or wurtzite structure, however, for this particular application there may be advantages to growing the

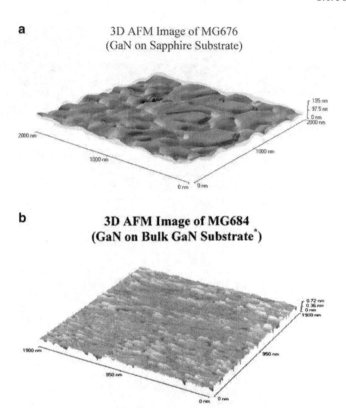

Fig. 12.4 (a) AFM image showing the surface morphology of a GaN film grown on sapphire, where defects emerging at the surface produce a rough surface. (b) AFM image showing the surface morphology of a GaN film grown on low defect density bulk GaN showing by comparison an atomically flat surface

metastable (001) oriented cubic or zinc-blende structure (see below). We have previously demonstrated that using arsenic as a surfactant, it is possible to grow the metastable form of (001) oriented zinc-blende GaN on GaAs or GaP substrates [18], as shown by the fact that the diffraction peak occurs at 40° in the Ω-2Θ plot (this peak is unique to zinc-blende material as shown in Fig. 12.5). This method can also be applied to other III-Nitride semiconductors as we have subsequently demonstrated and may be applicable to the growth of In-rich InGaN-based solar cells, especially to avoid the detrimental influence of internal electric fields (see below).

The growth of InGaN poses additional problems compared with the growth of GaN. In particular because the vapour pressure of In over InN is low, it is not possible to desorb excess indium before the films decompose. However, we have previously established that it is possible to grow InGaN films over the whole composition range from InN to GaN with good control over composition [19]

12 InGaN Technology for IBSC applications

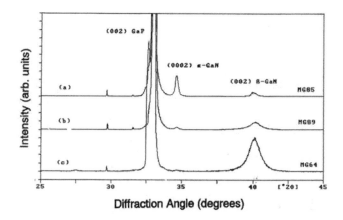

Fig. 12.5 Influence of arsenic on the growth of GaN, a film grown without arsenic present on the surface (MG85) crystallises in the hexagonal or wurtzite structure, but with increasing amounts of arsenic there is a transition to a mixed hexagonal and cubic film (MG89) to a pure cubic or zincblende film (MG64) (Reprinted with permission from [18]. Copyright 1995. American Institute of Physics)

Table 12.1 The primary optical phenomena, parameters, and theoretical background needed for describing the optical behaviour of light propagating through the solar cell structure (the absorptance A here corresponds to a single pass of light rays over a distance l_{opt}

Nominal composition	Peak position[a] (2Θ)	Calc. composition[b] (%)	EMPA composition (%)
AlN	36.042° (Theoretical 36.056°)	–	–
$Al_{0.75}Ga_{0.25}N$	35.645°	Al 73% : Ga 27%	Al 74% : Ga 26%
$Al_{0.5}Ga_{0.5}N$	35.268°	Al 48% : Ga 52%	Al 46% : Ga 54%
$Al_{0.25}Ga_{0.75}N$	34.854°	Al 20% : Ga 80%	Al 22% : Ga 78%
GaN	34.576° (Theoretical 34.570°)	–	–
$In_{0.25}Ga_{0.75}N$	33.727°	In 22% : Ga 78%	In 24% : Ga 76%
$In_{0.5}Ga_{0.5}N$	32.754°	In 48% : Ga 52%	In 50% : Ga 50%
$In_{0.75}Ga_{0.25}N$	32.160°	In 65% : Ga 35%	In 71% : Ga 29%
InN	31.379° (Theoretical 31.027°)	–	–

[a]Peak profiles fitted to give the peak position
[b]Calculated assuming Vergard's Law

(see Table 12.1) and with constant composition as a function of growth temperature as shown in Fig. 12.6. This ability to accurately adjust the In composition is not, however, trivial and requires close control over the both the In:Ga and group III (In+Ga):N flux ratios for the reasons discussed above. Thus, the margins for the successful growth of high-quality InGaN films are narrower than for GaN.

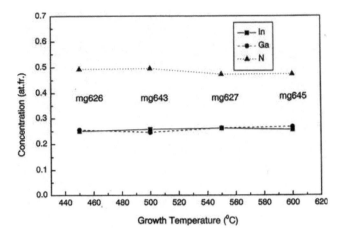

Fig. 12.6 Composition of InGaN films grown by PA-MBE as a function of growth temperature

The III–V nitrides also have much larger spontaneous polarisation and piezoelectric constants compared with other III-V materials [20]. The direction of the electric fields depends on the growth direction in wurtzite materials and may therefore assist or make more difficult the separation of charge carriers in a solar cell structure. The group III face will have a negative charge at the surface and should be doped p-type, whereas the N-face will accumulate a positive charge and should be doped n-type. Control over polarity during growth will thus be essential for structures based on wurtzite films. This problem may be entirely eliminated by growth of (001) oriented metastable cubic material (see above) where due to symmetry all electric fields are eliminated.

For In-rich InGaN, the large internal fields combined with the fact that the conduction band minimum lies well below the charge neutrality level lead to strong electron accumulation at the InN surface [21] and in InGaN for In concentrations $> \sim 40\%$ [22]. Thus, while p-doping may be achieved in In-rich InGaN, making a simple P–i–N diode for IBSC applications may not be as straightforward as in other III–V materials.

Other practical issues also make this task more difficult, in particular when growing In-rich InGaN films, due to the high carrier concentrations found in this materials system, a very large temperature rise occurs unless special precautions are taken. Recent measurements for In-rich InGaN films grown on a GaAs substrate show increases in temperature as large as 200°C when measured using the BandiT system, which in turn can lead to phase separation for In-rich samples [23]. Careful control of temperature using appropriate temperature ramps can eliminate this problem [23], but in much previous work on this topic, this problem has not been recognised due to the lack of temperature measurements.

12 InGaN Technology for IBSC applications

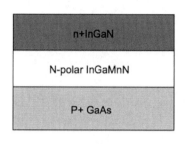 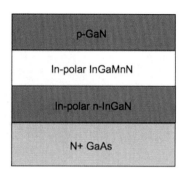

Fig. 12.7 Proposed InGaN IBSC structures. For N-polar samples, due to the influence of electric fields, the structure shown in (**a**) should be appropriate. For In(Ga) polar samples, the structure shown in (**b**) is more appropriate

12.4 Proposed IBSC Structures and Progress So Far

Taking all of the above factors into consideration, two different IBSC structures are proposed as shown in Fig. 12.7. Initial studies have concentrated on the InGaN wurtzite materials. Films of InGaN have been grown on semi-insulating (111) oriented GaAs substrates at temperatures from 400–550°C, with and without Mn fluxes to study the incorporation of Mn into InGaN. The results show that the solubility of Mn increases with decreasing growth temperature as is expected for InGaN, which will have a melting point significantly below that of GaN. In films where Mn is above a temperature-dependent critical value, x-ray studies show that the films phase separate into InN-rich and GaN-rich regions, but below this critical value single phase material is obtained. Films of this type will form that basis for InGaN IBSC structures. The resulting films are currently being studied for evidence of absorption below the InGaN band gap, which would indicate the formation of the impurity band needed for a successful IBSC device.

The structure shown in Fig. 12.7b has also been grown, without Mn in the InGaN region as a first test for the complete structure. The InGaN layer is grown on an n^+-GaAs substrate after removal of the native oxide. The growth temperature is approximately 500°C, after growth of the InGaN, a p-type GaN top contact is grown starting at 500°C and raising the temperature to 650°C, which previous studies have shown to be effective for p-type doping [24]. The transparent top GaN is doped with Mg and shows p-type conductivity using the Seebeck effect. Preliminary I–V characteristics, shown in Fig. 12.8, show clear evidence for having successfully obtained the required $P-i-N$ device characteristic, although the doping level in the intrinsic region remains to be determined. Similar structures are now being prepared with and without Mn in the active region and will be investigated in due course.

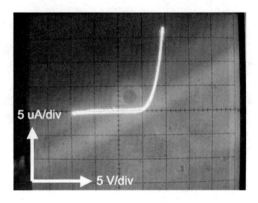

Fig. 12.8 I–V characteristic for the InGaN P–i–N test structure grown without Mn in the i region

12.5 Future Plans and Options

The optimum value for the In content in InGaN has yet to be determined, for the moment we are concentrating on samples with a fixed In content, but we will look in the future at other In/Ga compositions. The InGaN samples have to be grown at relatively low temperature and this may result in less than optimum quality of the resulting material. To further improve the quality of the InGaN, we will look at alternative growth techniques such as migration enhanced epitaxy [25], which is known to improve the properties of films grown at low temperature. One of the key problems with all nitride semiconductor structures is the high density of defects. It is well known that this can be improved significantly by growth on bulk substrates, which are now available commercially. One future possibility, therefore, is to consider growth of InGaN-based IBSC structures on bulk GaN substrates.

Finally, Mn is the first possible dopant considered for IBSCs based on InGaN [8]; however, other dopants need to be considered and investigated experimentally. Other possibilities include carbon (C), which is known to be a deep level in GaN and can be incorporated in large quantities into GaN.

12.6 Summary and Conclusions

In this chapter, we have discussed the possible application of InGaN layers doped with Mn to form the basis for Intermediate Band Solar Cell devices. We have discussed the reasons for choosing to study this promising system; the plasma-assisted molecular beam epitaxy method needed to grow such structures; the specific problems associated with the growth of InGaN and finally the progress made so far in achieving this overall goal. We have also discussed briefly other possible dopants e.g. C which might be used in InGaN for the same application.

References

1. V.Y. Davydov, A.A. Klochikhin, V.V. Emtsev, D.A. Kurdyukov, S.V. Ivanov, V.A. Vekshin, F. Bechstedt, J. Furthmuller, J. Aderhold, J. Graul, A.V. Mudryi, H. Harima, A. Hashimoto, A. Yamamoto, E.E. Haller, Phys. Stat. Sol. B-Basic Sol. Stat. Phys. **234**(3), 787 (2002)
2. J.W. Orton, C.T. Foxon, Rep. Progr. Phys. **61**(1), 1 (1998)
3. S. Nakamura, T. Mukai, M. Senoh, Appl. Phys. Lett. **64**(13), 1687 (1994)
4. S. Nakamura, G. Fasol, S.J. Pearton, *The Blue Laser Diode: The Complete Story*, 2nd edn. (Springer, Berlin, 2000)
5. U.K. Mishra, L. Shen, T.E. Kazior, Y.F. Wu, Proc. IEEE **96**(2), 287 (2008)
6. C.J. Humphreys, MRS Bull. **33**(4), 459 (2008)
7. A. Luque, A. Martí, Phys. Rev. Lett. **78**(26), 5014 (1997)
8. A. Martí, C. Tablero, E. Antolín, A. Luque, R.P. Campion, S.V. Novikov, C.T. Foxon, Sol. Energ. Mater. Sol. Cell. **93**(5), 641 (2009)
9. R.P. Campion, K.W. Edmonds, L.X. Zhao, K.Y. Wang, C.T. Foxon, B.L. Gallagher, C.R. Staddon, J. Crystal Growth **251**(1-4), 311 (2003)
10. S.V. Novikov, K.W. Edmonds, A.D. Giddings, K.Y. Wang, C.R. Staddon, R.P. Campion, B.L. Gallagher, C.T. Foxon, Semicond. Sci. Technol. **19**(3), L13 (2004)
11. M. Mesrine, N. Grandjean, J. Massies, Appl. Phys. Lett. **72**(3), 350 (1998)
12. A.V. Blant, O.H. Hughes, T.S. Cheng, S.V. Novikov, C.T. Foxon, Plasma Sources Sci. Technol. **9**(1), 12 (2000)
13. T. Li, R.P. Campion, C.T. Foxon, S.A. Rushworth, L.M. Smith, J. Cryst. Growth **251**(1–4), 499 (2003)
14. D. Cherns, L. Meshi, I. Griffiths, S. Khongphetsak, S.V. Novikov, N. Farley, R.P. Campion, C.T. Foxon, Appl. Phys. Lett. **92**(12), 121902 (2008)
15. C.T. Foxon, S.V. Novikov, J.L. Hall, R.P. Campion, D. Cherns, I. Griffiths, S. Khongphetsak, J. Cryst. Growth **311**(13), 3423 (2009)
16. B. Heying, I. Smorchkova, C. Poblenz, C. Elsass, P. Fini, S. Den Baars, U. Mishra, J.S. Speck, Appl. Phys. Lett. **77**(18), 2885 (2000)
17. C.T. Foxon, T.S. Cheng, D. Korakakis, S.V. Novikov, R.P. Campion, I. Grzegory, S. Porowski, M. Albrecht, H.P. Strunk, Mrs Internet J. Nitride Semicond. Res. **4**, (1999)
18. T.S. Cheng, L.C. Jenkins, S.E. Hooper, C.T. Foxon, J.W. Orton, D.E. Lacklison, Appl. Phys. Lett. **66**(12), 1509 (1995)
19. A.V. Blant, T.S. Cheng, C.T. Foxon, J.C. Bussey, S.V. Novikov, V.V. Tretyakov, III–V Nitrides **449**, 465 (1997)
20. F. Bernardini, V. Fiorentini, D. Vanderbilt, Phys. Rev. B **56**(16), 10024 (1997)
21. P.D.C. King, T.D. Veal, C.F. McConville, F. Fuchs, J. Furthmuller, F. Bechstedt, P. Schley, R. Goldhahn, J. Schormann, D.J. As, K. Lischka, D. Muto, H. Naoi, Y. Nanishi, H. Lu, W.J. Schaff, Appl. Phys. Lett. **91**, 092101 (2007)
22. P.D.C. King, T.D. Veal, H. Lu, P.H. Jefferson, S.A. Hatfield, W.J. Schaff, C.F. McConville, Phys. Stat. Sol. B-Basic Sol. Stat. Phys. **245**(5), 881 (2008)
23. J.L. Hall, A.J. Kent, C.T. Foxon, R.P. Campion, J. Cryst. Growth **312**(14), 2083 (2010)
24. C.T. Foxon, T.S. Cheng, N.J. Jeffs, J. Dewsnip, L. Flannery, J.W. Orton, I. Harrison, S.V. Novikov, B.Y. Ber, Y.A. Kudriavtsev, J. Cryst. Growth **189**, 516 (1998)
25. Y. Horikoshi, M. Kawashima, Jpn. J. Appl. Phys. Part 1-Regular Papers Short Notes and Review Papers **28**(8), 200 (1989)

Chapter 13
Ion Implant Technology for Intermediate Band Solar Cells

Javier Olea, David Pastor, María Toledano Luque, Ignacio Mártil, and Germán González Díaz

Abstract This chapter describes the creation of an Intermediate Band (IB) on single crystal silicon substrates by means of high-dose Ti implantation and subsequent Pulsed Laser Melting (PLM). The Ti concentration over the Mott limit is confirmed by Time-of-Flight Secondary Ion Mass Spectroscopy (ToF-SIMS) measurements and the recovery of the crystallinity after annealing by means of Glancing Incidence X Ray Diffraction (GIXRD) and Transmission Electron Microscopy (TEM). Rutherford Backscattering Spectroscopy (RBS) measurements show that most of the atoms are located interstitially.

Analysis of the sheet resistance and mobility measured using the van der Pauw geometry shows a temperature-dependent decoupling between the implanted layer and the substrate. This decoupling and the high laminated conductivity of the implanted layer could not be explained except if we assume that an IB has been formed in the semiconductor. A specific model for the bilayer electrical behaviour has been developed. The fitting of this model and also the simulation of the sheet resistance with the ATLAS code allow to determine that the IB energetic position is located around 0.36–0.38 eV below the conduction band. Carriers at the IB have a density very similar to the Ti concentration and behave as holes with mobilities as low as $0.4\,\text{cm}^2\,\text{Vs}^{-1}$.

13.1 Introduction

In the last decade, an enormous effort has been channeled towards the improvement of the efficiency in the solar cell technology [1]. One of the most promising concepts is the Intermediate Band Solar Cell (IBSC) [2], which is a forward-looking approach

G. González Díaz (✉)
Departamento de Física Aplicada III, Electricidad y Electrónica. Facultad de Ciencias Físicas, Universidad Complutense de Madrid, 28040 Madrid, Spain
e-mail: germang@fis.ucm.es

that can overcome the theoretical solar conversion efficiency limit for single junction solar cells [3]. This new technology is based on the idea of placing a new band of allowed states between the traditional valence and conduction bands that enable photons of energy lower than the semiconductor bandgap to be absorbed, yielding a higher efficiency. Hence, apart from the optical transition from the valence band to the conduction band, two new transitions could show up: one from the valence band to the Intermediate Band (IB) and the other from the IB to the conduction band. This capacity, when exploited without voltage degradation, permits to waste less incident photons, which translates in a higher efficiency conversion.

Among other possibilities like quantum dots [4] or highly mismatched alloys [5], ion implantation of deep level impurities can fulfill this task if the dose is high enough to exceed the Mott limit [6]. The use of classical group V donors or group III acceptors to produce an impurity band on Si or other semiconductors have been extensively studied from the fifties until now [7–10]. In those cases, the impurities are located substitutionally in the Si lattice and their solid solubility is high enough to overcome the limit to form a band. Due to the low ionization energy for these elements, the impurity band is only observed at low temperatures. The band has an energy width that depends on the spacing between the impurities but is in any case very narrow. Tight binding theory for narrow bands predicts that the mobility should be very low, a direct function on the impurity concentration and not depending on the temperature [7].

In the case of deep impurities, the technological problem associated with the formation of an IB seems to be very different. First of all, the solid solubility is orders of magnitude lower than the limit for metal-insulator transition. In this chapter, we will deal with Ti as a deep impurity which has a solid solubility in Si in the order of 10^{14} cm^{-3} [11], which is certainly lower than the required $5.9 \cdot 10^{19}$ cm^{-3} [12]. To overcome the solubility limit, we used highly non-equilibrium techniques, i.e., ion implantation and subsequent Pulsed-Laser Melting (PLM) annealing [13]. In fact, PLM technique has been extensively used to implement (IB) materials [14]. Ab-initio theoretical calculations show that Ti is a promising candidate for deep level IB. P. Wahnon and coworkers [15] have proved that interstitial Ti on Si form a half-filled band inside the gap and that the interstitial incorporation on the Si lattice is energetically more favourable than substitutional incorporation. Electrical transport properties of semiconductors with IB could be, in principle, quite different from semiconductors with a shallow impurity band. The effects of the semi-metallic band in the last case are observable only at very low temperatures due to the low ionization energies for the III or V doping elements. Also, at high doping and moderate temperatures, the impurity band merges with the valence or conduction band producing a degenerate semiconductor. In the case of deep impurities, the activation energy is higher and the effects will be present for a more extensive temperature range. However, ion implantation technology can create an IB region only in a thin superficial layer of the Si wafer and this fact makes the analysis more complex because the substrate will interfere in the measurements.

The study of the Hall effect on double layers was pioneered by Petritz [16] and continues now with different improvements [17]. Similar electrical behaviour appears frequently in III–V or II–VI semiconductors, and its anomalous Hall effect has to be modelled as two parallel bands [18, 19].

In this paper, we analyse the structural and electrical transport properties of single crystal Si in which an IB is formed by means of two non-equilibrium techniques: Ti ion implantation at doses well above the theoretical Mott limit and subsequent PLM of the implanted layer to recover the lattice damage.

13.2 Experimental

300 μm silicon (111) n-type samples ($\mu = 1,450\,\text{cm}^2\,\text{V}^{-1}\text{s}^{-1}$; $n = 2.2 \cdot 10^{13}\,\text{cm}^{-3}$ at room temperature) were implanted in an Ion Beam Services (IBS) refurbished VARIAN CF3000 Ion Implanter at 30 KeV with Ti at high doses ($10^{14}, 10^{15}, 5 \cdot 10^{15}, 10^{16}$ and $5 \cdot 10^{16}\,\text{cm}^{-2}$). Then the implanted Si samples were annealed by means of the PLM method to recover the sample crystallinity. The PLM annealing process was performed by J. P. Sercel Associates Inc. (New Hampshire, USA). Samples were annealed with one 20 ns long pulse of a KrF excimer laser (248 nm) at energy densities from 0.2 to 0.8 J cm^{-2}. PLM is a highly non-equilibrium processing technique, which is able to melt and recrystallize the Si surface up to about 100 nm depth in very short times (10^{-8}–10^{-6} s). This rapid recrystallization time allows the incorporation of Ti atoms to the Si at concentrations well above the solubility limit for this element [11]. Also, the PLM processing of the Ti implanted Si layer prevents secondary phase formation (i.e., titanium silicide) even when the equilibrium solubility limit has been greatly exceeded [20].

Time-of-Flight Secondary Ion Mass Spectroscopy (ToF-SIMS) measurements were performed with a ToF-SIMS IV model manufactured by ION-TOF, using a 25 keV pulsed Ga$^+$ beam at 45° incidence and O$_2$ flow. The raw profiles were calibrated by standard procedures, i.e., the concentration values were calculated from the ToF-SIMS counts by matching the Ti content in the implanted but not annealed sample to the nominal amount of Ti and the depth was calculated by optical profilometry measurement of the crater depth and assuming a constant erosion rate. Samples were also characterized by X-ray diffraction and Transmission Electron Microscopy (TEM). Glancing-Incidence X-Ray Diffraction (GIXRD) patterns were obtained by a Panalytical XPert PRO MRD diffractometer working with a Cu K$_\alpha$ source. The GIXRD patterns were collected at different incidence angles from 0.1 to 3° to study the crystal structure of the samples at different depths. Cross-sectional TEM images were obtained by a JEOL JEM-2000FX working at 200 keV. Simultaneously with the TEM measurements, Electron Diffraction (ED) patterns with a selected area of diffraction of about 50 nm were obtained. The ED images provided information on the crystalline morphology of the implanted layer. High-Resolution TEM (HRTEM) images were obtained by a JEOL JEM-4000 EX working at 400 keV. Channelling and random Rutherford Backscattering

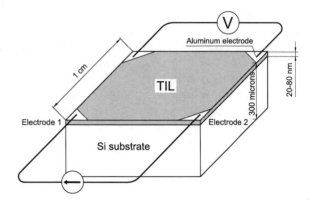

Fig. 13.1 Experimental set-up for the electrical van der Pauw characterization of the implanted samples (Reprinted with permission from [21]. Copyright 2009. Institute of Physics)

Spectroscopy (RBS) measurements were done to determine how the Ti atoms are located into the Si lattice (i.e., substitutional or interstitial). Measurements were conducted in a Cockcroft–Walton accelerator built by High Voltage Energy Europe. The equipment worked with 2 MeV He ions. Finally, Si implanted samples were electrically characterized by resistivity and Hall effect measurements in the van der Pauw configuration at variable temperature using a Keithley SCS 4200 model with 4 Source and Measure Units. A schematic view of the electrical set-up is showed in Fig. 13.1. Electrical contacts were Al vacuum deposited triangles at the four corners of a square sample of 1 cm side. In all the cases, the contacts were made at the Ti Implanted Layer (TIL). The samples were placed inside a homemade cryostat attached to a vacuum pump to avoid moisture condensation. The temperature was varied between 90 and 380 K. In some cases, samples were measured down to 7 K using a Janis closed cycle cryostat. We have measured the 4 van der Pauw configurations and also we have changed the polarity of the current source and the direction of the magnetic field to avoid any possibility to have spurious thermogalvanomagnetic effects.

13.3 Results

13.3.1 Structural Characterization

13.3.1.1 ToF-SIMS Profiles

Figure 13.2 shows the Ti depth profiles for implanted samples at 10^{14}, 10^{15}, $5 \cdot 10^{15}$, 10^{16} and $5 \cdot 10^{16}$ cm^{-2} after a PLM process at 0.8 J cm^{-2}. As comparison we have plotted also the profile for the non-annealed samples with doses of 10^{15} and 10^{16} cm^{-3}. As it can be seen, Ti profiles have been pushed to the Si surface, becoming steeper and with a higher and sharp maximum, although this effect is more pronounced for lower doses. The second peak that appears clearly for 10^{14} and

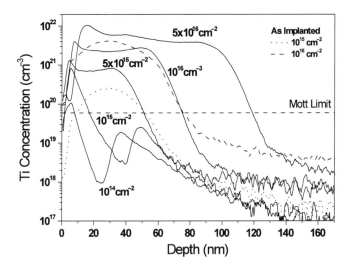

Fig. 13.2 ToF-SIMS profiles of Ti-implanted samples with 10^{14}, 10^{15}, $5 \cdot 10^{15}$, 10^{16} and $5 \cdot 10^{16}$ cm^{-2} doses after PLM annealing at 0.8 J cm^{-2}. Also the implanted but non-annealed samples with 10^{15} and 10^{16} cm^{-2} are also shown for reference (Reprinted with permission from [22]. Copyright 2010. American Institute of Physics)

10^{15} cm^{-2} doses at 40 nm and 50 nm, respectively, marks the border between melted and non-melted semiconductor zone. Clearly, the melted depth is related with the degree of amorphization which in turn is directly dependent on the implanted dose.

The comparison of the area under the SIMS profiles for the annealed and non-annealed samples shows that only about 60% of the Ti remains in the silicon sample after the PLM. For doses higher than 10^{15} cm^{-2}, we have a thickness of 20, 60, 80 or 120 nm with Ti concentration over the Mott limit. For the dose of 10^{14} cm^{-2}, the Ti volume concentration reaches this limit only in a very thin layer very close to the surface, where the ToF-SIMS data are very prone to errors.

13.3.1.2 GIXRD

Figure 13.3 shows the X-ray diffractograms at 0.4° glancing angle for the samples PLM annealed at 0.8 J cm^{-2} and implanted with different doses. For the samples implanted with the highest doses (10^{16} and $5 \cdot 10^{16}$ cm^{-2}), two diffraction peaks at 47.1 and 55.9° can be observed. For the sample with a dose of 5×10^{15} cm^{-2}, the peak placed at 47.1° disappears, but the peak at 55.9° is still present. However, both diffraction peaks vanish for the sample implanted with the 10^{15} cm^{-2} dose. These peaks are attributed to the (220) and (113) silicon reflections. It is worth highlighting that no peaks associated with Ti-Si phases are present. These results point out the formation of a silicon polycrystalline structure for the samples implanted with the

Fig. 13.3 GIXRD diffractograms at an incidence angle of 0.4° for the samples implanted with different doses and annealed at 0.8 J cm^{-2} (Reprinted with permission from [22]. Copyright 2010. Institute of Physics)

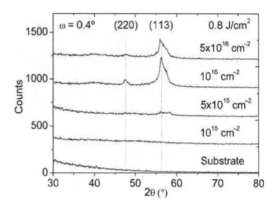

Fig. 13.4 GIXRD diffractograms at different incidence angles for the sample implanted with a 10^{16} cm^{-2} dose and annealed at 0.8 J cm^{-2} (Reprinted with permission from [22]. Copyright 2010. Institute of Physics)

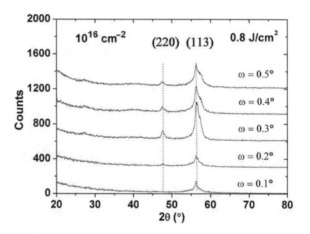

highest doses. To further study the crystal structure at different depths, the X-ray diffractograms were collected at different glancing angles.

Figure 13.4 shows the GIXRD at different glancing angles for the sample implanted with a 10^{16} cm^{-2} and annealed at 0.8 J cm^{-2}. The intensities of the (220) and (113) silicon reflections increase with increasing the glancing incidence angle up to 0.3°, and decrease with higher incidence angles. Analogous behaviour was observed for the other samples. This trend hints at the formation of a silicon polycrystalline phase in the implanted layer for the samples with the highest dose. However, no diffraction peaks were observed at any incidence angle for the sample implanted with the 10^{15} cm^{-2}, pointing to a good crystallization of the implanted layer. It is important to highlight that a virgin substrate has not any diffraction peak at this configuration. From these results, it is concluded that excellent recrystallization is obtained at 10^{15} cm^{-2} implanted dose.

13.3.1.3 TEM

The final confirmation of the different crystalline structure among the samples comes from the TEM images and ED measurements. Figure 13.5 shows the cross-sectional TEM images and the ED pattern of samples implanted with $5 \cdot 10^{16}$ down to 10^{15} cm^{-2} doses, and annealed at 0.8 J cm^{-2}. For the three first images, Fig. 13.5a–c, we can observe a polycrystalline layer on top of a single crystal silicon substrate. It is clearly appreciable that the grain size increases from Fig. 13.5a–c. The sample implanted with the highest dose (Fig. 13.5a) presents a nanocrystalline structure, with a mean grain size of about 5 nm. When the implanted dose decreases the grain size presents a dimension comparable to the implanted layer thickness. For the sample implanted with a implantation dose of $5 \cdot 10^{15}$ cm^{-2} (Fig. 13.5c), the thickness of the defective implanted layer is reduced down to about 30 nm. For the sample implanted with a 10^{15} cm^{-2} dose (Fig. 13.5d), no contrast difference appears between the implanted layer and the substrate, thus, the defective layer is suppressed. Except for some punctual defects, the crystal seems to be perfect.

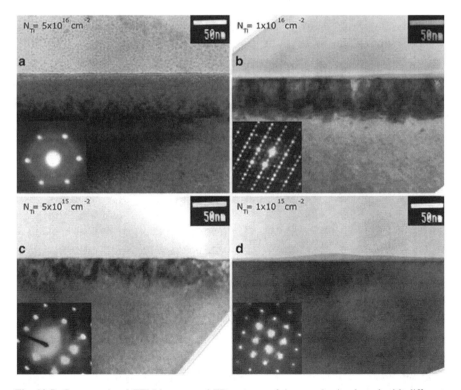

Fig. 13.5 Cross-sectional TEM images and ED patterns of the samples implanted with different doses, and annealed at 0.8 J cm^{-2} (Reprinted with permission from [22]. Copyright 2010. Institute of Physics)

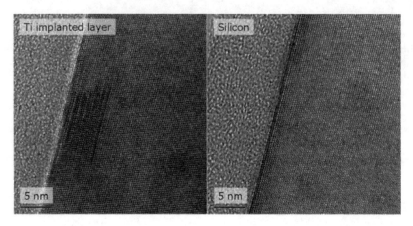

Fig. 13.6 a: High-Resolution TEM images of the sample implanted with a dose of 10^{15} cm^{-2} and annealed at 0.8 J cm^{-2} (*left*). b: Un-implanted sample (*right*) (Reprinted with permission from [22]. Copyright 2010. Institute of Physics)

Regarding the ED pattern, Fig. 13.5a shows that the sample implanted with the highest dose ($5 \cdot 10^{16}$ cm^{-2}) is polycrystalline with small grains. This is indicated by the diffraction dots which are grouped in concentric ring around the direct beam. ED diagram of Fig. 13.5b belonging to the sample implanted with a 10^{16} cm^{-2} dose presents an unusual pattern: the bright spots correspond to the [111] zone axis of the silicon, but intermediate spots appears between the main dots. This effect is due to the highly defective lattice reconstruction. For the $5 \cdot 10^{15}$ and 10^{15} cm^{-2} doped samples, the ED patterns of the implanted layer and the silicon substrate did not show difference, as it can be seen in Fig. 13.5c, d. This result points to a good reconstruction of the lattice, and strengthens the GIXRD results.

Figure 13.6a shows the High-Resolution TEM (HRTEM) image of the Ti implanted sample with 10^{15} cm^{-2} dose, while Fig. 13.6b shows for comparison the image of an un-implanted silicon substrate. Areas with a different contrast were observed seldom-distributed in the implanted layer. This effect is likely due to dangling bonds that produces stacking faults. This result points to an excellent crystalline reconstruction after the PLM process for this sample, as it was previously stressed by GIXRD data of Fig. 13.3. Also, it should be remarked that with the 10^{15} cm^{-2} dose, we obtain an implanted layer with a Ti concentration higher than the Mott limit (see Fig. 13.2), with an almost perfect lattice reconstruction, i.e., a material of choice for IBSC.

13.3.1.4 RBS

RBS profiles have been obtained for doses of 10^{15}, $5 \cdot 10^{15}$ and $5 \cdot 10^{16}$ cm^{-2} both in random and channeling configuration. The Ti peak is clearly observed at energies close to 1,650 keV, which is far from the maximum backscattered energy due to the

Si surface (1,330 keV). The area of this peak is proportional to the implanted dose and shows the broadening corresponding to the implantation depth that increases with the dose. Ti peaks obtained in random and channeled configurations are almost equal and the comparison of its area for the different doses shows that only a maximum of about 8% of the Ti is located at substitutional positions being therefore the most part of the implanted atoms located interstitially. Anyway, the uncertainty of these measures does not permit to obtain a real value of the substitutional percentage, which is in any case very low if any.

As a partial conclusion of the structural characterization, we can resume that for the 10^{15} cm^{-2} dose we have an almost perfect crystal with a layer of about 20 nm having a Ti interstitial concentration above the Mott limit. Higher doses produce deeper profiles but with a poorer crystalline structure.

13.3.2 Electrical Transport Properties

13.3.2.1 Sheet Resistance and Hall Mobility

Figure 13.7a shows the sheet resistance as a function of measured temperature for a reference Si substrate and for samples with implanted doses of 10^{14}, 10^{15}, $5 \cdot 10^{15}$ and 10^{16} cm^{-2}. All the implanted samples were PLM annealed with 0.8 J cm^{-2} energy density. Hollow symbols are the experimental measurements while lines with symbols correspond to a simulation we will explain later. We also checked that unimplanted substrates have the same electrical characteristics before and after a PLM process. We did this experiment to confirm that the annealing process has not any influence on the substrate sheet resistance or mobility. Figure 13.7b shows the mobility modulus versus temperature for the same samples quoted in the previous figure. In any case, the samples were n type, that is to say, they have negative mobility. Both the sheet resistance and the mobility for samples implanted with a 10^{14} cm^{-2} dose and their corresponding for the substrate are almost indistinguishable.

For samples from 10^{15} to 10^{16} cm^{-2} doses, sheet resistance shows an uncommon feature. For temperatures lower than 180–190 K, the sheet resistance is higher than the substrate resistance while for higher temperatures is lower. As we assume that we are dealing with a bilayer (implanted layer and substrate), this implies these layers are in parallel at high temperature and isolated at low temperature. For a parallel scheme, there is no way to have a sheet resistance higher than one of the branches, and for this reason we have to think in a decoupling at low temperatures. That means that for low temperatures we are measuring just the implanted layer while at high temperatures we are measuring both the implanted layer and the substrate in parallel.

The decoupling would be straightforward if we had opposite majority carriers in each layer at low temperature (We mean a rectifying behaviour) and the same type at high temperature, but there is no way to change from p type at low temperatures to n type at high temperatures for an usual semiconductor. We have to note that the substrate remains n type for all the temperature range.

In the following and in order to explain the results, we will assume that a half-filled IB has been formed at the implanted layer and there is a delocalization of the impurity electron wavefunctions in the thickness, where Ti concentration is overcoming the Mott limit as it is shown in Fig. 13.2; that is to say, 20 nm for 10^{15} cm^{-2} dose, 60 nm for $5 \cdot 10^{15}$ cm^{-2} dose and 80 nm for 10^{16} cm^{-2} dose.

This assumption is congruent with the different behaviour observed for the sample with 10^{14} cm^{-2} dose where it is supposed that the Mott limit has not been reached yet and there is not enough Ti to obtain the overlapping of the wavefunctions.

According to the IB theory, the band diagram of the TIL/n-Si double sheet should be like the one showed in Fig. 13.8. In the TIL side, the Fermi level (FL) is pinned at the IB energy (E_C–E_{IB}) due to the high density of states at this energy and is almost constant with the temperature. Electrons at the conduction band and holes

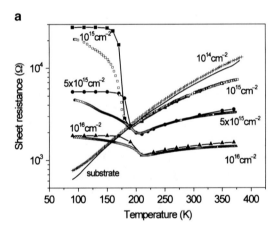
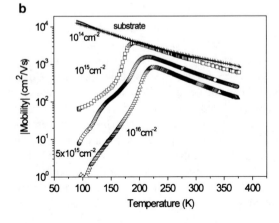

Fig. 13.7 (a) Sheet resistance as a function of measured temperature for a n-Si substrate (–) and for double sheet TIL/n-Si substrate for different implantation doses: 10^{14} cm^{-2} (*plus* experimental); 10^{15} cm^{-2} (*square* experimental; *filled square* ATLAS simulation); $5 \cdot 10^{15}$ cm^{-2} (◦ experimental; *bullet* ATLAS simulation) and 10^{16} cm^{-2} (*triangle* experimental; *filled triangle* ATLAS simulation). (b) Mobility absolute value as a function of the measured temperature for a n-Si substrate (–) and for double sheet TIL/n-Si substrate for different implantation doses: 10^{14} cm^{-2} (*plus*); 10^{15} cm^{-2} (*square*); $5 \cdot 10^{15}$ cm^{-2} (*circle*) and 10^{16} cm^{-2} (*triangle*) (Reprinted with permission from [21]. Copyright 2009. Institute of Physics)

13 Ion Implant Technology for Intermediate Band Solar Cells

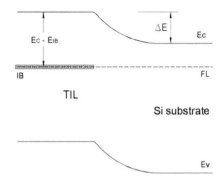

Fig. 13.8 Generic band diagram of the TIL/n-Si double sheet. This drawing is not at scale and correspond to the expected band diagram at $T = 90$ K (Reprinted with permission from [21]. Copyright 2009. Institute of Physics)

at the valence band are determined through standard Maxwell–Boltzman statistics and their densities are strongly dependent on the temperature in spite of the constant position of the Fermi level ($n = N_C \exp(-(E_C - E_{IB})/kT)$ for electrons and $p = N_V \exp(-(E_{IB} - E_V)/kT)$ for holes). Nevertheless, the carrier concentration at the IB will not agree with this statistics and they should be calculated as in a degenerated band, in agreement with the theory for semiconductors with IB [2].

Carriers at the IB could behave as electrons or holes depending on the sign of its effective mass, that is to say, the concavity or convexity of the band at the energy where it is crossed by the Fermi level. Of course the statistics now should be the Fermi Dirac one. Anyway, and independently of its type, carriers at the IB could not cross to the substrate because of the lack of continuity between the IB and the substrate. Tunneling is avoided by the low doping of the substrate that implies a wide depletion thickness, mostly developed in this substrate. Therefore at every temperature, IB carriers are electrically confined in the TIL plane, with low mobility as corresponding to a narrow band.

The sheet resistance at low temperatures only comes from the TIL because of the decoupling effect. To compute it, we have to take into account the electrons at the conduction band, holes at the valence band and carriers at the IB, irrespective of its sign. The only way to discriminate between electrons and holes at this band are the mobility measurements that determine that these carriers are holes as we will see later. Anyway, for the decoupling explanation its nature is irrelevant.

If we assume that the Fermi level (or the IB position which is the same at the TIL) is closer to the conduction band than to the valence band, the only carriers to worry about are electrons at the conduction band and, of course, holes at IB. At low temperature, if the IB energetic position is deep enough, the electron density should be very low at the TIL conduction band. In the substrate, the electron concentration is almost temperature independent (from 90 to 300 K) as corresponding to a semiconductor with a n-type shallow doping of $2.2 \cdot 10^{13}$ cm^{-3}. The substrate Fermi level position can be easily calculated using the Maxwell–Boltzmann statistics. At 90 K the FL in the substrate is about 0.1 eV below the conduction band. In the Fig. 13.8, we have assumed that the IB is deeper than this energy and as a consequence, the electrical connection between the two layers becomes unidirectional; the substrate can inject electrons to the TIL if the net applied voltage is high enough for the electrons to surmount the barrier energy but the TIL cannot inject electrons to the substrate because the electron density at TIL is negligible at this temperature. Due to the Fermi level position represented in Fig. 13.8, holes at TIL valence band or substrate valence band are also negligible. In a parallel double sheet, the fact of having an unidirectional contact between the layers means an effective decoupling of them because any current line coming and returning from TIL to the substrate has to cross twice the junction in opposite senses.

When the temperature goes up, the electron density at the TIL conduction band increases as it corresponds to the thermal generation from a level with an activation energy of $E_C - E_{IB}$. This activation energy is smaller than the Si intrinsic one and electrons density increases relatively fast. There should be a temperature at which the electron density in the TIL becomes similar to the substrate electron density, which remains constant. Electrons can now flow freely back and forth from one layer to another and both layers are now electrically coupled. At this temperature both conduction bands at the TIL and the substrate should be leveled and there is not any current limitation mechanism between them.

According to the previous paragraph, our bilayer limits the electron flux going from the implanted layer to the substrate but only at low temperatures. As holes flux does not cross from TIL to the substrate at the IB, this border behaves as a junction having the P side at the implanted layer and the N side at the substrate. This unidirectional behaviour disappears at high temperature. In this sense, and only in this sense, the implanted layer behaves as a P semiconductor at low temperatures. It is important to highlight that for a semiconductor with an IB the issue of the definition of the type is not as clear as for a normal semiconductor.

As the carrier density at the IB have only influence on the sheet resistance of the implanted layer and they are not involved in the coupling/decoupling mechanism, their nature is irrelevant to determine the current limitation mechanism at the TIL/substrate junction.

Assuming the previous hypothesis, the densities of carriers are easy to calculate using the classical semiconductor equations. Let us suppose that the IB is located 0.36 eV below the conduction band. As we will see in the next paragraph, this is the value that best fits the experimental results, it is not temperature dependent and it is worth to say that it is close to the Ti level referred in the literature [11]. With this hypothesis, at low temperature (90 K) the electrons carrier density in the TIL should be extremely low and the barrier height (ΔE, see Fig. 13.8) with the substrate is 0.26 eV. Conversely at 300 K the electron density in the TIL increases up to values similar to the electron concentration in the substrate. In that case, there is no appreciable barrier for electrons and they can flow freely in both directions, depending on the external voltage polarity.

13.3.2.2 ATLAS Sheet Resistance Simulation

To fully confirm the qualitative explanation of the previous paragraph, we have performed some simulations using the ATLAS code framework [23]. As in this code it is not possible to define a semiconductor with an IB, we have defined the TIL sheet as a "new" semiconductor with the following characteristics:

- *Conduction band:* Is the same than Si, having the same equivalent density of states, mobility, affinity and so on.
- The gap of this "new" semiconductor is the $E_C - E_{IB}$ energy, avoiding the VB at the TIL.
- *Density of States (DOS) and hole concentration in the IB:* as stated before, RBS measurements showed that Ti is located mostly at interstitial positions. As Ti electronic configuration is $1s^2\ 2s^2\ 2p^6\ 3s^2\ 3p^6\ 4s^2\ 3d^2$, each atom is able to give one to four electrons to the IB. If we assume as a first approach that each Ti atom gives an electron to the IB, the DOS should be 2× [Ti] (where [Ti] is the Ti concentration), in agreement with standard semiconductor theory of band formation, and the hole density will be the Ti concentration [Ti]. Later in this chapter, we have some considerations about the possibility of having more electrons per Ti atom. These assumptions are in accordance with the theoretical calculations of Wahnon and coworkers [15] that have shown through ab initio calculations that the Ti interstitial atoms in a host Si lattice should produce a half filled intermediate band.
- From the statistical point of view, the IB is supposed to be narrow enough to be considered as a single level.
- It is worth noting that, for an energy band narrow enough, NV (equivalent density of states) and DOS should be roughly the same. Hence, we will use for NV the DOS quoted above, i.e., $N_V = 2 \times [Ti]\,\text{cm}^{-3}$.
- Hole mobility at the IB (μ_{IB}) has to be guessed but it is supposed to have a very low value as corresponding to a narrow energy band. Careful Hall measurements at low temperatures limits this variation between about 0.1–0.6 cm^2 V^{-1} s^{-1} as we will see further on.
- TIL thickness will be obtained from the ToF-SIMS measurements at the point where the concentration becomes equal to the Mott limit. For easier calculation, we assume a squared-box Ti concentration obtained as: $[Ti] = 0.6 \times D/t$ where D is the implanted dose, t is the thickness obtained through ToF-SIMS measurement and the 0.6 figure takes into account the Ti losses during the PLM annealing as explained before.

The band diagram, resulting from ATLAS code, is depicted in Fig. 13.9 for two temperatures, 90 and 300 K. Notice that the TIL "normal" VB has disappeared in this model in relation to the model proposed in Fig. 13.8 because, as stated before, holes in both "normal" valence bands do not have any significant role in the model due to its very low carrier density.

We have simulated the bilayer sheet resistance for the 10^{15}, $5 \cdot 10^{15}$ and 10^{16} cm^{-2} doses, all of them with 0.8 J cm^{-2} annealing. For this simulation, we have used

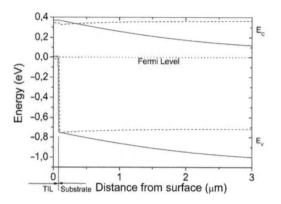

Fig. 13.9 Band diagram, obtained from ATLAS simulation, of the double sheet at $T = 90$ K ($-E_C$ and $-E_V$) and $T = 300$ K ($-E_C$ and $-E_V$) (Reprinted with permission from [21]. Copyright 2009. Institute of Physics)

a tri-dimensional structure like the one represented in Fig. 13.1. A comparison between sheet resistance experimental data and the ATLAS simulation values is presented in Fig. 13.7a. For the fitting, only two parameters have been scanned: the IB energetic position and the IB hole mobility. Best fit is obtained with $E_C - E_{IB} = 0.36$ eV for the three samples and hole mobility $\mu_p = 0.4 \text{ cm}^2 \text{ V}^{-1}\text{s}^{-1}$ for doses of 10^{15} cm^{-2} and $5 \cdot 10^{15}$ cm^{-2} and 0.6 cm^2 V^{-1}s^{-1} for dose of 10^{16} cm^{-2}. No fit is done neither in the IB hole concentration which remains equal to Ti volume concentration nor in the N_V value at the IB that is $2\times$ [Ti]. The temperature of the sheet resistance minimum is very sensitive to the pseudo-gap chosen, i.e. the energetic position of the IB, being the sheet resistance plateau at low temperatures dependent on the hole mobility times the IB hole density. For this reason, we cannot know exactly how many carriers are given for each Ti atom; it goes from one electron per Ti atom with a mobility of 0.4 cm^2 Vs^{-1} for the corresponding hole to four electrons with 0.1 cm^2 Vs^{-1} hole mobility.

The electron mobility at the conduction band of the implanted layer is a point of concern because it is well known that it should be very dependent on the crystallinity. Nevertheless, there is not appreciable variation on the fitting results if we scan its value at room temperature from 0 to $1{,}500$ cm^2 V^{-1} s^{-1}. That simply means that the product $p\mu_p$ at the intermediate band is always higher than $n\mu_n$ at the TIL conduction band for all the temperatures. p is the hole concentration at the IB, n is the electron concentration at the conduction band of the TIL and μ_p and μ_n the mobilities of these carriers.

At high temperatures the fitting is always excellent and the position of the minimum and its shape is very well modelled. For temperatures below the one corresponding to the resistance minimum, the simulated sheet resistance increases faster than the experimental results. The difference could be related with a possible uncontrolled current leak through the TIL-substrate interface. As this is similar to the behaviour of a reverse polarized P–N junction without special technology to avoid surface leaks, we suppose that here the most probable region to have superficial states is the rim of the sample, which is a highly damaged zone. This extra conduction path has not been simulated by the ATLAS code.

13 Ion Implant Technology for Intermediate Band Solar Cells 335

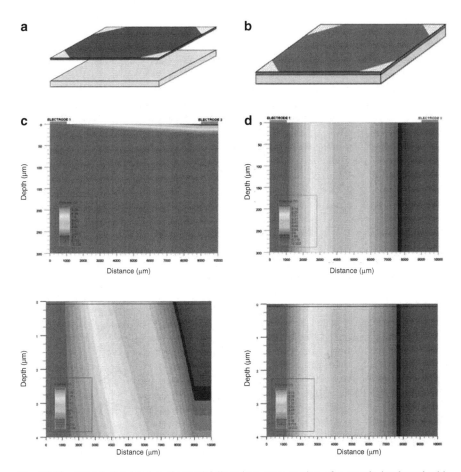

Fig. 13.10 ATLAS simulated equipotential lines in a cross section of a sample implanted with 10^{15} cm^{-2} dose. (**a**) is the whole sample at 90 K and (**b**) at 300 K, (**c**) and (**d**) are enlargements of the TIL area at 90 K and 300 K, respectively (Reprinted with permission from [21]. Copyright 2009. Institute of Physics)

Figure 13.10 shows the simulated equipotential lines in a cross section from the injection current electrode at left (electrode 1 in Fig. 13.1) to ground at right (electrode 2) at 90 (Fig. 13.10a, c) and 300 K (Fig. 13.10b, d). In both cases, the external potential is 10V positive at the left electrode. Figure 13.10b presents, for 300 K, the equipotential lines in the whole sample showing that they are perpendicular to the surface.

Figure 13.10d is an enlargement of the TIL area showing that there are no differences between the TIL and the substrate because the current is carried by both layers in parallel. At low temperature almost all the substrate has the potential of the positive electrode, as is depicted in Fig. 13.10a. All this behaviour is congruent with the idea that the junction between the TIL and the substrate is blocking if TIL is negative and not blocking with the opposite polarity. Figure 13.10c depicts the enlarged TIL area for low temperature showing perpendicular equipotential lines meaning that the current is supported only by this part of the bilayer. The potential differences between TIL and substrate causes a depletion zone at the substrate with a triangular shape which become thicker as we approach to the right (ground electrode) because the potential difference is becoming higher. The depletion zone, which is clearly shown in Fig. 13.10a, is essentially developed at the substrate and not in the TIL layer due to the very different carrier concentration between both layers.

Unfortunately, we were unable to simulate the Hall mobility with the ATLAS code because convergence problems.

13.3.2.3 An Analytical Model for the Sheet Resistance and for the Hall Mobility

To confirm the ATLAS results and also to simulate the mobility behaviour of our samples, we have developed an analytical model to explain the sheet resistance and mobility characteristics of our bilayers. As far as we know, this model has not been proposed in the literature and for this reason we will explain it from a general point of view below.

General Model for the Sheet Resistance of a Bilayer

We will assume a bilayer composed by layer 1 and layer 2 as in Fig. 13.11. Both layers are connected through resistors $R_t/2$, which model the current limitation for the current flux from one layer to the other.

Currents injected on layer 1 and 2, I_1 and I_2 could be obtained as:

$$I_1 = \frac{V}{R_{C1}} \quad I_2 = \frac{V}{(R_t + R_{C2})}, \tag{13.1}$$

where V is the voltage drop at the current source terminals. Obviously, $I = I_1 + I_2$. R_{C1} and R_{C2} are the resistances between two consecutive electrodes (a and b on Fig. 13.12) for layer 1 and (a' and b') for layer 2.

As we are interested in sheet resistance R_S and not in the resistance between two contacts R_C, it will be mandatory to obtain the relation between both magnitudes. We will call $\alpha = R_C/R_S$. This figure is solely dependent on the relative size of the electrical contact respective to the sample size. We will assume triangular contacts.

The resistance R_C has to be calculated numerically as there is not analytical model for it. We can use a code as ATLAS to determine the relation between this resistance and the parameter α quoted above but it is easy and more instructive to use the popular PSPICE code [24]. We will simulate the layer as an array of resistors with 19 rows and 19 columns. Resistor at the limits should have twice value because

13 Ion Implant Technology for Intermediate Band Solar Cells

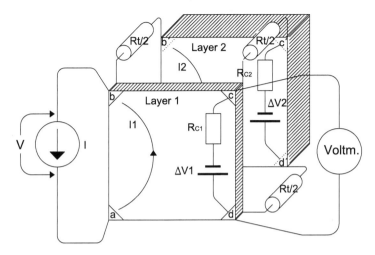

Fig. 13.11 Exploded view of the bilayer showing the current limiting resistances at the four corners, currents on each layer introduced at corners *a* and *b* and the voltages developed at the opposite corners *c* and *d* through a R_C resistance

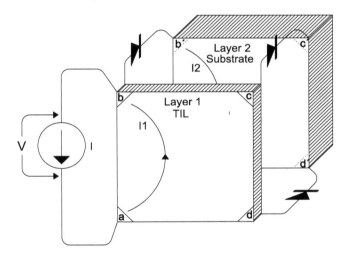

Fig. 13.12 Exploded view of the bilayer showing the diodes at the four corners that simulate the rectifying behaviour of the TIL-substrate interface

there is not current path outside these limits [25]. For easy calculation, we will use 1 ohm resistor and we will inject a 1 ampere current between two consecutive contacts and we will register the potential on the current source (V) and the potential difference at the opposite corners (ΔV) for different contact size. In the Table 13.1, we give the results of this simulation.

Column 1 is the contact size, i.e. the number of nodes that have been short circuited at the corners to simulate the equipotential triangular contact. The relative

Table 13.1 Potential at measurement corners (ΔV), sheet resistance (Rs), α factor (Rc/Rs) and sheet resistance error for different contact sizes

Cont. size	ΔV	Rs	α factor	Error(%)
1 × 1	0.222	1.006	5.174	0.610
2 × 2	0.222	1.004	2.390	0.429
3 × 3	0.220	0.999	1.901	−0.115
4 × 4	0.217	0.983	1.542	−1.746
5 × 5	0.210	0.951	1.255	−4.919

Table 13.2 α factor for Hall configuration

Cont. size	α factor
1 × 1	5.394
2 × 2	2.610
3 × 3	2.140
4 × 4	1.711

size could be obtained having in mind that the mesh size is 19 × 19. The second one is the differential voltage at potential measurement corners, the third one is the sheet resistance obtained with the classical van der Pauw formula $R_S = \pi/ln(2)\,\Delta V/I$. It is important to realize that the theoretical sheet resistance is just the resistance of each one of the differential elements we used to represent the complete sample, that is to say 1 ohm. The α factor is simply the voltage developed at the current source because the current is 1 Ampere and the sheet resistance is 1 ohm. For our sample which has approximately 1.5 mm side triangular contacts the α factor is very close to 2. The error in the sheet resistance measurement is as low as 0.1%. Table 13.2 gives the same factor α when the electrodes are not consecutive but opposite as is the case for Hall measurements.

Rewriting expressions 13.1, $I_1 = V/\alpha R_{S1}$ and $I_2 = V/(R_t + \alpha R_{S2})$, where R_{S1} and R_{S2} are the sheet resistances of the layer 1 and 2, respectively. Due to the presence of the R_t, the developed voltage on the opposite corners (c and d for layer 1 and c' and d' for layer 2 on Fig. 13.11) are not the same for both layers (it is the same only if the injected current on each layer is in a inverse proportion to the sheet resistance). To obtain the measured voltage we have to bear in mind the internal impedance across which the voltage is developed [16]. Of course, the resistance between electrodes c and d is the same than the one observed between electrodes a and b, that is to say R_{C1}. The same happen for electrodes c' and d', which have a resistance R_{C2}.

Each one of the voltages ΔV could be expressed according van der Pauw as:

$$\Delta V_1 = I_1 \cdot R_{S1} \cdot ln(2)/\pi \qquad \Delta V_2 = I_2 \cdot R_{S2} \cdot ln(2)/\pi \qquad (13.2)$$

and the equivalent voltage measured on the electrodes of the top layer (electrodes d and c on Fig. 13.11) with a infinite impedance voltmeter will be:

$$\Delta V = \Delta V_1 - R_{C1} \cdot \frac{\Delta V_1 - \Delta V_2}{R_t + R_{C1} + R_{C2}} = \frac{\Delta V_1 G_{C1} + \Delta V_2 \cdot \frac{G_t G_{C2}}{G_t + G_{C2}}}{G_{C1} + \frac{G_t G_{C2}}{G_t + G_{C2}}}, \qquad (13.3)$$

where $G_{C1} = 1/R_{C1}$ and $G_{C2} = 1/R_{C2}$ are the conductances of layer 1 and layer 2, respectively, and G_t is $1/R_t$.

Equation 13.3 could be written in a more compact way if we introduce a function $F = G_t/(G_t + G_{C2})$:

$$\Delta V = \frac{\Delta V_1 G_{C1} + \Delta V_2 G_{C2} F}{G_{C1} + G_{C2} F}. \tag{13.4}$$

The equivalent sheet resistance, R_{sheet}, i.e., the one measured on top of the layer 1 is:

$$R_{\text{sheet}} = \frac{\pi}{\ln(2)} \frac{\Delta V}{I} = \frac{\pi}{\ln(2)} \frac{\Delta V}{\left(\frac{V}{R_{C1}} + \frac{V}{R_t + R_{C2}}\right)} = \frac{G_{C1} + \frac{G_t^2 G_{C2}}{[G_t + G_{C2}]^2}}{\alpha \left(G_{C1} + \frac{G_t G_{C2}}{G_t + G_{C2}}\right)^2}. \tag{13.5}$$

Introducing the function F in the sheet resistance formula, we can write:

$$R_{\text{sheet}} = \frac{G_{C1} + G_{C2} F^2}{\alpha (G_{C1} + G_{C2} F)^2}. \tag{13.6}$$

In (13.5) and (13.6), $R_{\text{sheet}} \to R_{S1}$ when G_t or $F \to 0$ and to $R_{S1} \,//\, R_{S2}$ when $G_t \to \infty$ and $F \to 1$

General Model for the Hall Mobility of a Bilayer

Same procedure can be used to find the Hall mobility obtaining an effective mobility as:

$$\mu_{\text{eff}} = \frac{\mu_1 G_{C1} + \mu_2 (G_t/(G_t + G_{C2}))^2}{G_{C1} + G_{C2}(G_t/(G_t + G_{C2}))^2} = \frac{\mu_1 G_{C1} + \mu_2 G_{C2} F^2}{G_{C1} + G_{C2} F^2}, \tag{13.7}$$

where μ_1 and μ_2 are layer 1 and layer 2 mobilities. These mobilities have to be considered with its corresponding signs according if they are related to p type carriers or n type ones. Now the α coefficient has to be changed from 2 to 2.2, which is the correct value for R_C/R_S when the current is injected at the opposed electrodes in a square sample with relative electrode size as the one we have, as it is obtained from Table 13.2.

Model Application to Our Case

An electrical equivalent circuit for our bilayer is depicted in Fig. 13.12. We will assume that both layers are connected with four non-ideal diodes (one on each corner) that represent the electrical TIL-substrate junction. The characteristics of these diodes have been calculated in reference [26], where we showed that the reverse current depends exponentially with the temperature with an activation energy which is the difference between the IB energetic position and E_C. Some experimental data of the rectifying behaviour of these diodes could be found in [21].

This model is, strictly speaking, non linear but we will assume that the reverse characteristics of the non-ideal diode could be represented with a temperature depending resistor Rt. We have to bear in mind that the limiting diodes are the ones inversely polarized and that the main variation is not the voltage but the temperature. Direct polarized diodes will be substituted by short circuits.

The limiting resistor has to have a temperature dependence, which is the inverse of the junction saturation current, that is to say:

$$R_t = \frac{1}{G_t} = A \cdot e^{\frac{(E_C - E_{IB})}{KT}}, \qquad (13.8)$$

where the pre-exponential factor A includes all the parameters quoted in reference [26] and also the contact area.

According to the model, the TIL will be layer 1 and the substrate layer 2. Electron density at the conduction band of TIL will be n_1 and its mobility μ_{n_1}. Hole density at the IB will be p_1 and its mobility μ_{p_1}. Substrate electron parameters are n_2 and μ_{n_2}. As explained before, we will not take into account holes in the valence band neither at the TIL nor at the substrate. We will define the following magnitudes:

- $G_{S1} = 1/R_{S1} = q(n_1\mu_{n_1} + p_1\mu_{p_1})t_1$, TIL sheet conductance
- $G_{S2} = 1/R_{S2} = qn_2\mu_{n_2}t_2$, substrate sheet conductance,

where t_1 and t_2 are the TIL and substrate thicknesses, respectively.

- $G_{C1} = 1/R_{C1} = G_{S1}/\alpha$, TIL conductance between two consecutive electrodes
- $G_{C2} = 1/R_{C2} = G_{S2}/\alpha$, substrate conductance between two consecutive electrodes

According to (13.6), the sheet resistance is:

$$R_{\text{sheet}} = \frac{G_{C1} + G_{C2}F^2}{\alpha(G_{C1} + G_{C2}F)^2}.$$

With $F = G_t/(G_t + G_{C2})$ being G_t defined in (13.8).

According to the model and having in mind that we are dealing with electrons at the CB and holes at IB in the TIL layer and only electrons at the substrate, the effective mobility will be:

$$\mu_{\text{eff}} = \frac{-n_1\mu_{n_1}^2 t_1 + p_1\mu_{p_1}^2 t_1 - n_2\mu_{n_2}^2 t_2 F^2}{n_1\mu_{n_1}t_1 + p_1\mu_{p_1}t_1 + n_2\mu_{n_2}t_2 F^2}, \qquad (13.9)$$

where F has the same meaning than in the previous equations.

13 Ion Implant Technology for Intermediate Band Solar Cells

Table 13.3 IB energetic position, IB hole mobility and diode pre-exponential factor for the three doses

	Analytical model			ATLAS		
DOSE (cm^{-2})	10^{15}	$5 \cdot 10^{15}$	10^{16}	10^{15}	$5 \cdot 10^{15}$	10^{16}
E_C-E_{IB} (eV)	0.38	0.38	0.38	0.36	0.36	0.36
μ_{IB} (cm^2 Vs^{-1})	0.4	0.4	0.6	0.4	0.4	0.6
$A(\Omega)$	$2 \cdot 10^{-7}$	$3 \cdot 10^{-6}$	$5 \cdot 10^{-6}$	–	–	–

Now, remembering the meaning of all the variables we have in the previous equations and its dependence with the temperature we can adjust our model to the experimental results. As with the ATLAS code, we consider that the hole concentration at the IB in the TIL is the Ti density. To fit the model to the experimental measurements, we have to scan just three parameters: the IB energy (we assume is the same for all the samples), hole mobility at the IB and the A parameter of the diode saturation current. All the other parameters are determined by direct measurements (TIL thickness) or are a consequence of the model (hole concentration at the IB). As with the ATLAS simulation, the electron mobility at the conduction band of the TIL has no influence on the fitting when its value ranged from 0 to 1,500 cm^2 V^{-1} s^{-1}.

Table 13.3 shows the best fit parameters and Fig. 13.13a, b give us the fitted plots for the sheet resistance and the mobility for the same samples represented on Fig. 13.7a, b. Parameters are remarkably similar to the ones used for ATLAS fits, except a difference of 20 mV in the IB energetic position. In Table 13.3, we give also the parameters used in the ATLAS fitting for comparative purposes.

Hall Mobility at Low Temperatures

One of the most exciting aspects of this work is the nature (electrons or holes) of the carriers at the IB. The bilayer decoupling is independent on its nature and it is only dependent on the barrier current limitation for the electrons. The impossibility of the IB carriers to go to the substrate is irrespective whether they are electrons or holes. The only difference we can find between electrons or holes is its mobility sign. According to the model, at low temperatures $F \to 0$, layers are decoupled and we are measuring only the mobility at the TIL, which is:

$$\mu_{\text{eff}} = \frac{-n_1 \mu_{n1}^2 t_1 + p_1 \mu_{p1}^2 t_1}{n_1 \mu_{n1} t_1 + p_1 \mu_{p1} t_1}. \tag{13.10}$$

According to our model, holes mobility and its concentration at IB do not depend on the temperature while electrons terms do. Samples with low electron mobility should change the polarity at higher temperatures, while for samples with higher electron mobility, the polarity change should be observed at lower temperatures. Samples presented in the Fig. 13.7b do not change the mobility sign

Fig. 13.13 (a) Sheet resistance and (b) mobility as a function of measured temperature for an-Si substrate (—) and for double sheet TIL/n-Si substrate for different implantation doses: 10^{15} cm^{-2} (*square* experimental; *filled square* analytical model); $5 \cdot 10^{15}$ cm^{-2} (*circle* experimental; *bullet* analytical model) and 10^{16} cm^{-2} (*triangle* experimental; *filled triangle* analytical model) (Reprinted with permission from [21]. Copyright 2009. Institute of Physics)

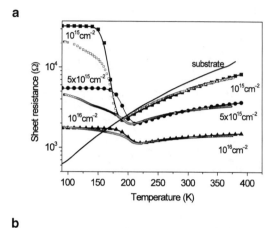

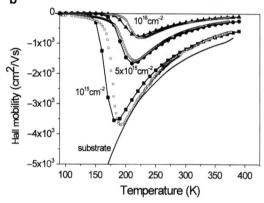

(at less at temperatures higher than 90 K), although it value decreases very quickly. Figure 13.14 shows the sheet resistance and Hall mobility of a sample implanted with a dose of 5×10^{15} cm^{-2} and annealed at 0.6 J cm^{-2}. This sample has presumably low electron mobility because it is annealed at low energy. The polarity change at $T = 190$ K is clearly shown in the inset. In the Fig. 13.15, we plot the Hall mobility of a sample implanted with 10^{16} cm^{-2} dose and annealed at a energy of 0.8 J cm^{-2}. In this sample, the electron mobility should be higher than the sample of the Fig. 13.14 because the higher annealing energy. Hall measurements were conducted down to a temperature of 7 K being the polarity change clearly evidenced at $T = 40$ K. Results of Figs. 13.14 and 13.15 are in fully agreement with the prediction of (13.10). Also, the value of the mobility at low temperatures is in good accordance with the values showed in Table 13.3, which were obtained through the fitting of the sheet resistance and mobility at higher temperatures instead of direct measurements.

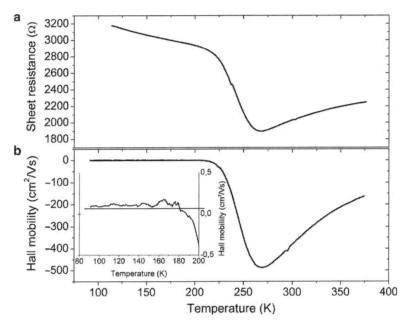

Fig. 13.14 Sheet resistance (**a**) and Hall mobility (**b**) for a representative sample in which p-type conduction has been measured ($5 \cdot 10^{15}$ cm^{-2} implantation dose; PLM at 0.6 J cm^{-2}). The inset shows the Hall mobility in the low temperature zone (Reprinted with permission from [21]. Copyright 2009. Institute of Physics)

Fig. 13.15 Hall mobility for a sample implanted with a dose of 10^{16} cm^{-2} and PLM at 0.8 J cm^{-2} measured down to 7 K and showing p-type conductivity at low temperatures. Inset shows an enlargement of the mobility at low temperature

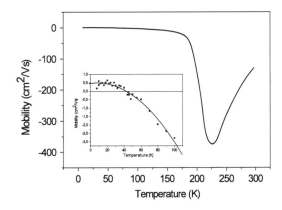

13.4 Discussion

The fitting obtained with the ATLAS code and with the analytical model are remarkably similar in spite of the different starting hypothesis. In the first case, we assume a heterojunction between the TIL and the substrate and in the second one the limitation is due to a generic temperature-dependent resistance. In this second case, we use a lumped equivalent circuit where the current limitation between both layers is concentrated at the corners, while ATLAS uses a true tri-dimensional grid. In both cases, we preserve the idea that all the Ti atoms contribute to the sheet conductivity of the implanted layer, which means that the IB is formed. The possibility of having a "normal" doping, that is to say, that the Ti atoms behave as donor or acceptor type shallow impurities is non-congruent with the Hall mobility sign change and also the TIL layer would be in parallel or isolated for all the temperatures. Anyway, the electronic configuration of the Ti electrons preclude the possibility of behave as a typical III or V dopant if it is substitutional to Si in the lattice. RBS measurements show that the Ti atoms are interstitially located and ab initio calculations show that in that case we have a half-filled intermediate band. This band is located closer to the conduction band than to the valence band and is fully compatible with the experimental data.

The IB hole mobility used in both models presents some controversial aspects. It is clear that we have not enough precision to determine its value but, anyway, it seems that it increases with the dose. As the dose goes up, TIL layer becomes more damaged and even polycrystalline and it seems that the mobility should go down in contradiction with the experimental data that show an increase from 0.4 to 0.6 cm^2 V^{-1} s^{-1}. This increase could be related to the IB width because as the Ti concentration increases, so should the bandwidth; this implies a reduction in the carrier effective mass, and hence an increase in the mobility [27]. Other very important argument to discuss is the absence of any IB effect in the 10^{14} cm^{-2} sample. This sample has a very high Ti concentration as it is shown in Fig. 13.2 but under the Mott limit except at the maximum where is just reaching this concentration. We assume that the Ti concentration is not enough to develop an IB, and the Ti electrons remain in the atom vicinity without any electrical activity. Finally, another fact that supports our hypothesis of the IB formation is the increase in the carrier lifetime as the doping is increased. In this experiment, which has been previously published [28], we used samples with doses of 10^{15}, $5 \cdot 10^{15}$ and 10^{16} cm^{-2} and we measured the carrier lifetime using a quasi-steady-state photoconductance technique. As Ti is a very well-known killer center, the lifetime should diminish when the dose increases. Nevertheless, the result is the opposite and the lifetime increases. This fact supports the theory that the formation of an IB which in turn implies the delocalization of the electrons wavefunctions avoids the non-radiative SRH recombination.

13.5 Conclusions

From GIXRD, TEM and ED results, it can be concluded that the Ti-implanted Si layer with a dose of 10^{15} cm^{-2} (average volume concentration of $3 \cdot 10^{20}$ cm^{-3} in the implanted thickness) and further PLM annealed at the highest energy density (0.8 J cm^{-2}) presents an excellent reconstruction of the crystal structure. Higher doses have a polycrystalline structure with decreasing grain size. RBS measurements show that most of the Ti is incorporated at interstitial positions. Sheet resistance and Hall mobility measurements prove the existence of an IB energetically located about 0.36–0.38 eV below the conduction band. In this IB, carriers behave as holes with high concentration, in the same order than Ti concentration, and low mobility. Holes remains confined in the implanted layer giving a strong laminar conductivity. IB physical thickness goes from the surface to the deep when the Ti profile has the value of the Mott concentration. Sheet resistance has been simulated with the ATLAS code and a specifically developed analytical model that corroborates the ATLAS simulation. The analytical model fits also the Hall mobility. Both the sheet resistance and mobility shows a very good accordance with the experimental data. All these results mean that Ti-implanted Si with the structural and electrical characteristics could be a material of choice for future IBSC.

Acknowledgements Authors would like to acknowledge the Nanotechnology and Surface Analysis Services of the Universidad de Vigo C.A.C.T.I. for ToF-SIMS measurements, the Center for Microanalysis of Materials of the Universidad Autónoma de Madrid for RBS measurements, C.A.I. de Difraccin de Rayos X of the Universidad Complutense de Madrid for GIXRD measurements, C.A.I. de Microscopa de la Universidad Complutense de Madrid for TEM analysis and C.A.I. de Técnicas Físicas of the Universidad Complutense de Madrid for ion implantation experiments. This work was made possible thanks to the FPI (Grant No. BES-2005-7063) of the Spanish Ministry of Education and Science. This work was partially supported by the Projects NUMANCIA-2 (No. S2009/ENE1477) funded by the Comunidad de Madrid GENESIS-FV (No. CSD2006-00004) funded by the Spanish Consolider National Program and by U.C.M.-C.A.M. under Grant CCG07-UCM/TIC-2804.

References

1. A. Martí, A. Luque, *Next Generation Photovoltaics: High Efficiency Through Full Spectrum Utilization* (Institute of Physics Publishing, Bristol, 2004)
2. A. Luque, A. Martí, Phys. Rev. Lett. **78**(26), 5014 (1997)
3. W. Shockley, H.J. Queisser, J. Appl. Phys. **32**(3), 510 (1961)
4. A. Luque, A. Martí, N. López, E. Antolín, E. Cánovas, C. Stanley, C. Farmer, L.J. Caballero, L. Cuadra, J.L. Balenzategui, Appl. Phys. Lett. **87**(8), 083505.1
5. K.M. Yu, M.A. Scarpulla, R. Farshchi, O.D. Dubon, W. Walukiewicz, Nuclear Instruments Methods Phys. Res. Section B-Beam Interactions with Mater. Atoms **261**(1–2), 1150 (2007)
6. N.F. Mott, Rev. Mod. Phys. **40**(4), 677–683 (1968)
7. E.M. Conwell, Phys. Rev. **103**(1), 51 (1956)
8. R.O. Carlson, Phys. Rev. **100**(4), 1075 (1955)
9. S. Liu, K. Karrai, F. Dunmore, H.D. Drew, R. Wilson, G.A. Thomas, Phys. Rev. B **48**(15), 11394 (1993)
10. A. Gaymann, H.P. Geserich, H. Vonlohneysen, Phys. Rev. Lett. **71**(22), 3681 (1993)
11. S. Hocine, D. Mathiot, Appl. Phys. Lett. **53**(14), 1269 (1988)
12. A. Luque, A. Martí, E. Antolín, C. Tablero, Phys. B-Condens. Matter **382**(1–2), 320 (2006)
13. M. Hernández, J. Venturini, D. Zahorski, J. Boulmer, D. Debarre, G. Kerrien, T. Sarnet, C. Laviron, M.N. Semeria, D. Camel, J.L. Santailler, Appl. Surface Sc. **208**, 345 (2003)
14. K.M. Yu, W. Walukiewicz, J.W. Ager, D. Bour, R. Farshchi, O.D. Dubon, S.X. Li, I.D. Sharp, E.E. Haller, Appl. Phys. Lett. **88**(9), 092110.1 (2006)

15. K. Sánchez, I. Aguilera, P. Palacios, P. Wahnon, Phys. Rev. B **79**(16), 165203 (2009)
16. R.L. Petritz, Phys. Rev. **110**(6), 1254 (1958)
17. D.C. Look, J. Appl. Phys. **104**(6), 063718 (2008)
18. D.C. Look, D.C. Walters, M.O. Manasreh, J.R. Sizelove, C.E. Stutz, K.R. Evans, Phys. Rev. B **42**(6), 3578 (1990)
19. D.C. Look, D.C. Reynolds, J.W. Hemsky, J.R. Sizelove, R.L. Jones, R.J. Molnar, Phys. Rev. Lett. **79**(12), 2273 (1997)
20. C.W. White, S.R. Wilson, B.R. Appleton, F.W. Young, J. Appl. Phys. **51**(1), 738 (1980)
21. J. Olea, G. Gonzalez-Diaz, D. Pastor, I. Martil, J. Phys. D-Appl. Phys. **42**(8), 085110 (2009)
22. J. Olea, M. Toledano-Luque, D. Pastor, E. San-Andrés, I. Martil, González-Díaz, J. Appl. Phys. **107**, 103524 (2010)
23. ATLAS, Device Simulator Framework distributed by Silvaco Data Systems Inc., 4701 Patrick Henry Device, Bldg 6, Santa Clara, CA 95054
24. PSPICE, Cadence Designs Systems Inc., 2655 Seely Avenue, San Jos, CA 95134
25. D.W. Koon, Rev. Sci. Instruments **77**(9), 094703 (2006)
26. G. Gonzalez-Díaz, J. Olea, I. Martil, D. Pastor, A. Martí, E. Antolín, A. Luque, Sol. Energ. Mater. Sol. Cell. **93**(9), 1668 (2009)
27. F.J. Blatt, *Physics of Electronic Conduction in Solids* (Mac Graw Hill, New York, 1968)
28. E. Antolín, A. Martí, J. Olea, D. Pastor, G. González-Díaz, I. Martil, A. Luque, Appl. Phys. Lett. **94**(4), 042115 (2009)

Index

Ab initio calculations, 287, 333
Absorber, 103
Absorptance, 97
Absorption, 144, 240, 241
Absorption coefficient, 97, 116
Absorption enhancement, 142, 143, 145, 146, 150
Accelerated ageing tests, 54
Acceptance angle losses, 68
Acoustic phonon scattering, 246
AC-system efficiencies, 1
Activation energy, 340
Advanced characterization techniques, MJSC, 45
AERONET database, 15
Alexander Maish, 76
AlGaN, 313
AM1.5G, 281
Amorphization, 325
Amorphous silicon thin film solar cells, 150
AM1.5 solar spectrum, 105
Analytical model, 336
Anderson, 287
Angular distribution function, ADF, 98, 119
Antiferromagnetic, 294
Antireflective coating, 96, 113, 114, 125
Antisite point defects, 281
(Al)GaInAs, 9
(Al)GaInP, 9
Aperture frame, 68
ARC, 114, 115
Aromatic hydrocarbon derivatives, 166
Arsenic, 315
a-Si:H, 101, 134, 144, 146, 147, 149, 280
a-Si:H top cell, 147, 148
ATLAS, 333
Atmospheric refraction corrections, 87

Atomic reservoirs, 290
Auger processes, 231
Auger related scattering, 246
Automatic error collection, 81
Autonomous tracking control, 76

Back electrical contact, 118
Backlash, 64
Back reflector, 96, 105, 113, 116–118, 125
Balance of system, BOS, 192
Banbdgap, 122
Band-anticrossing model, 283
Band-like absorption spectrum, 167
Band mixing, 233
Barrier height, 332
Bi-exciton recombination, 246
Bilayer, 329
Bilayer decoupling, 341
Binary chalcogenides, 288
Black body, 279
Bound state, 295
BSQ Solar, 73
Buffer, 11

Calibration model, 79
Carissa Plains plant, 76
Carrier cooling, 193
Carrier lifetime, 344
CdS, 201, 297
CdSe, 197, 201
CdSe/CdTe, 197
CdTe, 193, 197, 280–282
Cell performance, 106
CGS/CIGS, 122
Chalcogen, 288

Index

Chalcopyrite, 284
Chalcopyrite-based nanowires, 297
Characterization methods, 100
Charge-carrier generation rate profile, 101
Chemical potential, 288
CIGS, 101, 121
Closed-loop, 74
Clustered multi-start, 81
Coherence of the excitation light, 165
Coherent excitation, 166
Coherent propagation, 97
Collimated beam, 170
Colloidal, 191
Colloidal suspensions, 294
Complex refractive index, 97, 100
Concentration, 26
Concentrator solar cell, 13
Concentrix, 72
Confined states, 294
Conversion efficiency, 103
Copper phthalocyanine, CuPc, 150
Core-shell, 296
Cost, 24, 192
Coulomb coupling, 199
Croot mean square roughness, 98
Cross-section, 137
$Cu(In,Ga)(S,Se)_2$, 282
$Cu(In,Ga)Se_2$, 285
$(Cu,Ag)(Al,Ga,In)(S,Se,Te)_2$, 281
Cubic film, 315
Cu-containing chalcopyrite, 280
$CuGa_5S_8$, 289
$CuGaS_2$, 282, 283, 290
$CuGaSe_2$, 282, 283, 291
$CuIn_xGa_{1-x}Se_2$, 193
$CuInS_2$, 282, 283
$CuInSe_2$, 283, 297
Current matching, 106
Current voltage characteristics, 188
Current–matching, 3
Curvature of a wafer, 12
CuS, 289
Cu_2S, 289
Cu_2Se, 289

Decay time, 182
Decoupling, 329, 332
Defect maps, 50
Degradation, 54
Delayed fluorescence, 177, 179
Delayed sensitizer fluorescence, 182
Delocalization, 330
Densities of carriers, 332

Density functional theory, 292
Density of states (DOS), 237
Dependence on the beam diameter, 177
Depletion zone, 336
Detailed balance, 193, 229, 277
Detailed balance calculations, 2
Device-architecture, 186
Dichroic mirror, 123
Diffractive gratings, 110
Diffusion controlled, 177
Diffusive dielectric material, 113, 119
Difracting grids, 215
Diode saturation current, 341
–9,10-diphenylanthracene (DPA), 161
Dipole approximation, 240
Direct (specular) light, 100
Dislocation densities, 11
Dislocations, 10
Dispersion relation, 133, 141
Dispersion relation of a PSPP, 139, 140
Distributed Bragg reflector, 116
3D distributed model, 31, 42
Doctor blade technique, 172
Doping levels, 164
Double textures, 108
Down-conversion (DC), 184
Down-shifting (DS), 184
DSSC, 186
DSSC + UCd, 187
Dual-junction solar cells, 2, 31, 48
DX-like centres, 281
Dye, 201

Effective mass, 295
Efficiencies of triple-junction solar cell, 3
Efficiency, 106, 123, 230
Electrical network model, 8
Electrochemical potential, 284
Electroluminescence, EL, 47, 49
Electron cooling, 246
Electron-hole pair, 230
Electronic structure, 292
Electron-phonon scattering, 193, 231
Emergency stowage, 76
Emitter singlet state, 168
Encapsulation, 99
Energetically-conjoined TTA-UC, 177
Energetic schema, 159
Energy bandgaps, 101
Energy gap, 106, 121
Energy harvesting efficiencies, 17
Energy harvesting model, 15
Energy lost, 168

Index

Enthalpy of formation, ΔH_f, 288
Epitaxy, 24
EPS-Tenerife, 78
Equipotential lines, 335
EtaOpt, 3, 4
EUCLIDES^{TM}, 78
Euler angles, 80
Excitation pathways, 163
Exponential smoothing, 78
External quantum efficiency, 104, 109, 119, 169
Extinction cross-section, 137

Fabry-Perot resonance, 146, 147, 149
FEMOS, 101, 111
Fermi level, 230, 293, 331
Fermi's golden rule, 246
Ferromagnetic, 294
Finite-depth-well effective mass approximation, 296
Finite difference time domain, 108
Finite element method (FEM), 101
Flexure floor, 73
Foundation, 68
Fraunhofer ISE, 2
Frölich interaction Hamiltonian, 244
Front surface field (FSF), 5
Function F, 339

Ga, 312
GaAs, 5, 26
GaAs single-junction concentrator solar cell, 2
GaAs solar cells, 150
GaInNAs, 17
GaInP, 2
GaInP / GaAs dual-junction solar cell, 7, 31
GaInP / GaInAs / Ge triple-junction solar cell, 14, 43
GaInP/GaAs dual-junction solar cell, 23
GaMnN, 311
GaN, 309
Ga$_2$S$_3$, 289
GaS, 289
GaSe, 289
Ga$_2$Se$_3$, 289
Ge bottom cell, 9
Geotechnical analysis, 68
Ge substrate, 25, 28
Glancing-Incidence X-Ray Diffraction, GIXRD, 323
Glass transition temperature, 171
Global absorption, 142

Glove-box, 170
Grain boundaries, 281, 282
1D grating, 111
Grating equation, 110
Grid parity, 132
Groove period, 110
Guided mode, 134, 138, 139, 141, 144

Hall-effect, 287
Hall mobility, 339
3D-harmonic oscillator, 296
Haze, 98, 108, 109, 119
Heptamer (OF7), 161
Heterojunction, 343
High-concentration photovoltaics, 1
Hot carriers, 192
Hot excitons, 192
Hybrid multi-terminal configuration, 123
Hybrid sun tracking controllers, 77

IBSC. *See* Intermediate band
Ideal efficiency of a multi-junction solar cell, 14
Ideality factors, 283, 285
I-III-VI$_2$, 281
II-VI QDs, 200
III-V multi-junction solar cells, 2, 23
III-V QDs, 200
III-V semiconductor, 193
Impact ionization, 193
Impurity approach, 287
Impurity bands, 287
InAlGaAs space layers, 253
InAs, 197, 201
InAs/GaAs QDs, 234
Incomplete selectivity of absorption coefficients, 286
InGaN, 309
InN, 314
InP, 197, 201
Intensity depence, 176
Intensity dependence of the UC, 160
Interband, 196
Interband tunneling, 5
Interdiffusion, 296, 298
Interference, 97, 116
Intermediate band, 209, 251
    ab-initio, 225
    bandwidth, 214
    bulk systems, 215
    chalcopyrite semiconductors, 253
    deep centres, 220

dilute II-VI oxide semiconductors, 253
GaNAs, 222
highly mismatched alloys, 222, 253
In_2S_3, 223
InGaN, 220, 253
limiting efficiency, 209
metallization of solar cells, 268
model, 213
molecular approach, 223
optimum gaps, 252
photon selectivity, 215
quantum dot, 215, 216, 251
quantum dot die manufacture, 265
quantum efficiency, 217, 270
quantum wells, 217
quasi-Fermi levels, 214
role of the emitters, 212
silicon, 253
thin films, 220
two photon absorption, 272
voltage preservation, 211, 272
ZnTe, 222, 253
Intermediate band formation, 235
Intermediate band solar cell, 230, 321
Internal energy conversion channel, 168
Interstitial, 329
Inter-system crossing (ISC), 159
Intraband, 196
Intrinsic defects, 281
Inverse Auger, 193
Inverted methamorphic, 4, 25
Ion implantation, 322
IPCE curves, 184
IRL, intermediate reflecting layer, 147, 149
Isofoton, 67
IV-VI QDs, 200

Kinetic, 298
Knudsen cells, 312
Koopman spiral, 82
kppw code, 234, 249
$\mathbf{k} \cdot \mathbf{p}$ theory, 231, 232
Kretschmann configuration, 140, 150
KrF excimer laser, 323

Lambertian (cosine) function, 119
LA phonons, 245
Lateral current, 35, 39
Lattice-matched, 23, 25
Lattice-matched triple-junction solar cell, 2, 4, 17
Learning coefficient, 29

Least squares, 81
LED-like approach, 41
Levenberg-Marquardt, LM, 81
Light absorption, 106
Light emitting diode, LED, 53, 280
Light management, 124, 215
Light profile, 31, 33, 41
Light scattering, 96, 103, 106
Light trapping, 96, 103, 111, 132, 133, 144, 147, 152
Limiting efficiency, 211
Linear function, 176
Local absorption, 142
Local mobility, 165
Local temperature, 177
Local viscosity, 171
LO-phonon, 244
LSC, 189
LSPP, localized surface plasmon polariton, 132, 133, 135, 136, 138, 141, 143, 144, 148
LTSpice, 5, 8
Luminiscence, 283
Lumped equivalent circuit, 344

Magnetic properties, 294
Magnetoresistance, 287
Makov-Payne correction, 247
Material dispersion, 135
Maximum service wind speed, MSWS, 64
Maxwell equations, 101
MBE. *See also* Molecular beam epitaxy, 311, 312
μc-Si, 134, 149
μc-Si bottom cell, 147
μc-Si:H, 101
Mean offsets, 88
Mean Time to Failure, MTTF, 53, 55
Metallated porphyrin macrocycles (MOEP), 159
Metallisation, 14
Metamorphic, 4
Metamorphic triple-junction solar cell, 12, 17
Method of discretisation, 100
Micromorph, 121
Minibands, 238, 294
Minimum enclosing circle, MEC, 71
Minority carrier, 283
Mn, 310
Model based calibrated approach, 77
Model free predictive approach, 77
Modulated photonic crystal, 117, 118
Molar concentration dependence, 179

Index   351

Molar ratio, 173, 179
Molecular beam epitaxy, 251, 253, 278
Molecular couples, 186
Momentum, 194
Mott, 287
Mott limit, 344
Mott transition, 209, 215
MOVPE, 25
Multicomponent organic systems, 174
Multiexcitons, 196
Multijunction devices, 96, 121, 125, 184
Multijunction solar cell, 105, 121
Multilayer structures, 97
Multiple Exciton Generation, MEG, 191

Nanocrystalline, 327
Nanodisc, 145, 146, 148, 149
Nanoparticle, 112, 125, 133, 134, 138, 144–147
Nanostructure, 192, 294
Nanostructuring approach, 294
Nanotubes, 199
Nanowires, 296
Near field, 136, 137, 140, 144, 146
Neutral complex, 281
Nitrogen, 312
Noble metal, 297
Non-coherent excitation, 166
Non-coherent photons, 158
Non-radiative channels, 179
Non-radiative recombination, 285
Non-radiative relaxation channels, 175
Non-radiative scattering, 242
Non-radiatively, 176
Non-volatile, 171
Non-volatile solvent, 172, 180
Numerical modelling, 4
Numerical simulators, 100

Off-stoichiometry, 281
Oligomer, 161, 171
One axis tracking, 62
Open-circuit photocurrent, 188
Open-circuit voltage, 103
Open-loop controllers, 76
Optical losses, 96, 103, 104, 106, 113
Optical matrix element, 240
Optical modelling, 100, 124
Optical models, 100
Optical path, 97, 103, 106, 111
Optical phonons, 244
Optoelectronic, 29

Ordered vacancy compounds, 281
Organic dye, 187
Organic solar cell, 133, 141, 150, 151, 158
Oscillator strength, 240
Otto configuration, 140, 150
Oxygen, 170

PbS, 201
PbSe, 195, 197
PbTe, 197
PdTBP, 175
π-conjugated, 185
Peak current, 35, 36, 38
Pedestal tracker, 62
Perimeter, 30, 32, 41, 42, 55
Periodic array of QD's, 231
Periodic Born-von Karman boundary conditions, 234
Periodic surface, 110
Periodic texturisation, 101
Permittivity, 135
PF film, 181
Phonon bottleneck, 244
Phonons, 244
Phosphorescence, 162
Photocharging, 197
Photocurrent, 103, 284
Photoluminescence, PL, 47, 197
Photon conversion, 185
Photon flux, 284
Photonic crystal, 96, 113, 116, 118
Photonic-crystal-like, 125
Photon management, 131–134, 136, 141, 142, 152
Photon-sorting, 240
Photoreflectance, 296
Piezoelectric field, 233
Plane wave (PW) methodology, 234
Plasma-assisted, 312
Plasma-enhanced, 312
Plasmons, 132, 215
Plastic foil substrates, 112
Pn-junctions, 14
Point defects, 281
Pointing vector, 80
Polar coupling, 244
Polar facets, 281
Polarity change, 341
Polarizability, 137
Polarizability of a sphere, 136
Polarons, 244
Polycrystalline, 327
Polymers, 201

Porous Si-oxide, 115
Positioning resolution, 64
Position Sensitive Device, PSD, 84
Poynting vector, 142
Primary axis, 79
Proportional-integral (PI) controller, 79
Pseudo-gap, 334
PSPICE, 336
PSPP, propagating surface plasmon polariton, 132, 133, 138, 140, 141, 150
Pulsed-Laser Melting, PLM, 322
Pump-probe, 196
PV modules, 99
Pyramidal-type SnO_2:F superstrate, 110

Q-band, 167, 175, 187
Q-band absorption, 173
QD. *See* Quantum dot
QD array, 231
QD array Brillouin Zone, 237
QD array wavenumber, 237
Quadruple-junction solar cell, 17
Quantum dot QD, 191
   capped, 259
   capping, 257
   InAs, 252, 256
   multi-layer stacks, 260
   photoluminescence, 260
   seeded, 259
Quantum-dot solar cell, 294
Quantum efficiency, QE, 45, 106, 111
Quantum well, 192, 296
Quantum wire, 192
Quantum yields, 169, 170, 193
Quenched, 163
Q.Y., 181

Radiative lifetime, 243
Radiative limit, 243, 278
Radiative recombination, 242
Random, 108
r_{anode}, 38
Rare earth, 287
Rare-earth (RE) doped phosphors, 158
Ray-tracing, 101
Re-absorption, 175, 177
Recombination, 278
Recombination rate, 280
Record efficiency, 23
Recrystallization, 326
RE-doped crystalline glasses, 188
Red shifted, 242

Reflectance, 115
Reflection, 97
Reflection high energy electron difraction. *See* RHEED
Refractive index grading, 114
Refractive indices, 114
Reliability, 52–54
Residual oxygen, 170
Resonance wavelength, 135, 137, 138, 144, 146–149
Resonant frequency, 112
RHEED, 255
Root-mean-square-roughness, 98, 108
Rutherford backscattering spectrometry, RBS, 328

Sandia National Laboratories, 64
Sapphire, 312
Saturation current, 285, 286
Scalar scattering theory, 109
Scattered (diffused) light, 100
Scattering cross-section, 137, 144
Scattering order, 110
Schottky, 202
Schrödinger equations, 234
Secondary axis, 79
Semiconductor QD, 231
Sensitizer, 163
Sensitizer triplet, 166
Sentaurus TCAD, 5, 7
Series resistance, 27, 29, 43, 54
Sesonant tunnelling, 5
Sheet resistance, Rs, 329, 336
Shockley-Queisser, 192, 279
Short-circuit current density, 106
Si, 193, 197
SiC, 312
Six-junction solar cell, 17
Size scaling, 188
SMARTS, 15
Snell's law, 107
Solar One plant, 76
Solar spectrum, 103
Solfocus, 72
Solid state thin polymer films, 165
Solubility, 294
Solubility limits, 288, 322
Solvothermal methods, 294
Soret-band, 175, 179
Space dimensions, 100
Spectral power density, 158
Spectral response, SR, 45
Spectrum splitting, 123, 125

Index

SPICE, 8
Spin-orbit interaction, 233
SPP, surface plasmon polariton, 132, 134, 152
Step-graded buffer, 9
Step-wise UC, 165
Stiffness, 67
Stow position, 64
Strain, 9, 12, 233
Strain compensated tunnel diodes, 13
Strain relaxation, 12
Structural bending, 66
Structural flexure, 68
Styrene matrix, 171
Sub-bandgap absorption, 123
Sub-linear, 177
Substitutional, 329
Substitutional impurities, 288, 294
$SunDog^®$, 82
Sun ephemeris, 63
Sunlight concentration, 157
Sunlight focusing, 188
Sunlight-UC, 169
Sunlight UC organic solar cells, 183
Sun pointing sensor, 63
Sun sensor, 74
SunShine, 100, 115
$SunSpear^®$, 84
Superlattice, 296
Superstrate, 99
Surface features, 108
Surface plasmon absorption, 116
Surface plasmon polaritons, 112
Surface reconstruction, 281
Surface treatments, 198
Surface-textured interfaces, 125

Tandem, 123
Tandem $a\text{-}Si\!:\!H/\mu c$–Si:H, 109
Tandem cell, 147
TCO substrate, 105
TE polarization, 241
Tetraanthraporphyrins, 185
Tetranaphthoporphyrins, 185
Textured interfaces, 96, 107
Texturisation, 97, 114
Thermalisation, 167
Thermalisation losses, 103, 121
Thermal stress, 185
Thermochemical, 289
Thickness, 103
Thin-film intermediate band solar cells (TF-IBSC), 277
Thin-film solar cells, 95, 124, 134, 150, 280

Threshold energy, 194
Tilt-roll tracker, 62
Time of Flight Secondary Ion Mass Spectroscopy, ToF-SIMS, 323
Time-dependent perturbation theory, 246
Ti-Si phases, 325
Total reflection, 107
Tracking accuracy, 66
Tracking Accuracy Sensor, TAS, 84
Tracking error measurements, 77
Transient absorption, 196
Translational symmetry, 234
Transmission, 97
Transparency window, 167, 173, 175
Transparent conductive oxides, 96
Transport properties, 248
Trions, 198
Triple junction solar cell, 24, 45
Triplet harvesting, 175, 176
Triplet-triplet transfer (TTT), 161, 166
TTA-schema, 177
TTA-UC, 159
TTA-UCd, 188
Tunnel diode, 13
Tunnel diode models, 5
Tunneling, 331
Tunnel junction, 35, 45
Two axis tracking, 62
Two-photon absorption, 280

UC-couple, 177
UCd, 173, 186, 188
UC–pathway, 161
UC-sunlight concentrators, 157
UC-$TypeI$, 163
UC-$TypeII$, 163
Ultimate efficiency limit, 229
Ultra high concentration, UHCPV, 27, 39
Up-conversion (UC), 184
US National Climatic Center, 64
USNO MICA software, 87
Utilisation of the solar spectrum, 105, 106

Van der Pauw, 324
Van Hove singularities, 239
Vergardś Law, 315
Virtual energetic levels, 167
Viscosity, 177, 180
Volatile organic solvent, 170

Wavelength-selective intermediate reflector, WSIR, 121, 123

Whisker, 297
White diffusive dielectrics, 116
White paint, 96, 119, 125
Wide gap semiconductors, 280
Wigner-Weisskopf description, 245
W-textured superstrate, 110

Wurtzite, 312

Zinc-blende film, 315
ZnTe, 283
ZnTe:O, 283

Printed by Publishers' Graphics LLC USA
MO20120327-110
2012